Methods in Molecular Biology

Series Editor
John M. Walker
School of Life Sciences
University of Hertfordshire
Hatfield, Hertfordshire, AL10 9AB, UK

For further volumes:
http://www.springer.com/series/7651

Chromosomal Mutagenesis

Second Edition

Edited by

Shondra M. Pruett-Miller

Department of Genetics,
Genome Engineering and iPSC Center,
Washington University School of Medicine,
St. Louis, MO, USA

Editor
Shondra M. Pruett-Miller
Department of Genetics
Genome Engineering and iPSC Center
Washington University School of Medicine
St. Louis, MO, USA

ISSN 1064-3745 ISSN 1940-6029 (electronic)
ISBN 978-1-4939-1861-4 ISBN 978-1-4939-1862-1 (eBook)
DOI 10.1007/978-1-4939-1862-1
Springer New York Heidelberg Dordrecht London

Library of Congress Control Number: 2014955158

Printed on acid-free paper

Humana Press is a brand of Springer
Springer is part of Springer Science+Business Media (www.springer.com)

Dedication

For Ryan and Jace.

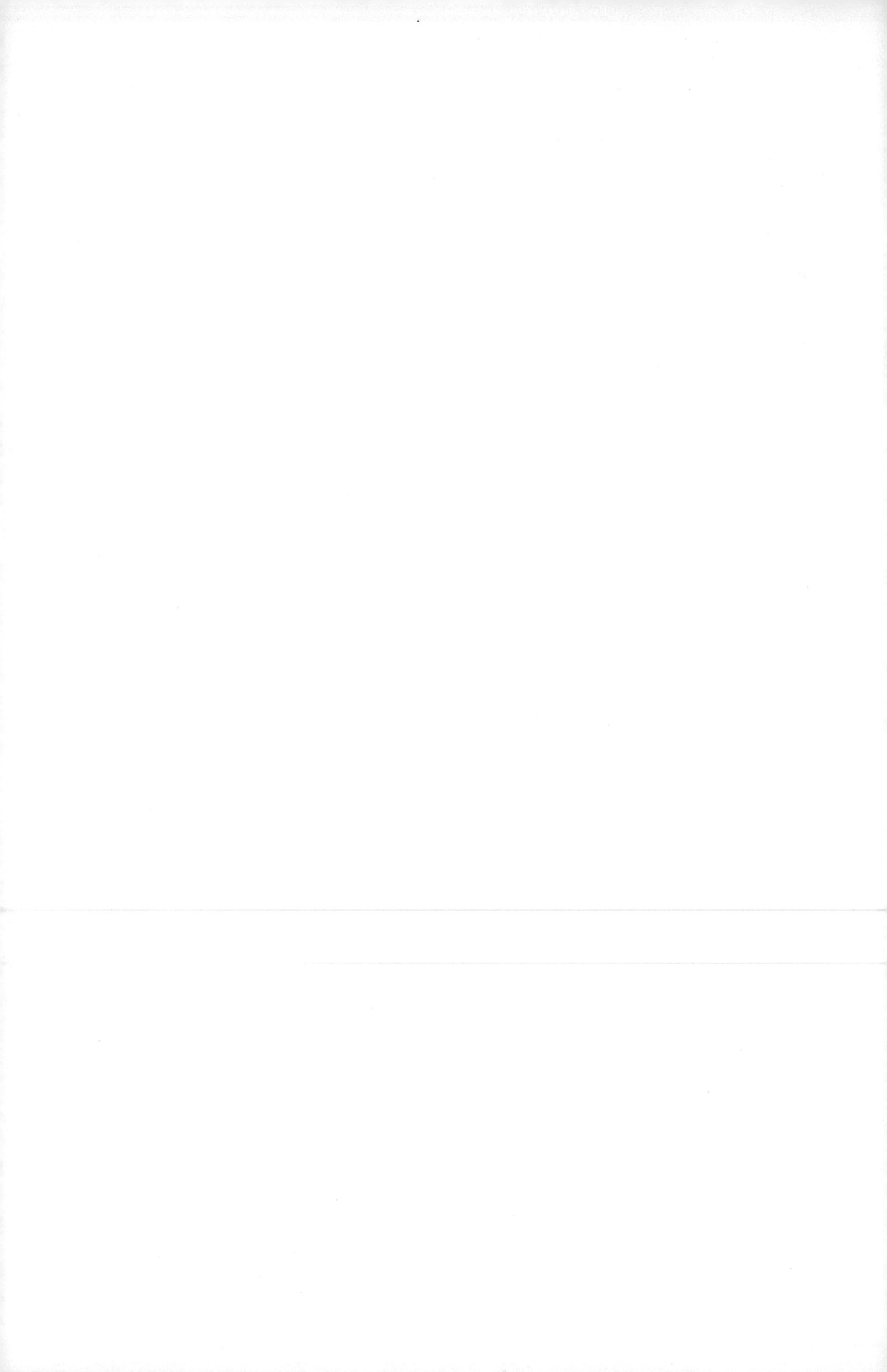

Preface

The era of genomics has ushered in exciting new opportunities for scientists to not only better understand biology but also to enhance biology. Although it has been over a decade since the human genome was fully sequenced, the tools needed to precisely and effectively make targeted genomic modifications have only recently become efficient and available. Through chromosomal mutagenesis, we now have the potential to create cell and animal-based models that have and will continue to become essential not only in the field of basic research but also in drug development, personalized medicine, and even food production. In this edition of *Chromosomal Mutagenesis*, we provide comprehensive coverage and detailed protocols on current and emerging mutagenesis methods focusing specifically on mammalian systems and commonly used model organisms.

The field of genome engineering has rapidly evolved over the last decade. Since the last edition of this book, only 6 years ago, major advances and discoveries have made chromosomal mutagenesis a widely used technique and one that is available to any molecular biology laboratory. This edition contains detailed protocol chapters but also review and case study chapters from leaders in the field. The book starts out with a review from ZFN pioneer, Dana Carroll, discussing the history and major advances in nuclease-mediated chromosomal mutagenesis.

One limitation to chromosomal mutagenesis in mammalian cells has been the inability to make homozygous modifications in the diploid genome. In the chapter by Kosuke Yusa and colleagues, they describe one method to overcome this limitation by using haploid murine embryonic stem cells along with piggyBac transposon-based insertional mutagenesis to generate a genome-wide mutant library that can be used for forward genetic screening. Another limitation for successful chromosomal mutagenesis has been the difficulty in generating targeted modifications in a single-site specific manner. Jonathan Geisinger and Michele Calos contributed a chapter describing the simultaneous use of phiC31 and Bxb1 integrase which permits greater control and accuracy over the site of integration.

Other strategies to make specific, targeted modifications rely on homologous recombination or homology-directed repair. One such approach is presented in the chapter contributed by Faisal Reza and Peter Glazer, in which they describe a protocol for using triplex-forming molecules in combination with recombinagenic DNA to create precise, genomic modifications. The use of the adeno-associated virus (AAV) also relies on the endogenous cellular recombination machinery to create site-specific modifications. In the chapter by Rob Howes and Christine Schofield, they describe how to design and use AAV gene targeting vectors and viruses to create targeted, isogenic cell lines.

The rate of homologous recombination is fairly low in mammalian somatic cells. However, it has been shown that the rate of homologous recombination can be increased by over 1,000-fold by introducing a targeted double-strand break near the desired genomic modification site. There are several flavors of engineered endonucleases and the field has advanced at an incredible rate in the last few years with the advent of transcription activator-like effector nucleases (TALENs) and clustered regularly interspaced short palindromic repeats (CRISPR)/Cas9 nucleases. The first targetable nucleases to be used successfully in genome engineering were meganucleases. In general, meganucleases have not become standard reagents used in a wide variety of labs because of the initial difficulty in targeting meganucleases to unique and novel genomic sites. However, in the chapter by Barry Stoddard and colleagues, they describe an approach for efficiently engineering meganucleases to specifically cleave a wide variety of DNA sequences that can be performed in an academic lab setting.

One reason meganucleases have historically been difficult to redirect is due to the architecture of this type of nuclease: the DNA binding and the DNA cleavage domains are structurally integrated. This architecture makes it difficult to redirect the binding without affecting the cleavage. Zinc finger nucleases (ZFNs) and TALENs are another type of engineered endonuclease that can be engineered to make targeted double-stranded DNA breaks. Both ZFNs and TALENs have structurally distinct DNA binding and DNA cleavage domains making this type of nuclease easier to target to specific regions of genomic DNA without affecting the cleavage activity. Unlike their predecessors, ZFNs and TALENs are relatively easy to design and target. Several strategies for TALEN design and assembly have been developed, and two such strategies are described here. In the chapter by Daniel Voytas and colleagues, they describe the most well-cited strategy for TALEN assembly: the Golden Gate Platform. In the chapter by Veit Hornung and colleagues, they describe a high-fidelity strategy for TALEN assembly that is amenable to high-throughput assembly with reduced labor and cost.

As mentioned earlier, specificity is key when making genomic modifications, and even more so when making such modifications using nucleases, which literally cut genomic DNA. Double-strand DNA breaks are the most dangerous type of DNA lesion for the cell, and therefore mammalian cells are very efficient at repairing such breaks. However, in efforts to rejoin these dangerous DNA lesions, the endogenous cellular repair machinery sometimes makes mistakes. Therefore, any unintended DNA breaks caused by an engineered endonuclease could lead to unwanted chromosomal mutagenesis. To this point, Sandrine Boissel and Andrew Scharenberg contribute a chapter detailing the assembly and characterization of Mega-TALs, which are a fusion of a meganuclease and a TAL effector DNA binding domain. These monomeric, engineered proteins can be hyperspecific and highly efficient genome editing tools.

The latest genome editing tool to enter the field is the CRISPR/Cas9 system. Unlike meganucleases, ZFNs, or TALENs, the CRISPR/Cas9 system does not require protein engineering every time a new target sequence is identified. The CRISPR/Cas9 system has been engineered into a two component system consisting of a guide RNA and a Cas9 protein. The guide RNA directs the cleavage activity of the Cas9 protein through complementary base pairing with the target genomic site. Le Cong and Feng Zhang contribute a chapter detailing the design, assembly, and functional validation of CRISPR/Cas9 nucleases.

When using a nuclease to create user-defined targeted genomic modifications, a donor template containing the specific desired modification must also be introduced. There are numerous types of modifications that can be introduced, and with each, a unique set of considerations. The chapter written by Greg Davis and I describes important parameters and details for designing donor plasmids for small modifications made to the genome using ZFNs. The chapter by Malkov and colleagues details important considerations and gives a case study for creating a donor plasmid for use in endogenously tagging genes with fluorescent proteins. Tony Gutschner contributes a chapter containing a detailed protocol for using nucleases and donor plasmids to silence long noncoding RNAs.

Recently, it was shown that donor templates used with nucleases can not only be in the traditional donor plasmid form, but could also be single-stranded oligodeoxynucleotides (ssODNs). Several key factors make using ssODNs a great alternative for certain types of modifications. For example, ssODNs are small and can therefore be put into cells in molar excess compared with donor plasmids, which are typically a limiting factor for modification rates. Also, ssODNs can easily be ordered from an oligo synthesis company and received within a matter of days. The chapter by Greg Davis and colleagues describes the use of single-stranded oligo donors as repair templates after a nuclease-induced double-strand break. Kunitoshi Chiba and Dirk Hockemeyer contribute a chapter describing the use of both donor plasmids and single-stranded oligo donors along with ZFNs, TALENS, and/or CRISPR nucleases in induced pluripotent stem cells (iPSCs). No matter the type of nuclease used to create the targeted, double-strand break, and no matter what type of repair template, if any, is used, the ultimate goal in genome editing is to be able to identify a correctly edited cell. In the chapter by Matthew Porteus, he describes and reviews several approaches to increasing the efficiency of genome editing in order to achieve this goal.

This edition focuses on chromosomal mutagenesis in mammalian cells, but we thought it important to include a few chapters on creating modified animal models as well. The zebrafish has long been an excellent model organism especially for developmental biology because of their rapid generation time, transparent embryos, and ease of forward genetics. However, until recently, reverse genetic approaches in zebrafish have been limited and time-consuming. In the chapter by Victoria Bedell and Stephen Ekker, they describe how to use engineered endonucleases to create both targeted knockout and knockin zebrafish models via direct embryo injection. Although reverse genetic approaches have been available for mouse models for several decades, available techniques have traditionally required the use of murine embryonic stem cells (mESCs) and can become time-consuming to produce. However, with advances in genome engineering, both mouse and rat models can be created via direct embryo injection of engineered endonuclease. By completely skipping the requirement for culturing and modifying the embryonic stem cells, this technique for rodent model development can significantly reduce the generation time. In the chapter by Takehito Kaneko and Tomoji Mashimo, they detail the process of creating genetically modified rodent models via direct embryo injection of ZFNs, TALENs, and/or CRISPRs. With the advent of numerous, novel animal models, it is important to also develop techniques to efficiently and effectively preserve them. Takehito Kaneko contributes a chapter on conserving animal strains via sperm preservation via freeze-drying.

As a whole, this edition of *Chromosomal Mutagenesis* includes detailed protocols, case studies, and reviews from thought-leaders in the field. It is my hope that this work will speed scientific discovery and aid in the next advances in the field. I want to thank each of the authors for their intellectual contributions as well as their enthusiasm in providing each chapter. I also want to thank John Walker at the University of Hertfordshire for his help and support throughout the compilation of this work.

St. Louis, MO, USA *Shondra M. Pruett-Miller*

Contents

Contributors

VICTORIA M. BEDELL • *Department of Biochemistry and Molecular Biology, Mayo Clinic, Rochester, MN, USA*
SANDRINE BOISSEL • *Center for Immunity and Immunotherapies, Seattle Children's Research Institute, Seattle, WA, USA*
MICHELE P. CALOS • *Department of Genetics, School of Medicine, Stanford University, Stanford, CA, USA*
DANA CARROLL • *Department of Biochemistry, School of Medicine, University of Utah, Salt Lake City, UT, USA*
TOMAS CERMAK • *Department of Genetics, Cell Biology & Development and Center for Genome Engineering, University of Minnesota, Minneapolis, MN, USA*
FUQIANG CHEN • *Sigma-Aldrich Biotechnology, St. Louis, MO, USA*
KUNITOSHI CHIBA • *Department of Molecular and Cell Biology, University of California, Berkeley, CA, USA*
MICHAEL CHOI • *Division of Basic Sciences, Fred Hutchinson Cancer Research Center, Seattle, WA, USA*
LE CONG • *Broad Institute of MIT and Harvard, Cambridge, MA, USA; McGovern Institute for Brain Research, Department of Brain and Cognitive Sciences, Department of Biological Engineering, Massachusetts Institute of Technology, Cambridge, MA, USA*
GREGORY D. DAVIS • *Sigma-Aldrich Biotechnology, St. Louis, MO, USA*
STEPHEN C. EKKER • *Department of Biochemistry and Molecular Biology, Mayo Clinic, Rochester, MN, USA*
JOHN FETTER • *Cell-Based Assays/Reporter Cell Lines, Sigma-Aldrich Research Biotech, St. Louis, MO, USA*
JONATHAN M. GEISINGER • *Department of Genetics, School of Medicine, Stanford University, Stanford, CA, USA*
PETER M. GLAZER • *Department of Therapeutic Radiology, School of Medicine, Yale University, New Haven, CT, USA; Department of Genetics, School of Medicine Yale University, New Haven, CT, USA*
TONY GUTSCHNER • *Department of Genomic Medicine, The University of Texas M.D. Anderson Cancer Center, Houston, TX, USA*
DIRK HOCKEMEYER • *Department of Molecular and Cell Biology, University of California, Berkeley, CA, USA*
VEIT HORNUNG • *Institute of Molecular Medicine, University Hospital, University of Bonn, Bonn, Germany*
ROB HOWES • *MedImmune, Cambridge, UK*
TAKEHITO KANEKO • *Institute of Laboratory Animals, Graduate School of Medicine, Kyoto University, Kyoto, Japan*

DMITRY MALKOV • *Cell-Based Assays/Reporter Cell Lines, Sigma-Aldrich Research Biotech, St. Louis, MO, USA*
TOMOJI MASHIMO • *Institute of Laboratory Animals, Graduate School of Medicine, Kyoto University, Kyoto, Japan*
STEPHEN J. PETTITT • *Breakthrough Breast Cancer Research Centre and CRUK Gene Function Laboratory, Institute of Cancer Research, London, UK*
MATTHEW PORTEUS • *Department of Pediatrics, Stanford Medical School, Stanford, CA, USA*
SHONDRA M. PRUETT-MILLER • *Genome Engineering and iPSC Center, Washington University, St. Louis, MO, USA*
FAISAL REZA • *Department of Therapeutic Radiology, School of Medicine, Yale University, New Haven, CT, USA*
ANDREY SAMSONOV • *Cell-Based Assays/Reporter Cell Lines, Sigma-Aldrich Research Biotech, St. Louis, MO, USA*
ANDREW M. SCHARENBERG • *Department of Immunology, Center for Immunity and Immunotherapies, Seattle Children's Research Institute, University of Washington, Seattle, WA, USA*
JONATHAN L. SCHMID-BURGK • *Institute of Molecular Medicine, University Hospital, University of Bonn, Bonn, Germany*
TOBIAS SCHMIDT • *Institute of Molecular Medicine, University Hospital, University of Bonn, Bonn, Germany*
CHRISTINE SCHOFIELD • *Horizon Discovery Ltd., Waterbeach, Cambridge, UK*
COLBY G. STARKER • *Department of Genetics, Cell Biology & Development and Center for Genome Engineering, University of Minnesota, Minneapolis, MN, USA*
BARRY L. STODDARD • *Division of Basic Sciences, Fred Hutchinson Cancer Research Center, Seattle, WA, USA*
RYO TAKEUCHI • *Division of Basic Sciences, Fred Hutchinson Cancer Research Center, Seattle, WA, USA*
E-PIEN TAN • *Wellcome Trust Sanger Institute, Cambridge, UK*
DANIEL F. VOYTAS • *Department of Genetics, Cell Biology & Development and Center for Genome Engineering, University of Minnesota, Minneapolis, MN, USA*
KOSUKE YUSA • *Wellcome Trust Sanger Institute, Cambridge, UK*
NATHAN ZENSER • *Cell-Based Assays/Reporter Cell Lines, Sigma-Aldrich Research Biotech, St. Louis, MO, USA*
FAN ZHANG • *Cell-Based Assays/Reporter Cell Lines, Sigma-Aldrich Research Biotech, St. Louis, MO, USA*
FENG ZHANG • *Broad Institute of MIT and Harvard, Cambridge, MA, USA; McGovern Institute for Brain Research, Department of Brain and Cognitive Sciences, Department of Biological Engineering, Massachusetts Institute of Technology, Cambridge, MA, USA*
HONGYI ZHANG • *Cell-Based Assays/Reporter Cell Lines, Sigma-Aldrich Research Biotech, St. Louis, MO, USA*

Chapter 1

Genome Editing by Targeted Chromosomal Mutagenesis

Dana Carroll

Abstract

The tools for genome engineering have become very powerful and accessible over the last several years. CRISPR/Cas nucleases, TALENs and ZFNs can all be designed to produce highly specific double-strand breaks in chromosomal DNA. These breaks are processed by cellular DNA repair machinery leading to localized mutations and to intentional sequence replacements. Because these repair processes are common to essentially all organisms, the targetable nucleases have been applied successfully to a wide range of animals, plants, and cultured cells. In each case, the mode of delivery of the nuclease, the efficiency of cleavage and the repair outcome depend on the biology of the particular system being addressed. These reagents are being used to introduce favorable characteristics into organisms of economic significance, and the prospects for enhancing human gene therapy appear very bright.

Key words Genome engineering, Zinc-finger nucleases (ZFNs), Transcription activator-like effector nucleases (TALENs), CRISPR/Cas nucleases, DNA repair, Nonhomologous end joining (NHEJ), Homologous recombination

1 Introduction

To a large extent, progress in science travels on the back of technological advances. The invention of the microscope opened up a previously invisible world. The invention of the bubble chamber allowed physicists to detect subatomic particles. In the realm of molecular biology, where would we be without restriction enzymes, DNA cloning, DNA sequencing, high-speed computing? The chapters in this volume of Chromosomal Mutagenesis reflect the rapid development of new tools for making designed and highly specific changes in the genomes of a wide variety of cells and organisms. This progress has, in turn, facilitated the detailed analysis of gene function in model organisms, and it promises to speed the production of improved traits in livestock and crop plants. Applications to human therapy have been initiated as well.

Comparison of this volume with its predecessor, now only 6 years old, shows how rapidly targetable nucleases have come to dominate the field of genome engineering. In this introductory

Shondra M. Pruett-Miller (ed.), *Chromosomal Mutagenesis*, Methods in Molecular Biology, vol. 1239,
DOI 10.1007/978-1-4939-1862-1_1, © Springer Science+Business Media New York 2015

chapter, I review the history of these reagents and describe some of the systems to which they have been applied. A number of more detailed recent reviews are also available [1–4].

2 The Nucleases

2.1 Meganucleases

Double-strand breaks (DSBs) in chromosomal DNA stimulate homologous recombination (HR) and inaccurate repair by nonhomologous end joining (NHEJ). Important experiments demonstrating that these processes occur at unique DSBs were performed with the meganucleases, HO and I-*Sce*I. These are natural proteins from yeast that have long recognition sites—24 bp and 18 bp, respectively. When their sites were introduced at unique locations in the genomes of yeast [5, 6] and mammalian cells [7, 8], the breaks induced upon expression of the corresponding nucleases greatly stimulated these repair processes. Other enzymes in the large family of meganucleases (also called homing endonucleases) have been engineered to recognize and cleave sequences other than their natural targets [9], but most of the success in nuclease-based targeting has been obtained with novel reagents that have DNA-recognition and DNA-cleavage domains that can be manipulated separately.

2.2 Zinc-Finger Nucleases

Cys_2His_2 zinc fingers (ZFs) are DNA-binding modules found in sequence-specific transcription factors in all eukaryotic organisms [10]. Two cysteine and two histidine residues coordinate a single zinc atom in each finger. Amino acid side chains at specific positions make direct contact with 3 (or 4) DNA base pairs in the major groove; different amino acids specify different base pairs. It takes at least three fingers to provide sufficient energy for stable association with DNA, and adding more fingers increases affinity.

The *Fok*I restriction endonuclease recognizes the sequence 5′-GGATG-3′ and cuts 9 and 13 base pairs away on the two DNA strands, regardless of the actual sequence at that position, leaving a 4-nt 5′ overhang. Chandrasegaran and colleagues demonstrated that the binding and cleavage activities could be physically isolated in separate domains of the protein [11]. They also showed that cleavage specificity could be altered by fusing the nonspecific cleavage domain to alternative DNA-recognition domains, including ZFs [12, 13]. The cleavage domain must dimerize to be active, and because the dimer interface is weak, two sets of fingers are required, each linked to a monomer of the cleavage domain [14]. The ZF-*Fok*I fusions were originally called chimeric restriction enzymes, but they are now called zinc-finger nucleases (ZFNs).

To change the binding specificity of a ZFN monomer is, in principle, a simple matter of changing the identity of the fingers to bind new DNA triplets. The first novel ZFNs were produced by assembling new combinations of fingers identified as being specific for particular triplets in other contexts. In practice, however, only a minority of such constructs work well, due to influences of neighboring fingers and/or local DNA configuration [15–18]. Schemes have been devised to select new ZF combinations for new targets and to identify fingers that work well together [19–21], but these can be rather laborious. The best archives of ZFs and finger combinations are held by Sangamo Biosciences and marketed through Sigma-Aldrich.

Early experiments showed that ZFNs were capable of finding and cleaving their targets in various types of eukaryotic cells [22–25], and completely novel ZF combinations led to the mutagenesis and replacement of endogenous genomic sequences in whole organisms [26–28]. The range of organisms that have now been targeted with ZFNs is quite impressive [1, 4, 29].

2.3 TALENs

Transcription activator-like effectors (TALEs) are proteins produced by plant-pathogenic bacteria to regulate host gene expression [30]. The DNA-recognition domain of each TALE is made up of repeating modules of about 34 amino acids. Two of these amino acids, called repeat-variable di-residues (RVDs) vary in concert with the sequence of the target DNA, and two groups discovered that there is, in fact, a simple recognition code, albeit with some variants [31, 32]. In the one-letter amino acid shorthand, the RVD NI recognizes A, HD recognizes C, NG recognizes T, and NN is used for G, but can also bind A. Once this code was established, the ZFN precedent made it clear how to use these modules in specific nucleases by linkage to the *Fok*I cleavage domain [33].

Because each TALE module recognizes a single base pair, designing TALENs for new targets is quite straightforward. Constructing them presents a small challenge: the framework sequence of each module is essentially the same, so simple PCR approaches lead to deletions during construction. A number of alternative procedures have been developed, and all of them seem to work [34–39].

For most purposes the four standard TALE modules work well, but some situations benefit from alternatives. The modules with NK or NH in the recognition positions are more specific for G, but they provide less binding energy than NN [40–42]. They can be used to improve discrimination against related sequences, but inserting more than one or two in a TALEN monomer can significantly reduce overall affinity. In addition, the standard modules contribute quite differently to the overall binding strength [41, 42]. Methylation of C on carbon 5 interferes with recognition by the standard HD module. Replacement with N* (where * indicates deletion of the amino acid at position 13) effectively eliminates this inhibition [43], but also provides less affinity.

2.4 CRISPR/Cas Nucleases

Like the components of ZFNs and TALENs, the most recent targetable nucleases came from an unexpected source. Many bacteria and archaea carry a system of adaptive immunity against invading genomes (viruses, plasmids) based on RNA-guided DNA cleavage [44]. The cumbersome full name of the components of this system, clustered regularly interspaced short palindromic repeats, is euphoniously abbreviated as CRISPR. Among several variants, the Type II CRISPR system depends on a single protein, Cas9, and two short RNAs to execute DNA cleavage. One of these RNAs, the crRNA, selects the cleavage target by base pairing to it. For purposes of genome engineering, the two RNAs are commonly fused into one, called a guide RNA (gRNA) or single guide RNA (sgRNA) [45].

To accomplish targeted cleavage, one needs simply to express the Cas9 protein and deliver the gRNA to the cell or organism of interest. CRISPR nucleases have two major advantages over ZFNs and TALENs: (1) only a single, constant protein is needed, so once its delivery has been engineered, it can be used repeatedly; and (2) multiple gRNAs can be delivered, simultaneously or in parallel, to target multiple genes. gRNAs typically have 20 nucleotides of homology to the target sequence, although somewhat shorter and longer matches have been used. In addition, Cas9 requires a short sequence in the target just outside the RNA-DNA hybrid regions, called the protospacer adjacent motif (PAM) [44]. For the commonly used Cas9 from *Streptococcus pyogenes*, the PAM is 5′-NGG, where N is any base; 5′-NAG is also recognized, but with lower efficiency of cleavage [46–48]. Cas9 from other species require different PAMs, and the range of targetable sequences grows as more are characterized [49]. The CRISPR reagents are so new on the scene that optimizing their design and use is still very much under way.

3 Consequences of Targeted Double-Strand Breaks

3.1 Double-Strand Break Repair

A break across both strands of chromosomal DNA constitutes potentially lethal damage to cells. Whole segments of the genome may be lost by degradation, mis-segregation, or chromosomal rearrangements. Therefore, cells have dedicated mechanisms to restore the integrity of their chromosomes, and these are typically assigned to one of two categories: homology-dependent and homology-independent repair [50]. In the former, intact sequences matching those surrounding the break are used as a template for repair. In normal circumstances, the source of the template is a sister chromatid or homologous chromosome. In homology-independent processes, the broken ends are stuck back together, often inaccurately, with no instructions from a template.

The targetable nucleases take advantage of both processes by introducing a break at a specific site, which is then repaired by homologous recombination (HR) with a template provided by the experimenter, or by nonhomologous end joining (NHEJ) leading to localized sequence changes. In many cell types, NHEJ appears to dominate the choice of repair pathway, so introduction of small insertions and deletions (indels) at the cut site occurs more frequently than use of a designed donor. (I have qualified the statement about NHEJ dominance because some repair processes—HR with the sister chromatid, accurate re-ligation—restore the original sequence at the target and are experimentally undetectable.) NHEJ is sufficient when a gene knockout is the desired outcome. Introduction of designed sequence changes from a donor DNA relies on HR. Disabling DNA ligase IV, a key component of the major NHEJ pathway, makes the balance more favorable for HR [51, 52], but this is not possible in most situations. Instead, one can design the donor DNA to ensure the best outcome.

3.2 Donor Design

Both long, double-stranded and short, single-stranded DNAs have been used successfully as donors. Long DNAs need to have a total of 1–2 kb of homology with the target, including sequences on both sides of the break [23, 53]. This requirement is less strict than what has been observed for "classical" gene targeting in cultured mammalian cells [54], presumably reflecting a difference in mechanism, as well as efficiency, when the target is activated with a break. Long donors can carry subtle sequence changes or gene-sized insertions or deletions relative to the natural locus. Both linear and circular donors are used. Circles are clearly preferable in some situations [51], and the choice depends on the cells and mode of delivery. It should be noted that DNAs delivered to cultured cells may have an altered configuration by the time they reach the nucleus [55].

A further issue concerns conversion tracts—i.e., how far from the nuclease-induced break are sequences incorporated from the donor. In cultured mammalian cells, these tracts are quite short, falling off significantly within 100–200 bp [56]. In Drosophila, the tracts are somewhat longer than this in the larval germ line, but are extend at least several kb in embryonic cells [53]. Clearly this parameter is cell-type specific. It would be useful to know what governs and limits conversion tracts in any of these situations, since long tracts allow introduction of multiple changes over a broad distance using a single nuclease.

Single-stranded oligonucleotide donors with as little as 30–40 bases of homology to the target have been used successfully [53, 57, 58]. Of course, only local sequence changes can be introduced from short donors, with one exception. Oligonucleotides have been designed with one stretch of homology close to the break and another to sequences some distance away. HR with such a donor creates a deletion between the homologies, and this can be as large as tens of kb [57].

3.3 Off-Target Effects

While we are delighted by the efficiencies with which the various nucleases lead to modification of their intended targets, there are concerns regarding discrimination against other genomic sites. In fact, the rest of the genome constitutes a very large pool of potential secondary targets. Cleavage at any site may lead to NHEJ mutagenesis, and such unwanted mutations could endanger the health of cells and organisms and confound analyses of gene function.

The first ZFNs for a genomic target showed significant lethality [27] due to excessive cleavage [59], and early experiments in mammalian cells also revealed some toxicity [22]. Often this toxicity is a property of one ZFN in a pair, presumably due to homodimerization at secondary sites. The situation was greatly improved by amino acid substitutions in the *Fok*I dimer interface that prevent homodimerization, while allowing the required heterodimers to form [60, 61]. The first-generation modifications displayed reduced cleavage efficiency in some contexts [62], and this was alleviated by second-generation adjustments [63].

TALENs appear to be inherently more specific, even with the wild-type cleavage domain. This may reflect the fact that, unlike the case with zinc fingers, there is considerable interaction between consecutive TALE modules when bound to DNA [64, 65]. Mismatch of one module may disrupt neighboring interactions. As a precaution, the obligate heterodimer cleavage domain is often used in TALEN designs.

Much noise was made recently regarding off-target cleavage by CRISPR nucleases [46–48, 66–68]. At the same time, solutions that improve specificity were rapidly found [48, 69, 70]. Although a 22-bp sequence (20 nt in the gRNA, plus the 2-bp PAM) should be unique in pretty much any genome, Cas9 tolerates some mismatches between the gRNA and target. This may, in fact, be adaptive for the organisms relying on CRISPR systems for defense. The next virus infecting a cell could be related, but not identical to the one that established the integrated sequence.

An effective approach to reducing off target cleavage by Cas9 has been to insist on simultaneous recognition with two gRNAs directed to sequences near each other on opposite DNA strands. One nuclease site in Cas9 is mutated, so it becomes a nicking enzyme. When nicks directed by the paired gRNAs are within about 100 bp of each other, both NHEJ mutagenesis and HR with an oligonucleotide donor proceed with an efficiency similar to that stimulated by a frank DSB [48, 69, 70]. Another approach has been to shorten the gRNA slightly, on the theory that a single mismatch will be more destabilizing in a shorter duplex [71]. As long as 17–19 nt still match, the on-target efficiency remains robust.

4 Applications of Targetable Nucleases

4.1 Targeting in a Range of Organisms

By now the genomes of approximately 50 different organisms have been successfully modified with the targetable nucleases, and the number keeps growing. It is not my intent to cite all of the many applications; they can be found in various other reviews [1–4] and in the primary literature.

ZFNs were first applied to several model organisms, including Drosophila [26, 27], *C. elegans* [28], Arabidopsis [72], and zebrafish [73, 74]. Early experiments in human cells showed that ZFNs were effective there [24, 22, 23]. Excitement was generated in the rat research community by the demonstration that ZFNs delivered to one-cell embryos were effective in generating targeted mutations [75]. The gene targeting approach devised for mice [76] was not useful in its sister organism, due to the absence of robust embryonic stem cells.

It didn't take long after the description of the TALE recognition code for the corresponding nucleases to be applied to a range of cells and organisms. TALENs were more reliably designed for new targets than ZFNs, and the efficiency of cleavage was typically higher [3]. The adoption of the CRISPR nucleases has been stunning, and for good reason. As described above, they are very easy to design and highly effective. The ability to multiplex their targets by simple gRNA design has led to applications not conceivable for ZFNs or TALENs [77–79].

Some of the most promising applications of the targetable nucleases are to crop plants, livestock, and humans. Early experiments were performed in tobacco [25, 80] and maize [81], species in which whole plants can be regenerated from engineered cells or calli. Quite a number of additional crops have been targeted with TALENs and CRISPRs [82–89]. In the realm of food animals, both embryo injection and somatic cell nuclear transfer have been used to generate stocks with targeted genome alterations [90–92].

The targetable nucleases are obviously well suited to making desirable changes in the human genome, including the correction of disease mutations. Correcting a gene at its normal locus is preferable to delivering a therapeutic gene to essentially random sites with viral or transposon vectors. Not only is insertional mutagenesis avoided, but all the natural regulatory features are intact. Uses of this sort have been demonstrated in cultured cells and in animal models [see reviews for examples [1, 4, 93]]. In situations in which disease mutations are distributed throughout a particular gene, a cDNA carrying all downstream exons can be inserted into a site near the 5′ end of the gene [94, 95].

The most accessible applications in humans are ones in which the genome modification can be done ex vivo. Nuclease efficacy has been demonstrated in human embryonic stem cells and in induced pluripotent stem cells with this approach in mind. Any treatment involving blood cell lineages can be envisioned using hematopoietic stem cells from the bone marrow [96]. A current clinical trial uses ZFNs to knock out the CCR5 gene in T cells [93] as a means to protect them from HIV-1 infection. After ex vivo mutagenesis, the cells are infused back into the patient that originally donated them, completing an autologous transplant.

4.2 Delivery of the Reagents

The case of human applications demonstrates an aspect of the targetable nucleases that affects many systems—i.e., delivery of the nucleases and donor DNA. Taking human disease as an example, only a limited number of conditions can reasonably be treated ex vivo. Delivery of the necessary components to intact people is quite challenging. A great deal of research has gone into delivery of therapeutic genes with viral vectors or with vector-free methods, but not all cells and organs can be accessed. As an example of the challenges, consider cystic fibrosis, a very common genetic disease. First, many organs are affected, most seriously the lungs, pancreas, and the male reproductive system. How can nucleases and donors be administered to all of these? Can therapeutically useful doses be delivered to endothelial cells deep in the alveoli of the lung?

Delivery is an issue that affects model organisms as well, and each experimental system requires its own solutions. For a number of years, nuclease mutagenesis in the popular nematode, *C. elegans*, had only been achieved in somatic cells [28]. Germ line modification was thought to be restricted by potent RNA interference. More recently, both TALENs and CRISPRs have proved quite effective in the germ line, with delivery by injection of DNA, mRNA, or proteins [97–106]. In zebrafish, targeted mutagenesis has been achieved with all of the nuclease platforms using mRNA injection into embryos [107, 108], but HR has been more recalcitrant [109, 110]. It is likely that this reflects the developmental stage of the organism. After fertilization, the embryo undergoes very rapid cell divisions, with little or no gene expression, until mid-blastula. During this period, there is little time to wait for an orderly process of repair by HR, so the more immediate solution of NHEJ is preferred, despite the inherent dangers. In all these cases, we are dependent on the biology of the system for accommodation of the reagents and processing of the breaks.

5 Concluding Remarks

Progress in genome engineering with targetable nucleases has been quite amazing. In comparing the various platforms, I think it is fair to say that TALENs and CRISPRs have replaced ZFNs as the choice for routine laboratory use. This is based on their ease of construction and the reliability of new designs for new targets.

The CRISPR system has a number of distinct advantages, as cited above, but also has some limitations. The requirement for a specific PAM sequence restricts target choice to some extent, albeit modestly. Identification of additional Cas9 proteins with alternative PAMs will help to relieve this constraint [49]. TALENs, in contrast, have very few constraints on target choice. The 5′ base in each binding site is designated to be a T, but even this rule has been violated without inactivating the nuclease [111]. Even with this limitation, the number of bases in each binding site and in the spacer between them can be varied over considerable ranges to accommodate the initial T.

Future applications of the nuclease technology may be limited only by the imagination and insight of researchers. Clearly human gene therapy with these reagents will be expanded. Precise modification of food sources, both plants and animals, should find more ready acceptance, both by the public and by regulatory agencies, since in most cases, no foreign genetic material is being introduced. In the pharmaceutical realm, strains of animals—rats and zebrafish come immediately to mind—can be engineered with human disease mutations or with sensitizing mutations to enhance drug screening. It's a whole new world, and it will be great fun to watch how it develops.

Acknowledgments

I am grateful to people who have worked in my lab on the targetable nucleases over more than 18 years, and to our many collaborators. Work in my lab has been supported by research grants from the National Institutes of Health, most recently R01GM078571, and by the University of Utah Cancer Center Support Grant.

References

1. Carroll D (2014) Genome engineering with targetable nucleases. Annu Rev Biochem 82: 409–439
2. Gaj T, Gersbach CA, Barbas CF 3rd (2013) ZFN, TALEN, and CRISPR/Cas-based methods for genome engineering. Trends Biotechnol 31:397–405
3. Joung JK, Sander JD (2013) TALENs: a widely applicable technology for targeted genome editing. Nat Rev Mol Cell Biol 14:49–55
4. Segal DJ, Meckler JF (2013) Genome engineering at the dawn of the golden age. Annu Rev Genomics Hum Genet 14:135–158
5. Plessis A, Perrin A, Haber JE et al (1992) Site-specific recombination determined by I-*SceI*, a mitochondrial group I intron-encoded endonuclease expressed in the yeast nucleus. Genetics 130:451–460
6. Rudin N, Sugarman E, Haber JE (1989) Genetic and physical analysis of double-strand break repair and recombination in *Saccharomyces cerevisiae*. Genetics 122:519–534
7. Choulika A, Perrin A, Dujon B et al (1995) Induction of homologous recombination in mammalian chromosomes by using the I-*Sce*I system of *Saccharomyces cerevisiae*. Mol Cell Biol 15:1968–1973
8. Rouet P, Smih F, Jasin M (1994) Introduction of double-strand breaks into the genome of mouse cells by expression of a rare-cutting endonuclease. Mol Cell Biol 14: 8096–8106

9. Stoddard BL (2011) Homing endonucleases: from microbial genetic invaders to reagents for targeted DNA modifications. Structure 19:7–15
10. Pabo CO, Peisach E, Grant RA (2001) Design and selection of novel Cys_2His_2 zinc finger proteins. Annu Rev Biochem 70:313–340
11. Li L, Wu LP, Chandrasegaran S (1992) Functional domains in *Fok*I restriction endonuclease. Proc Natl Acad Sci U S A 89:4275–4279
12. Kim Y-G, Cha J, Chandrasegaran S (1996) Hybrid restriction enzymes: zinc finger fusions to *Fok*I cleavage domain. Proc Natl Acad Sci U S A 93:1156–1160
13. Kim Y-G, Chandrasegaran S (1994) Chimeric restriction endonuclease. Proc Natl Acad Sci U S A 91:883–887
14. Smith J, Bibikova M, Whitby FG et al (2000) Requirements for double-strand cleavage by chimeric restriction enzymes with zinc finger DNA-recognition domains. Nucleic Acids Res 28:3361–3369
15. Carroll D, Morton JJ, Beumer KJ et al (2006) Design, construction and in vitro testing of zinc finger nucleases. Nat Protoc 1:1329–1341
16. Kim JS, Lee HJ, Carroll D (2010) Genome editing with modularly assembled zinc-finger nucleases. Nat Methods 7:91
17. Ramirez CL, Foley JE, Wright DA et al (2008) Unexpected failure rates for modular assembly of engineered zinc fingers. Nat Methods 5:374–375
18. Segal DJ, Beerli RR, Blancafort P et al (2003) Evaluation of a modular strategy for the construction of novel polydactyl zinc finger DNA-binding proteins. Biochemistry 42: 2137–2148
19. Gupta A, Christensen RG, Rayla AL et al (2012) An optimized two-finger archive for ZFN-mediated gene targeting. Nat Methods 9:588–590
20. Maeder ML, Thibodeau-Beganny S, Osiak A et al (2008) Rapid "Open-Source" engineering of customized zinc-finger nucleases for highly efficient gene modification. Mol Cell 31:294–301
21. Sander JD, Dahlborg EJ, Goodwin MJ et al (2011) Selection-free zinc-finger-nuclease engineering by context-dependent assembly (CoDA). Nat Methods 8:67–69
22. Porteus MH, Baltimore D (2003) Chimeric nucleases stimulate gene targeting in human cells. Science 300:763
23. Urnov FD, Miller JC, Lee Y-L et al (2005) Highly efficient endogenous gene correction using designed zinc-finger nucleases. Nature 435:646–651
24. Alwin S, Gere MB, Gulh E et al (2005) Custom zinc-finger nucleases for use in human cells. Mol Ther 12:610–617
25. Wright DA, Townsend JA, Winfrey RJ Jr et al (2005) High-frequency homologous recombination in plants mediated by zinc-finger nucleases. Plant J 44:693–705
26. Bibikova M, Beumer K, Trautman JK et al (2003) Enhancing gene targeting with designed zinc finger nucleases. Science 300:764
27. Bibikova M, Golic M, Golic KG et al (2002) Targeted chromosomal cleavage and mutagenesis in *Drosophila* using zinc-finger nucleases. Genetics 161:1169–1175
28. Morton J, Davis MW, Jorgensen EM et al (2006) Induction and repair of zinc-finger nuclease-targeted double-strand breaks in *Caenorhabditis elegans* somatic cells. Proc Natl Acad Sci U S A 103:16370–16375
29. Carroll D (2011) Genome engineering with zinc-finger nucleases. Genetics 188:773–782
30. Bogdanove AJ, Voytas DF (2011) TAL effectors: customizable proteins for DNA targeting. Science 333:1843–1846
31. Boch J, Scholze H, Schornack S et al (2009) Breaking the code of DNA binding specificity of TAL-Type III effectors. Science 326: 1509–1512
32. Moscou MJ, Bogdanove AJ (2009) A simple cipher governs DNA recognition by TAL effectors. Science 326:1501
33. Christian M, Cermak T, Doyle EL et al (2010) Targeting DNA double-strand breaks with TAL effector nucleases. Genetics 186:757–761
34. Briggs AW, Rios X, Chari R et al (2012) Iterative capped assembly: rapid and scalable synthesis of repeat-module DNA such as TAL effectors from individual monomers. Nucleic Acids Res 40:e117
35. Cermak T, Doyle EL, Christian M et al (2011) Efficient design and assembly of custom TALEN and other TAL effector-based constructs for DNA targeting. Nucleic Acids Res 39:e82
36. Reyon D, Khayter C, Regan MR et al (2012) Engineering designer transcription activator-like effector nucleases (TALENs) by REAL or REAL-Fast assembly. Curr Protoc Mol Biol Chapter 12, Unit 12 15
37. Reyon D, Tsai SQ, Khayter C et al (2012) FLASH assembly of TALENs for high-throughput genome editing. Nat Biotechnol 30:460–465
38. Sanjana NE, Cong L, Zhou Y et al (2012) A transcription activator-like effector toolbox for genome engineering. Nat Protoc 7:171–192

39. Schmid-Burgk JL, Schmidt T, Kaiser V et al (2013) A ligation-independent cloning technique for high-throughput assembly of transcription activator-like effector genes. Nat Biotechnol 31:76–81
40. Christian ML, Demorest ZL, Starker CG et al (2012) Targeting G with TAL effectors: a comparison of activities of TALENs constructed with NN and NK repeat variable diresidues. PLoS One 7:e45383
41. Meckler JF, Bhakta MS, Kim MS et al (2013) Quantitative analysis of TALE-DNA interactions suggests polarity effects. Nucleic Acids Res 41:4118–4128
42. Streubel J, Blucher C, Landgraf A et al (2012) TAL effector RVD specificities and efficiencies. Nat Biotechnol 30:593–595
43. Valton J, Dupuy A, Daboussi F et al (2012) Overcoming transcription activator-like effector (TALE) DNA binding domain sensitivity to cytosine methylation. J Biol Chem 287:38427–38432
44. Wiedenheft B, Sternberg SH, Doudna JA (2012) RNA-guided genetic silencing systems in bacteria and archaea. Nature 482:331–338
45. Jinek M, Chylinski K, Fonfara I et al (2012) A programmable dual-RNA-guided DNA endonuclease in adaptive bacterial immunity. Science 337:816–821
46. Hsu PD, Scott DA, Weinstein JA et al (2013) DNA targeting specificity of RNA-guided Cas9 nucleases. Nat Biotechnol 31:827–832
47. Jiang W, Bikard D, Cox D et al (2013) RNA-guided editing of bacterial genomes using CRISPR-Cas systems. Nat Biotechnol 31:233–239
48. Mali P, Aach J, Stranges PB et al (2013) CAS9 transcriptional activators for target specificity screening and paired nickases for cooperative genome engineering. Nat Biotechnol 31: 833–838
49. Esvelt KM, Mali P, Braff JL et al (2013) Orthogonal Cas9 proteins for RNA-guided gene regulation and editing. Nat Methods 10:1116–1121
50. Chapman JR, Taylor MR, Boulton SJ (2012) Playing the end game: DNA double-strand break repair pathway choice. Mol Cell 47:497–510
51. Beumer KJ, Trautman JK, Bozas A et al (2008) Efficient gene targeting in Drosophila by direct embryo injection with zinc-finger nucleases. Proc Natl Acad Sci U S A 105:19821–19826
52. Bozas A, Beumer KJ, Trautman JK et al (2009) Genetic analysis of zinc-finger nuclease-induced gene targeting in Drosophila. Genetics 182:641–651
53. Beumer KJ, Trautman JK, Mukherjee K et al (2013) Donor DNA utilization during gene targeting with zinc-finger nucleases. G3 (Bethesda) 3:657–664
54. Deng C, Capecchi MR (1992) Reexamination of gene targeting frequency as a function of the extent of homology between the targeting vector and the target locus. Mol Cell Biol 12:3365–3371
55. Wake CT, Vernaleone F, Wilson JH (1985) Topological requirements for homologous recombination among DNA molecules transfected into mammalian cells. Mol Cell Biol 5:2080–2089
56. Elliott B, Richardson C, Winderbaum J et al (1998) Gene conversion tracts from double-strand break repair in mammalian cells. Mol Cell Biol 18:93–101
57. Chen F, Pruett-Miller SM, Huang Y et al (2011) High-frequency genome editing using ssDNA oligonucleotides with zinc-finger nucleases. Nat Methods 8:753–755
58. Radecke S, Radecke F, Cathomen T et al (2010) Zinc-finger nuclease-induced gene repair with oligodeoxynucleotides: wanted and unwanted target locus modifications. Mol Ther 18:743–753
59. Beumer K, Bhattacharyya G, Bibikova M et al (2006) Efficient gene targeting in Drosophila with zinc-finger nucleases. Genetics 172:2391–2403
60. Miller JC, Holmes MC, Wang J et al (2007) An improved zinc-finger nuclease architecture for highly specific genome cleavage. Nat Biotechnol 25:778–785
61. Szczepek M, Brondani V, Buchel J et al (2007) Structure-based redesign of the dimerization interface reduces the toxicity of zinc-finger nucleases. Nat Biotechnol 25:786–793
62. Beumer KJ, Trautman JK, Christian M et al (2013) Comparing ZFNs and TALENs for gene targeting in Drosophila. G3 (Bethesda) 3:1717–1725
63. Doyon Y, Vo TD, Mendel MC et al (2011) Enhancing zinc-finger-nuclease activity with improved obligate heterodimer architectures. Nat Methods 8:74–79
64. Deng D, Yan C, Pan X et al (2012) Structural basis for sequence-specific recognition of DNA by TAL effectors. Science 335:720–723
65. Mak AN-S, Bradley P, Cernadas RA et al (2012) The crystal structure of TAL effector PthXo1 bound to its DNA target. Science 335:716–719
66. Cradick TJ, Fine EJ, Antico CJ et al (2013) CRISPR/Cas9 systems targeting beta-globin and CCR5 genes have substantial off-target activity. Nucleic Acids Res 41:9584–9592

67. Fu Y, Foden JA, Khayter C et al (2013) High-frequency off-target mutagenesis induced by CRISPR-Cas nucleases in human cells. Nat Biotechnol 31:822–826

68. Pattanayak V, Lin S, Guilinger JP et al (2013) High-throughput profiling of off-target DNA cleavage reveals RNA-programmed Cas9 nuclease specificity. Nat Biotechnol 31: 839–843

69. Cho SW, Kim S, Kim Y et al (2013) Analysis of off-target effects of CRISPR/Cas-derived RNA-guided endonucleases and nickases. Genome Res 24:132–141

70. Ran FA, Hsu PD, Lin CY et al (2013) Double nicking by RNA-guided CRISPR Cas9 for enhanced genome editing specificity. Cell 154:1380–1389

71. Fu Y, Sander JD, Reyon D et al (2014) Improving CRISPR-Cas nuclease specificity using a modified guide RNA architecture. Nat Biotechnol 32:279–284

72. Lloyd A, Plaisier CL, Carroll D et al (2005) Targeted mutagenesis using zinc-finger nucleases in *Arabidopsis*. Proc Natl Acad Sci U S A 102:2232–2237

73. Doyon Y, MaCammon JM, Miller JC et al (2008) Heritable targeted gene disruption in zebrafish using designed zinc-finger nucleases. Nat Biotechnol 26:702–708

74. Meng X, Noyes MB, Zhu LJ et al (2008) Targeted gene inactivation in zebrafish using engineered zinc-finger nucleases. Nat Biotechnol 26:695–701

75. Geurts AM, Cost GJ, Freyvert Y et al (2009) Knockout rats via embryo microinjection of zinc-finger nucleases. Science 325:433

76. Capecchi MR (2005) Gene targeting in mice: functional analysis of the mammalian genome for the twenty-first century. Nat Rev Genet 6:507–512

77. Shalem O, Sanjana NE, Hartenian E et al (2013) Genome-scale CRISPR-Cas9 knockout screening in human cells. Science 343: 84–87

78. Wang H, Yang H, Shivalila CS et al (2013) One-step generation of mice carrying mutations in multiple genes by CRISPR/Cas-mediated genome engineering. Cell Rep 153:910–918

79. Wang T, Wei JJ, Sabatini DM et al (2013) Genetic screens in human cells using the CRISPR/Cas9 system. Science 343:80

80. Townsend JA, Wright DA, Winfrey RJ et al (2009) High-frequency modification of plant genes using engineered zinc-finger nucleases. Nature 459:442–445

81. Shukla VK, Doyon Y, Miller JC et al (2009) Precise genome modification in the crop species Zea mays using zinc-finger nucleases. Nature 459:437–441

82. Feng Z, Zhang B, Ding W et al (2013) Efficient genome editing in plants using a CRISPR/Cas system. Cell Res 23:1229–1232

83. Jiang W, Zhou H, Bi H et al (2013) Demonstration of CRISPR/Cas9/sgRNA-mediated targeted gene modification in Arabidopsis, tobacco, sorghum and rice. Nucleic Acids Res 41:e188

84. Li T, Liu B, Spalding MH et al (2012) High-efficiency TALEN-based gene editing produces disease-resistant rice. Nat Biotechnol 30:390–392

85. Mao Y, Zhang H, Xu N et al (2013) Application of the CRISPR-Cas system for efficient genome engineering in plants. Mol Plant 6:2008–2011

86. Miao J, Guo D, Zhang J et al (2013) Targeted mutagenesis in rice using CRISPR-Cas system. Cell Res 23:1233–1236

87. Nekrasov V, Staskawicz B, Weigel D et al (2013) Targeted mutagenesis in the model plant Nicotiana benthamiana using Cas9 RNA-guided endonuclease. Nat Biotechnol 31:691–693

88. Shan Q, Wang Y, Chen K et al (2013) Rapid and efficient gene modification in rice and brachypodium using TALENs. Mol Plant 6:1365–1368

89. Shan Q, Wang Y, Li J et al (2013) Targeted genome modification of crop plants using a CRISPR-Cas system. Nat Biotechnol 31: 686–688

90. Carlson DF, Tan W, Lillico SG et al (2012) Efficient TALEN-mediated gene knockout in livestock. Proc Natl Acad Sci U S A 109: 17382–17387

91. Tan W, Carlson DF, Lancto CA et al (2013) Efficient nonmeiotic allele introgression in livestock using custom endonucleases. Proc Natl Acad Sci U S A 110:16526–16531

92. Tan WS, Carlson DF, Walton MW et al (2012) Precision editing of large animal genomes. Adv Genet 80:37–97

93. Urnov FD, Rebar EJ, Holmes MC et al (2010) Genome editing with engineered zinc finger nucleases. Nat Rev Genet 11:636–646

94. Li H, Haurigot V, Doyon Y et al (2011) In vivo genome editing restores haemostasis in a mouse model of haemophilia. Nature 475: 217–221

95. Lombardo A, Genovese P, Beausejour CM et al (2007) Gene editing in human stem cells

using zinc finger nucleases and integrase-defective lentiviral vector delivery. Nat Biotechnol 25:1298–1306

96. Li L, Krymskaya L, Wang J et al (2013) Genomic editing of the HIV-1 coreceptor CCR5 in adult hematopoietic stem and progenitor cells using zinc finger nucleases. Mol Ther 21:1259–1269
97. Chen C, Fenk LA, de Bono M (2013) Efficient genome editing in Caenorhabditis elegans by CRISPR-targeted homologous recombination. Nucleic Acids Res 41:e193
98. Chiu H, Schwartz HT, Antoshechkin I et al (2013) Transgene-free genome editing in Caenorhabditis elegans using CRISPR-Cas. Genetics 195:1167–1171
99. Cho SW, Lee J, Carroll D et al (2013) Heritable gene knockout in Caenorhabditis elegans by direct injection of Cas9-sgRNA ribonucleoproteins. Genetics 195:1177–1180
100. Dickinson DJ, Ward JD, Reiner DJ et al (2013) Engineering the Caenorhabditis elegans genome using Cas9-triggered homologous recombination. Nat Methods 10:1028–1034
101. Friedland AE, Tzur YB, Esvelt KM et al (2013) Heritable genome editing in C. elegans via a CRISPR-Cas9 system. Nat Methods 10:741–743
102. Katic I, Grosshans H (2013) Targeted heritable mutation and gene conversion by Cas9-CRISPR in Caenorhabditis elegans. Genetics 195:1173–1176
103. Lo TW, Pickle CS, Lin S et al (2013) Heritable genome editing using TALENs and CRISPR/Cas9 to engineer precise insertions and deletions in evolutionarily diverse nematode species. Genetics 195:331–348
104. Tzur YB, Friedland AE, Nadarajan S et al (2013) Heritable custom genomic modifications in Caenorhabditis elegans via a CRISPR-Cas9 system. Genetics 195: 1181–1185
105. Waaijers S, Portegijs V, Kerver J et al (2013) CRISPR/Cas9-targeted mutagenesis in Caenorhabditis elegans. Genetics 195:1187–1191
106. Wood AJ, Lo TW, Zeitler B et al (2011) Targeted genome editing across species using ZFNs and TALENs. Science 333:307
107. Blackburn PR, Campbell JM, Clark KJ et al (2013) The CRISPR system – keeping zebrafish gene targeting fresh. Zebrafish 10: 116–118
108. Kok FO, Gupta A, Lawson ND et al (2014) Construction and application of site-specific artificial nucleases for targeted gene editing. Meth Mol Biol (Clifton, NJ) 1101: 267–303
109. Bedell VM, Wang Y, Campbell JM et al (2012) In vivo genome editing using a high-efficiency TALEN system. Nature 491:114–118
110. Zu Y, Tong X, Wang Z et al (2013) TALEN-mediated precise genome modification by homologous recombination in zebrafish. Nat Methods 10:329–331
111. Miller JC, Tan S, Qiao G et al (2011) A TALE nuclease architecture for efficient genome editing. Nat Biotechnol 29:143–148

Chapter 2

piggyBac Transposon-Based Insertional Mutagenesis in Mouse Haploid Embryonic Stem Cells

Stephen J. Pettitt, E-Pien Tan, and Kosuke Yusa

Abstract

Forward genetic screening is a powerful non-hypothesis-driven approach to unveil the molecular mechanisms and pathways underlying phenotypes of interest. In this approach, a genome-wide mutant library is first generated and then screened for a phenotype of interest. Subsequently, genes responsible for the phenotype are identified. There have been a number of successful screens in yeasts, *Caenorhabditis elegans* and *Drosophila*. These model organisms all allow loss-of-function mutants to be generated easily on a genome-wide scale: yeasts have a haploid stage in their reproductive cycles and the latter two organisms have short generation times, allowing mutations to be systematically bred to homozygosity. However, in mammals, the diploid genome and long generation time have always hampered rapid and efficient production of homozygous mutant cells and animals. The recent discovery of several haploid mammalian cell lines promises to revolutionize recessive genetic screens in mammalian cells. In this protocol, we describe an overview of insertional mutagenesis, focusing on DNA transposons, and provide a method for an efficient generation of genome-wide mutant libraries using mouse haploid embryonic stem cells.

Key words DNA transposon, *PiggyBac*, Haploid, Mouse ES cells, Genetic screen, Insertional mutagenesis, Forward genetics

1 Introduction

1.1 DNA Transposons

DNA Transposons are mobile genetic elements that can excise from their original position and reintegrate elsewhere in the genome. Transposons comprise two essential components: DNA flanked by terminal inverted repeats (TIRs) and a transposase enzyme that recognizes and catalyzes the mobilization of these DNA segments by a "cut-and-paste" mechanism (Fig. 1a, b). Autonomous transposons contain the gene encoding the transposase. As genetic tools, DNA transposons are used as a nonautonomous system, where the transposon carries a gene of interest such as a selectable marker (cargo) and the transposase is supplied from a separate expression vector (Fig. 1a). Typical transposon

Shondra M. Pruett-Miller (ed.), *Chromosomal Mutagenesis*, Methods in Molecular Biology, vol. 1239,
DOI 10.1007/978-1-4939-1862-1_2, © Springer Science+Business Media New York 2015

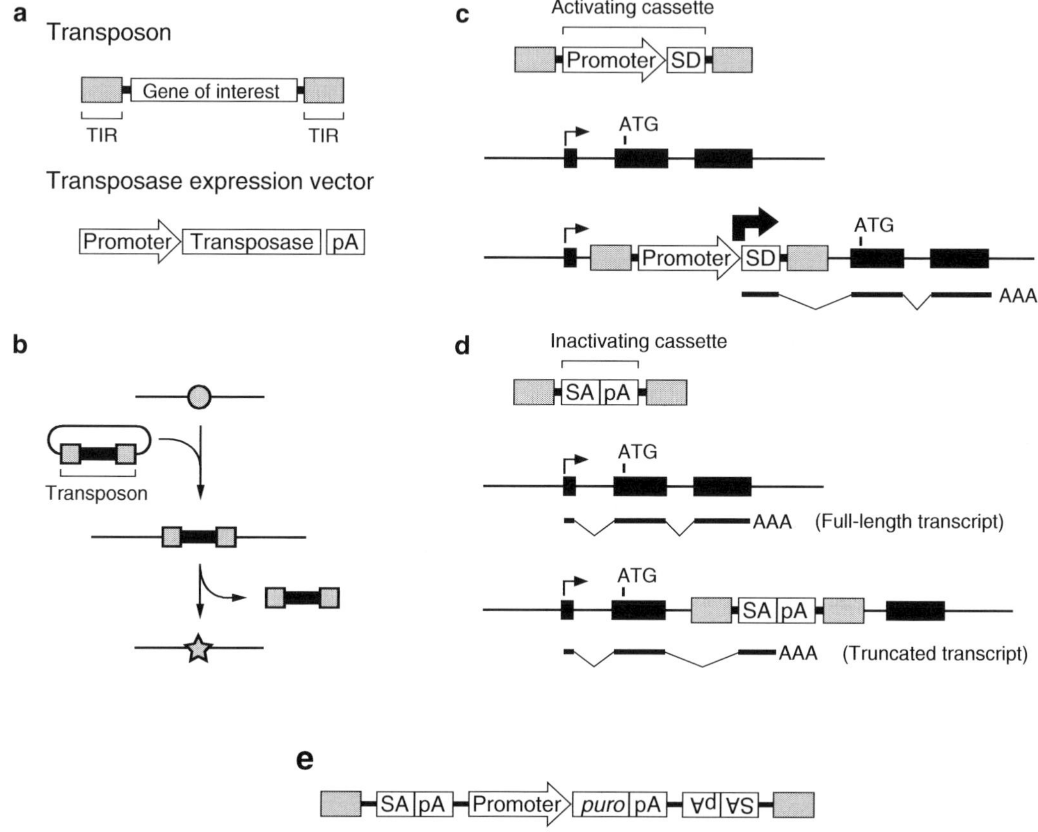

Fig. 1 DNA transposon and its applications for gene trap. (**a**) Nonautonomous DNA transposon system. TIR, terminal inverted repeat. (**b**). Schematic of transposition. Transposases bind to the TIR, excise the transposon from the transposon vector and insert the transposon into the target site (*circle*). After excision of the transposon, a footprint (*asterisk*) is commonly found, except for the *piggyBac* transposon. (**c**) Transposon with an activating cassette. Mutated genes will be overexpressed or ectopically expressed. SD, splice donor site; ATG, translation initiation codon; AAA, poly A tail. (**d**) Transposon with an inactivating cassette. The cassette disturbs generating a full-length transcript. SA, splice acceptor site; pA, polyadenylation signal sequence. (**e**) An inactivating transposon used in ref. [33]

cargos include an expression cassette for transgenesis [1, 2], an enhancer-trap cassette for *cis* element analyses [3, 4], or a gene-trap cassette for mutagenesis [5, 6].

Forward genetic screens provide a powerful means for a non-hypothesis-driven approach to gene function analysis. In this approach, genome-wide random mutants are first generated and then screened for a phenotype of interest. Subsequently, the causative genes in isolated mutants are identified. DNA transposons have been used as a favorable insertional mutagen in model organisms such as *Drosophila* [7, 8] and *Caenorhabditis elegans* [9, 10]. This is because (1) the highly efficient transposition enables (near-) saturated mutagenesis in a test organism and (2) the mutation (i.e., transposon integration site) can be readily identified. Since transposon sequences can be used as a molecular tag, integration sites can be isolated by analyzing transposon–genome junction sequences using well-established PCR-based methods such as splinkerette-PCR [11] or LAM-PCR [12]. These PCR-based methods have recently been adapted to high-throughput analyses using next generation sequencing technologies [13, 14].

Although DNA transposons were discovered in the 1950s, transposon systems for mammalian cells are relatively new. The first transposon system active in mammalian cells, *Sleeping Beauty*, was reported in 1997 [1]. This transposon system was reconstructed from fossil transposon sequences found in the Salmonid genome. An insect-derived transposon, *piggyBac*, was also found to be active when introduced into mammalian cells in 2005 [2]. Since then, a number of applications in mammals have been reported. For instance, these transposons have been utilized for germ-line mutagenesis in mice [3] and rats [15] and transgenesis in bigger mammals such as pig [16, 17]. An example of a clinically important application of DNA transposons is somatic mutagenesis for cancer gene discovery. The flexibility in designing transposon cargos allows researchers to develop sophisticated screening platforms and has resulted in the identification of a number of novel cancer genes [18–22]. In addition, clinical trials of transposon-mediated gene therapy are currently ongoing [23].

In parallel to these applications, efforts to improve transposition efficiency have been made. The transposase for both *Sleeping Beauty* and *piggyBac* systems have been engineered and hyperactive versions of both transposases are now available [24, 25]. These engineered transposases show a 10–100-fold increase in transposition efficiency compared to the original versions of the transposases.

Both systems show efficient transposition in mammalian cells; however, there are a few characteristic differences. For use in mutagenesis, the following two factors are important: (1) preference of integration sites and (2) footprint mutations. The genome-wide mapping of transposon integration sites found that the *piggyBac* transposon preferentially integrates into actively transcribed genes and regions with an open chromatin structure such as DnaseI hypersensitive sites and transcription factor binding sites [26]. In contrast, *Sleeping Beauty* integrates more evenly in the genome [27]. The *piggyBac* transposon is therefore more efficient in disrupting genes expressed in the target cells. Footprint mutations refer to sequence alterations at the site of transposon excision; most DNA transposons leave signature footprints. For instance, *Sleeping Beauty* transposon integrates into a 2-bp target site: TA. Duplication of this target site during insertion followed by incomplete excision changes the TA to TAG(T/A)CTA, a net insertion of 5 bp [1]. In contrast, *piggyBac* integrates into a 4-bp target site, TTAA, and does not change the sequence upon excision [28]. As a result of this, even if *piggyBac* undergoes multiple transpositions, the genome, in principle, remains intact apart from the final integration site. This eliminates the chance of inducing mutations that are not tagged by the transposon and are therefore difficult to map. This seamless excision of *piggyBac* has also led to its use as a fully removable vector in the generation of transgene-free induced pluripotent stem cells (iPSCs) [29] and the correction of inherited mutations in human iPSCs from patients with alpha-1 antitrypsin deficiency [30, 31].

1.2 Gene Trap Design

Two types of mutagenic transposon cargo (activating and inactivating) can be used, depending on whether a gain-of-function or loss-of-function screen is being performed. The activating elements consist of a promoter and a splice donor site (Fig. 1c). When this element is inserted upstream of the translation initiation codon of a gene, the promoter activates or increases gene expression. Hence, the activating element can be used for gain-of-function screens. On the other hand, the inactivating element consists of a splice acceptor site and a polyadenylation signal sequence (Fig. 1d). When this element is inserted in the intron of a gene, splicing between an upstream exon and the splice acceptor site occurs and a fusion transcript will be generated. This prevents the production of a full-length transcript. The inactivating element can therefore be used for loss-of-function screens. In cell culture systems, it is difficult to deliver the mutagenic element into every cell. As a result, in addition to the mutagenic elements, a drug selection cassette is typically inserted in transposon vectors and used to enrich cells harboring the transposon (Fig. 1e) [32, 33].

1.3 Haploid Mammalian Cells

Even with efficient insertional mutagens like DNA transposons, the diploid nature of the mammalian genome has always hampered loss-of-function screening. In the cell culture system, it is in principle impossible to convert randomly generated genome-wide heterozygous mutations into homozygotes. This issue was partially solved by increasing the frequency of loss-of-heterozygosity (LOH) using a *Blm* RecQ DNA helicase deficient background [34, 35]. However, since the frequency of LOH is only 10^{-4} events per cell per generation [34, 36], homozygous cells represent a very small fraction of a population consisting of at least 10,000 times more heterozygous mutants. It is therefore not an easy task to achieve sufficient coverage of genome-wide mutagenesis. In addition, this system was only available in mouse embryonic stem (ES) cells.

RNA interference (RNAi) has therefore served as a powerful tool for genome-wide loss-of-function screening in mammalian cell culture [37]. Since it suppresses gene expression at the mRNA level by degrading mRNA or inhibiting translation in a sequence-specific manner, ploidy is not an issue and any cell line can be used. Despite numerous successful screens, there are two well-known issues associated with RNAi: imperfect and/or transient suppression of gene function and off-target effects. "Validated" siRNAs and shRNAs are now available from a number of suppliers; however the actual efficiencies of gene inactivation of siRNAs/shRNAs in different target cell lines could differ. At the genome-wide scale, it is difficult to be sure of the true performance of every siRNA/shRNA in a set. Most successful screens use targeted libraries of RNAi reagents targeting a subset of genes. These are more accurately described as high throughput reverse genetic tools rather than a genuine forward genetic system.

The recent discovery of haploid mammalian cells has brought new possibilities for in vitro genetic screening. Thus far, four haploid cell lines have been reported: a near-haploid human chronic myeloid leukemia cell line [38] and its partially reprogrammed derivative [39], and mouse [40–42] and monkey [43] haploid ES cells. Since these cells contain only one copy of each gene, insertional mutagenesis is the most effective means of mutagenesis as discussed above. It is however reported that all haploid cell lines spontaneously diploidize and the haploid fraction is gradually lost from the culture. It is important to purify haploid fractions routinely and use highly pure haploid cell cultures for mutagenesis. As long as cells are mutagenized in the haploid state, subsequent diploidization is not an issue since mutated alleles become two copies, i.e., homozygous mutations.

Here, we describe experimental procedures for mouse haploid ES cell culture, purification of the haploid fraction using cell sorting, and *piggyBac*-mediated mutagenesis. Examples of successful screens can be found in refs. [33] and [42].

2 Materials

2.1 piggyBac Transposon Resources

The following plasmid DNAs are available from the Wellcome Trust Sanger Institute (http://www.sanger.ac.uk/form/-Rj-FxNGbR4yIRjwH0UOlnA).

1. Hyperactive mammalian-codon optimized *piggyBac* transposase (pCMV-hyPBase) [24].
2. Basic cloning vectors with *piggyBac* transposon DNA (pPB-LR5, pPB-RL5) [44].
3. Various mutagenic *piggyBac* transposons [6, 22, 33].

2.2 Cell Culture

1. Mouse haploid ES cells. This protocol has been developed with HAP-3 cells from the laboratory of Anton Wutz [40].
2. (Optional) Diploid ES cells purified from HAP-3 (*see* **Note 1**).
3. ES cell medium: Ndiff 227 (StemCells), 0.375 % BSA, 1 % Knockout serum replacement (Invitrogen), 1× nonessential amino acid, 3 μM CHIR99021 (Axon Medchem, Reston, VA,

USA), 1 μM PD0325901 (Axon Medchem), 1,000 U/ml Leukemia inhibitory factor (Millipore, Billerica, MA, USA) (*see* **Notes 2** and **3**).

4. Cell suspension medium: DMEM/F12, 0.375 % BSA.
5. Freezing medium: DMEM/F12, 20 % knockout serum replacement, 10 % DMSO.
6. PBS.
7. Accutase (Millipore).
8. Feeder cells (mouse embryonic fibroblasts inactivated by mitomycin C) (*see* **Notes 4** and **5**).
9. Feeder cell medium: DMEM, 10 % FBS, 2 mM GlutaMAX (Invitrogen, Carlsbad, CA, USA), 0.1 mM 2-mercaptoethanol, 1× nonessential amino acid.
10. Gene Pulser Xcell (Bio-Rad, Hercules, CA, USA).
11. 0.4 cm cuvette.

2.3 Fluorescence Activated Cell Sorting (FACS)

1. A flow sorter with a suitable laser for exciting Hoechst 33342 (e.g., 407 nm violet or a UV laser). This protocol was developed using a BD FACS Aria.
2. A flow cytometry analyzer (e.g., BD LSRFortessa).
3. Hoechst 33342 (Invitrogen).
4. PI staining solution: PBS, 5 μg/ml propidium iodide, 160 μg/ml RNAse A.
5. Cell strainer, 40 μm.

3 Methods

The most straightforward way to mutagenize haploid ES cells using *piggyBac* is to co-transfect ES cells with a transposon plasmid and a plasmid expressing the transposase. The transposon will be mobilized from the plasmid and inserted into the genome at a frequency much higher than random integration of the plasmid. Cells bearing chromosomal insertions of the transposon can be selected by drug selection using an appropriate selectable marker in the transposon cargo—in the protocol here, puromycin is used (Fig. 1e). A few days after transfection, the transposase expression plasmid will be lost and the transposon insertions are then stable. These stable mutants can subsequently be expanded as a pool and frozen to make libraries of mutants. Since mutants may grow at different rates, care should be taken to ensure that the cell number comfortably exceeds the number of mutants in the library whenever the pool is divided (e.g., when passaging cells, setting up screens or freezing). This reduces the chance of losing mutants from the pool due to poor sampling of mutants with low cell numbers.

Typically 10 μg of pCMV-hyPBase plasmid is used in a transfection of 1×10^7 ES cells. Varying the amount of transposon plasmid can result in lower numbers of mutants with single transposon insertions (100–500 ng transposon) or high numbers of mutants that are likely to have several insertions (above 5 μg transposon plasmid) [6]. The ratio of 10 μg transposase : 1 μg transposon used here provides a good balance—the majority of mutants have one or two insertions and can be easily mapped in subcloned mutants by splinkerette PCR and Sanger sequencing. A typical electroporation under these conditions will yield around 50,000 mutants using a *piggyBac* transposon with its own promoter driving a selectable marker. Using a promoter trap transposon will reduce this to about half since around 50 % of *piggyBac* insertions are in genes [2, 26].

Diploid cells in the culture prior to electroporation will result in undesirable heterozygous mutants. These diploid cells are also more efficiently transfected than haploids, so a small contaminating population of diploid cells may disproportionately reduce the number of true null mutants in the resulting libraries (Fig. 2). Therefore it is important to purify haploid cells before mutagenesis. This can be done directly by sorting Hoechst 33342-stained cells by FACS based on DNA content, as used in the initial establishment of haploid ES cell lines [40]. However, since this stain is toxic, the large

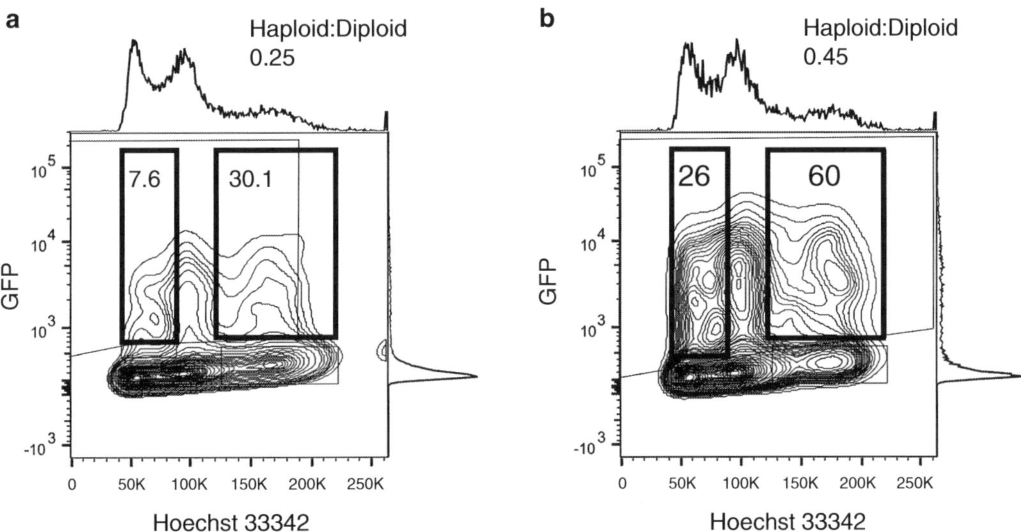

Fig. 2 Increased electroporation efficiency in haploid cells at high voltages. (**a**, **b**) A mixed population of haploid and diploid ES cells was transfected with a GFP expression plasmid using an electroporation voltage of 230 V (**a**) or 270 V (**b**) with a capacitance of 500 μF. After 48 h, cells were stained with Hoechst 33342 as described in the text and analyzed. Numbers show the percentage of cells of that phase of the cell cycle that are GFP-positive, the gates represent GFP+ G1-S haploid and GFP+ S-G2 diploid respectively. The use of higher electroporation voltages results in an increase in overall transfection efficiency as well as a decreased bias towards diploid cells (shown as the haploid–diploid ratio). A slight increase in cell death (not shown) is outweighed by the higher transfection efficiency, so more transfected live cells are obtained overall

numbers of cells required for transfection can be more efficiently obtained by sorting unstained cells based on forward/side scatter. Due to their smaller size, haploid ES cells have lower forward and side scatter and can be purified very efficiently on this basis. In this protocol an aliquot of the culture is stained with Hoechst 33342 to assess the composition of the culture and set up FACS gates, then the bulk of the culture is sorted without staining.

3.1 Cell Culture of Haploid Mouse ES Cells

1. Culture ES cells on feeder cells until they reach 70–80 % confluence. Replace the medium every day (*see* **Note 6**).
2. Wash the cells with PBS once, add Accutase (0.5 ml per well of 6-well plate) and incubate the cells at 37 °C for 5 min.
3. Add 2.5 ml of cell suspension medium to each well of 6-well plates and pipette to break up cell clumps. Transfer cell suspension to a 15 ml conical tube and centrifuge at 200 × *g* for 3 min at room temperature (15–25 °C).
4. Remove supernatant and resuspend cell pellets with ES cell medium. Plate the cells onto new feeder plates at a split ratio of 1:2–1:4.

3.2 Purification of the Haploid Fraction

1. Culture ES cells until they reach 70–80 % confluence. Refresh the medium every day.
2. Wash the cells with once, add Accutase (0.5 ml/a well of 6-well plate) and incubate the cells at 37 °C for 5 min.
3. Add 2.5 ml of cell suspension medium to each well of 6-well plates. Transfer 1×10^6 cells to a 15 ml conical tube and the rest of cells to another 15 ml conical tube. Centrifuge them at 200 × *g* for 3 min at room temperature.
4. Remove supernatant. Resuspend the 1×10^6 cell pellet in 1 ml of ES cell medium containing 20 μg/ml Hoechst 33342. Resuspend the cell pellet in the other tube with ES cell medium at a density of 5×10^6–1×10^7 cells/ml. This density may differ depending on cell sorting instruments.
5. Incubate both cell suspensions at 37 °C for 20 min with occasional agitation.
6. Pass the cells through a 40 μm cell strainer.
7. Analyze cell cycle profile using the stained cells and make a gate that represents the G1 phase of haploid cells, on the FSC-SSC (forward scatter-side scatter) plot (Fig. 3) (*see* **Note 7**).
8. Load the unstained sample and sort cells based on these FSC-SSC parameters into a collection tube containing 1 ml of ES cell medium.
9. Centrifuge the sorted cells at 200 × *g* for 3 min at room temperature and plate the cells onto an appropriate feeder plate according to the number of sorted cells (6–10×10^5 events per well of a 6-well plate).

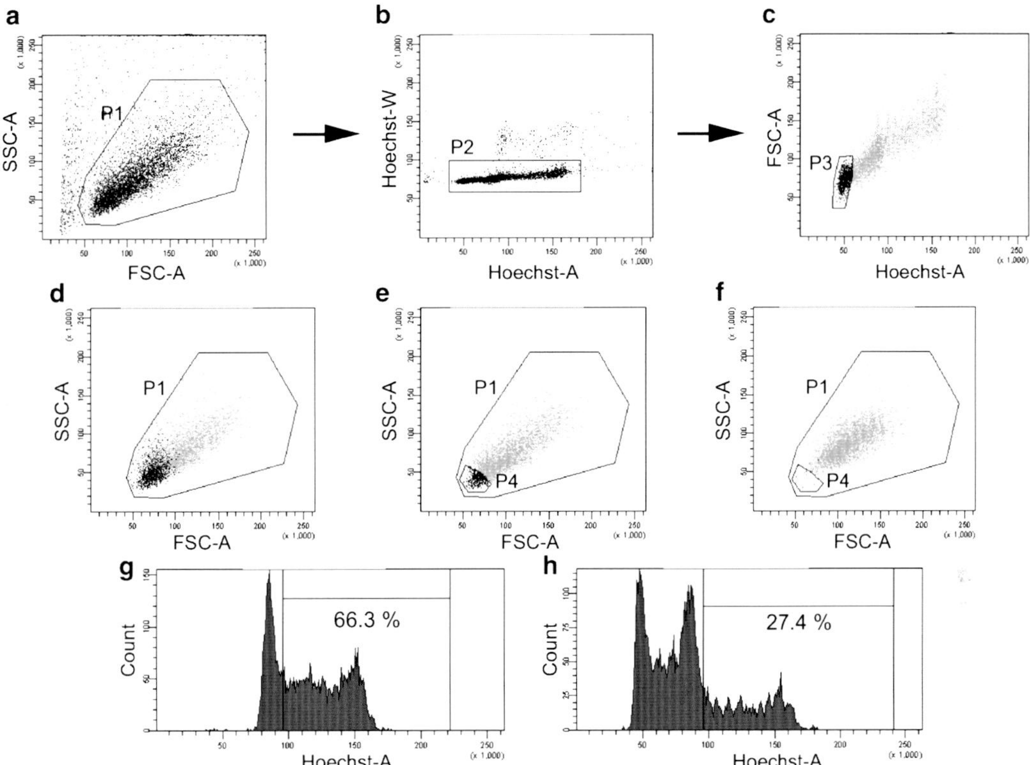

Fig. 3 Setting up the sorting gate. (**a–c**) Scatter plots from Hoechst 33342-stained haploid ES cells. (**a**) A FSC-SSC profile of all events. Gate P1 for all ES cell population. (**b**) A Hoechst Area—Hoechst Width profile of the events in Gate P1. Gate P2 contains singlets. (**c**) A HoechstA-FSC profile of the events in Gate P2. Gate P3 for haploid cells in the G1 phase. (**d**) A FSC-SSC profile of the events in Gate P2. Events in Gate P3 are shown in *black* and the rest of the events are shown in *grey*. (**e**) Gate P4 for sorting. Note that the Gate P4 is smaller than the area that represents haploid-G1 cells shown in (**d**). (**f**) A FSC-SSC profile of diploid ES cells. Some diploid cells are overlapping the area of haploid-G1 cells shown in (**d**), whereas there are only a few cells in Gate P4. (**g**, **h**) A cell cycle profile of diploid (**g**) and mixed haploid and diploid (**h**) ES cells. The S-G2 fraction of the diploid cells consists of approximately two third of the population. This ratio can be used to calculate the diploid cell fraction contaminated in haploid cell culture. In the case of (**h**), 41 % of the cells are diploid

3.3 piggyBac Transposon-Mediated Mutagenesis

1. Culture ES cells until they reach 70–80 % confluence. Aim to passage cells as little as possible prior to transfection to maintain a higher haploid fraction. Fewer than three passages are recommended (*see* **Note 8**).
2. Wash the cells with PBS once, add Accutase (0.5 ml per well of a 6-well plate) and incubate the cells at 37 °C for 5 min.
3. Add 2.5 ml of cell suspension medium to each well of a 6-well plate, transfer cell suspension to a 15 ml conical tube and centrifuge at $200 \times g$ for 3 min at room temperature (15–25 °C).
4. Resuspend 1×10^7 cells in 800 μl PBS and mix them with 1 μg transposon vector and 10 μg pCMV-hyPBase. Transfer the

mixture to an electroporation cuvette. Electroporate using Bio-Rad GenePulser (270 V, 500 μF) (*see* **Note 9**).

5. Plate cells on three 10-cm feeder plates. Plating 1/100 and 1/1,000 of the volume on six-well plates is also useful to estimate the number of mutants later (*see* **step 7**).
6. On the following day, replace the medium with ES cell medium containing 1 μg/ml puromycin. Continue puromycin selection for 7–10 days (*see* **Note 10**).
7. When puromycin-resistant colonies become big enough, count colony numbers on the 1/100 and 1/1,000 cell plates (Subheading 3.3, **step 5**) to estimate the complexity. Harvest cells from the three 10-cm dishes as described above and passage the cells to appropriate feeder plates. Fix an aliquot of 10^5–10^6 cells for later analysis of ploidy (*see* Subheading 3.4).
8. Freeze the cells (*see* **Note 11**).

3.4 Analysis of Ploidy in Fixed Cells by Propidium Iodide Staining

1. Fix cells by adding up to 10^6 cells in PBS (100 μl) to 1 ml 70 % ethanol at −20 °C. Fix overnight. Fixed cells can be stored at −20 °C for several weeks before analysis (*see* **Note 12**).
2. Centrifuge cells at 800 × *g* and resuspend in 1 ml PBS. Centrifuge again.
3. Resuspend in 300 μl PI staining solution.
4. Incubate in the dark at room temperature for 30 min.
5. Analyze by flow cytometry to obtain a cell cycle profile.

4 Notes

1. A FACS profile of the diploid ES cells is not necessary. As shown in Fig. 3f, however, it is very useful to make a tighter gate for haploid cells in the G1 phase with a minimum contamination of diploid cells in the G1 phase.
2. It is convenient to prepare 1,000× 2i mixture (3 mM CHIR99021 and 1 mM PD0325901 in DMSO). First, dissolve 5 mg of PD0325901 into 5.2 ml DMSO (Sigma-Aldrich, cat. no. D2650) and 15 mg of CHRI99021 into 5.4 ml DMSO, separately. These are 2,000× concentrated solutions. Subsequently, mix equal volumes of both solutions and aliquot 600 μl per cryovial. The aliquots can be stored at −20 °C for up to 1 year. Up to 5 rounds of freeze/thaw will not reduce activity of the inhibitors.
3. Adding knockout serum replacement at 1 % is beneficial for maintaining haploid fractions and proliferation of ES cells in this culture condition.
4. Puromycin-resistant mouse embryonic fibroblasts are ideal, but wild-type feeder cells can also be used for puromycin selection

since fibroblasts are less sensitive to puromycin than ES cells. Alternatively, the puromycin-resistant pSNL feeder cell line can be used.

5. Culture MEFs (at passage 3 or 4) in 15-cm dishes until they reach 100 % confluence. Replace old medium with 15 ml of MEF medium containing 15 μg/ml mitomycin C and incubate the dishes at 37 °C for 2.5 h. Next, wash the cells twice with PBS, add 15 ml of fresh MEF medium and incubate the cells overnight. On the following day, collect and count the inactivated MEFs. Re-plate the cells onto Gelatin/MEF-coated tissue culture plates at a density of 4.5×10^4 cells/cm^2 or freeze them at −80 °C at desirable concentrations in MEF medium containing 10 % (vol/vol) DMSO.
6. Differentiation of mouse ES cells is inhibited by the 2i inhibitor cocktail. This allows feeder-free culture of mouse ES cells while maintaining a high fraction of naïve pluripotent ES cells. However, ES cells occasionally do not adhere well onto gelatin-coated plates in these culture conditions, probably because of batch-to-batch differences in culture medium. Feeder cells greatly help ES cells to adhere to culture plates.
7. As shown in Fig. 2d–f, a larger gate surrounding the entire haploid-G1 cells cause diploid cell contamination. A tighter gate shown in Fig. 2e has to be used. If the Hoechst cell cycle profile is not well-defined, the incubation time can be increased. It is critical to stain cells in ES cell medium at the correct pH and at 37 °C, since the Hoechst dye is actively transported.
8. To obtain 1×10^7 cells within two passages after cell sorting, approximately 1×10^6 haploid-G1 cells need to be sorted. This is likely to require at least 10^7 cells as input, depending on the purity of the starting culture. If the haploid fraction in the starting culture is very low, a few round of purification will be required.
9. We recommend fixing a small fraction of the cells and analyzing to estimate percentage of the haploid fraction at the time of mutagenesis (*see* Subheading 3.4).
10. If pSNL feeders are used, the puromycin concentration should be 3 μg/ml.
11. Freeze cells in aliquots of at least 20 times the library complexity (e.g., for a library of 1,000 mutants, freeze in aliquots of at least 20,000 cells). When the frozen cells are thawed, it is important to measure the viability and estimate how many viable cells are in the initial plating. A low viable cell count (less than three times the library complexity) risks distorting the representation of mutants in the pool and losing mutants.
12. It can be useful to routinely fix samples as cells are passaged to monitor the culture. These can be stored and analyzed in

batches as convenient. Knowing the percentage of haploid cells at the time of transfection (Subheading 3.3, **step 4**) is particularly important to be confident that the library will contain sufficient numbers of null mutants. An aliquot of cells can also be fixed immediately after sorting to check purity.

Acknowledgements

This work was supported by the Wellcome Trust (WT077187), Breakthrough Breast Cancer and Cancer Research UK.

References

1. Ivics Z, Hackett PB, Plasterk RH, Izsvak Z (1997) Molecular reconstruction of Sleeping Beauty, a Tc1-like transposon from fish, and its transposition in human cells. Cell 91:501–510
2. Ding S, Wu X, Li G, Han M, Zhuang Y, Xu T (2005) Efficient transposition of the piggyBac (PB) transposon in mammalian cells and mice. Cell 122:473–483
3. Kokubu C, Horie K, Abe K, Ikeda R, Mizuno S, Uno Y, Ogiwara S, Ohtsuka M, Isotani A, Okabe M, Imai K, Takeda J (2009) A transposon-based chromosomal engineering method to survey a large cis-regulatory landscape in mice. Nat Genet 41:946–952
4. Ruf S, Symmons O, Uslu VV, Dolle D, Hot C, Ettwiller L, Spitz F (2011) Large-scale analysis of the regulatory architecture of the mouse genome with a transposon-associated sensor. Nat Genet 43:379–386
5. Horie K, Yusa K, Yae K, Odajima J, Fischer SE, Keng VW, Hayakawa T, Mizuno S, Kondoh G, Ijiri T, Matsuda Y, Plasterk RH, Takeda J (2003) Characterization of Sleeping Beauty transposition and its application to genetic screening in mice. Mol Cell Biol 23: 9189–9207
6. Wang W, Bradley A, Huang Y (2009) A piggyBac transposon-based genome-wide library of insertionally mutated Blm-deficient murine ES cells. Genome Res 19:667–673
7. Bellen HJ, O'Kane CJ, Wilson C, Grossniklaus U, Pearson RK, Gehring WJ (1989) P-element-mediated enhancer detection: a versatile method to study development in Drosophila. Genes Dev 3:1288–1300
8. Spradling AC, Stern DM, Kiss I, Roote J, Laverty T, Rubin GM (1995) Gene disruptions using P transposable elements: an integral component of the Drosophila genome project. Proc Natl Acad Sci U S A 92:10824–10830
9. Greenwald I (1985) lin-12, a nematode homeotic gene, is homologous to a set of mammalian proteins that includes epidermal growth factor. Cell 43:583–590
10. Moerman DG, Benian GM, Waterston RH (1986) Molecular cloning of the muscle gene unc-22 in Caenorhabditis elegans by Tc1 transposon tagging. Proc Natl Acad Sci U S A 83:2579–2583
11. Devon RS, Porteous DJ, Brookes AJ (1995) Splinkerettes–improved vectorettes for greater efficiency in PCR walking. Nucleic Acids Res 23:1644–1645
12. Schmidt M, Schwarzwaelder K, Bartholomae C, Zaoui K, Ball C, Pilz I, Braun S, Glimm H, von Kalle C (2007) High-resolution insertion-site analysis by linear amplification-mediated PCR (LAM-PCR). Nat Methods 4:1051–1057
13. Uren AG, Mikkers H, Kool J, van der Weyden L, Lund AH, Wilson CH, Rance R, Jonkers J, van Lohuizen M, Berns A, Adams DJ (2009) A high-throughput splinkerette-PCR method for the isolation and sequencing of retroviral insertion sites. Nat Protoc 4:789–798
14. Koudijs MJ, Klijn C, van der Weyden L, Kool J, ten Hoeve J, Sie D, Prasetyanti PR, Schut E, Kas S, Whipp T, Cuppen E, Wessels L, Adams DJ, Jonkers J (2011) High-throughput semiquantitative analysis of insertional mutations in heterogeneous tumors. Genome Res 21:2181–2189
15. Kitada K, Ishishita S, Tosaka K, Takahashi R, Ueda M, Keng VW, Horie K, Takeda J (2007) Transposon-tagged mutagenesis in the rat. Nat Methods 4:131–133
16. Garrels W, Mates L, Holler S, Dalda A, Taylor U, Petersen B, Niemann H, Izsvak Z, Ivics Z, Kues WA (2011) Germline transgenic pigs by Sleeping Beauty transposition in porcine zygotes and targeted integration in the pig genome. PLoS One 6:e23573

17. Wu Z, Xu Z, Zou X, Zeng F, Shi J, Liu D, Urschitz J, Moisyadi S, Li Z (2013) Pig transgenesis by piggyBac transposition in combination with somatic cell nuclear transfer. Transgenic Res 22:1107–1118
18. Collier LS, Carlson CM, Ravimohan S, Dupuy AJ, Largaespada DA (2005) Cancer gene discovery in solid tumours using transposon-based somatic mutagenesis in the mouse. Nature 436:272–276
19. Dupuy AJ, Akagi K, Largaespada DA, Copeland NG, Jenkins NA (2005) Mammalian mutagenesis using a highly mobile somatic Sleeping Beauty transposon system. Nature 436: 221–226
20. Starr TK, Allaei R, Silverstein KA, Staggs RA, Sarver AL, Bergemann TL, Gupta M, O'Sullivan MG, Matise I, Dupuy AJ, Collier LS, Powers S, Oberg AL, Asmann YW, Thibodeau SN, Tessarollo L, Copeland NG, Jenkins NA, Cormier RT, Largaespada DA (2009) A transposon-based genetic screen in mice identifies genes altered in colorectal cancer. Science 323:1747–1750
21. Keng VW, Villanueva A, Chiang DY, Dupuy AJ, Ryan BJ, Matise I, Silverstein KA, Sarver A, Starr TK, Akagi K, Tessarollo L, Collier LS, Powers S, Lowe SW, Jenkins NA, Copeland NG, Llovet JM, Largaespada DA (2009) A conditional transposon-based insertional mutagenesis screen for genes associated with mouse hepatocellular carcinoma. Nat Biotechnol 27: 264–274
22. Rad R, Rad L, Wang W, Cadinanos J, Vassiliou G, Rice S, Campos LS, Yusa K, Banerjee R, Li MA, de la Rosa J, Strong A, Lu D, Ellis P, Conte N, Yang FT, Liu P, Bradley A (2010) PiggyBac transposon mutagenesis: a tool for cancer gene discovery in mice. Science 330: 1104–1107
23. Kebriaei P, Huls H, Jena B, Munsell M, Jackson R, Lee DA, Hackett PB, Rondon G, Shpall E, Champlin RE, Cooper LJ (2012) Infusing CD19-directed T cells to augment disease control in patients undergoing autologous hematopoietic stem-cell transplantation for advanced B-lymphoid malignancies. Hum Gene Ther 23: 444–450
24. Yusa K, Zhou L, Li MA, Bradley A, Craig NL (2011) A hyperactive piggyBac transposase for mammalian applications. Proc Natl Acad Sci U S A 108:1531–1536
25. Mates L, Chuah MK, Belay E, Jerchow B, Manoj N, Acosta-Sanchez A, Grzela DP, Schmitt A, Becker K, Matrai J, Ma L, Samara-Kuko E, Gysemans C, Pryputniewicz D, Miskey C, Fletcher B, VandenDriessche T, Ivics Z, Izsvak Z (2009) Molecular evolution of a novel hyperactive Sleeping Beauty transposase enables robust stable gene transfer in vertebrates. Nat Genet 41:753–761
26. Li MA, Pettitt SJ, Eckert S, Ning Z, Rice S, Cadinanos J, Yusa K, Conte N, Bradley A (2013) The piggyBac transposon displays local and distant reintegration preferences and can cause mutations at noncanonical integration sites. Mol Cell Biol 33:1317–1330
27. Liang Q, Kong J, Stalker J, Bradley A (2009) Chromosomal mobilization and reintegration of Sleeping Beauty and PiggyBac transposons. Genesis 47:404–408
28. Fraser MJ, Ciszczon T, Elick T, Bauser C (1996) Precise excision of TTAA-specific lepidopteran transposons piggyBac (IFP2) and tagalong (TFP3) from the baculovirus genome in cell lines from two species of Lepidoptera. Insect Mol Biol 5:141–151
29. Yusa K, Rad R, Takeda J, Bradley A (2009) Generation of transgene-free induced pluripotent mouse stem cells by the piggyBac transposon. Nat Methods 6:363–369
30. Yusa K, Rashid ST, Strick-Marchand H, Varela I, Liu PQ, Paschon DE, Miranda E, Ordonez A, Hannan NR, Rouhani FJ, Darche S, Alexander G, Marciniak SJ, Fusaki N, Hasegawa M, Holmes MC, Di Santo JP, Lomas DA, Bradley A, Vallier L (2011) Targeted gene correction of alpha1-antitrypsin deficiency in induced pluripotent stem cells. Nature 478:391–394
31. Yusa K (2013) Seamless genome editing in human pluripotent stem cells using custom endonuclease-based gene targeting and the piggyBac transposon. Nat Protoc 8:2061–2078
32. Yusa K, Takeda J, Horie K (2004) Enhancement of Sleeping Beauty transposition by CpG methylation: possible role of heterochromatin formation. Mol Cell Biol 24:4004–4018
33. Pettitt SJ, Rehman FL, Bajrami I, Brough R, Wallberg F, Kozarewa I, Fenwick K, Assiotis I, Chen L, Campbell J, Lord CJ, Ashworth A (2013) A genetic screen using the PiggyBac transposon in haploid cells identifies Parp1 as a mediator of olaparib toxicity. PLoS One 8: e61520
34. Yusa K, Horie K, Kondoh G, Kouno M, Maeda Y, Kinoshita T, Takeda J (2004) Genome-wide phenotype analysis in ES cells by regulated disruption of Bloom's syndrome gene. Nature 429:896–899
35. Guo G, Wang W, Bradley A (2004) Mismatch repair genes identified using genetic screens in Blm-deficient embryonic stem cells. Nature 429:891–895
36. Luo G, Santoro IM, McDaniel LD, Nishijima I, Mills M, Youssoufian H, Vogel H, Schultz RA,

Bradley A (2000) Cancer predisposition caused by elevated mitotic recombination in Bloom mice. Nat Genet 26:424–429

37. Boutros M, Ahringer J (2008) The art and design of genetic screens: RNA interference. Nat Rev Genet 9:554–566
38. Carette JE, Guimaraes CP, Varadarajan M, Park AS, Wuethrich I, Godarova A, Kotecki M, Cochran BH, Spooner E, Ploegh HL, Brummelkamp TR (2009) Haploid genetic screens in human cells identify host factors used by pathogens. Science 326:1231–1235
39. Carette JE, Raaben M, Wong AC, Herbert AS, Obernosterer G, Mulherkar N, Kuehne AI, Kranzusch PJ, Griffin AM, Ruthel G, Dal Cin P, Dye JM, Whelan SP, Chandran K, Brummelkamp TR (2011) Ebola virus entry requires the cholesterol transporter Niemann-Pick C1. Nature 477:340–343
40. Leeb M, Wutz A (2011) Derivation of haploid embryonic stem cells from mouse embryos. Nature 479:131–134
41. Yang H, Shi L, Wang BA, Liang D, Zhong C, Liu W, Nie Y, Liu J, Zhao J, Gao X, Li D, Xu GL, Li J (2012) Generation of genetically modified mice by oocyte injection of androgenetic haploid embryonic stem cells. Cell 149:605–617
42. Elling U, Taubenschmid J, Wirnsberger G, O'Malley R, Demers SP, Vanhaelen Q, Shukalyuk AI, Schmauss G, Schramek D, Schnuetgen F, von Melchner H, Ecker JR, Stanford WL, Zuber J, Stark A, Penninger JM (2011) Forward and reverse genetics through derivation of haploid mouse embryonic stem cells. Cell Stem Cell 9:563–574
43. Yang H, Liu Z, Ma Y, Zhong C, Yin Q, Zhou C, Shi L, Cai Y, Zhao H, Wang H, Tang F, Wang Y, Zhang C, Liu XY, Lai D, Jin Y, Sun Q, Li J (2013) Generation of haploid embryonic stem cells from Macaca fascicularis monkey parthenotes. Cell Res 23:1187–1200
44. Cadinanos J, Bradley A (2007) Generation of an inducible and optimized piggyBac transposon system. Nucleic Acids Res 35:e87

Chapter 3

Using Phage Integrases in a Site-Specific Dual Integrase Cassette Exchange Strategy

Jonathan M. Geisinger and Michele P. Calos

Abstract

ΦC31 integrase, a site-specific large serine recombinase, is a useful tool for genome engineering in a variety of eukaryotic species and cell types. ΦC31 integrase performs efficient recombination between its *attB* site and either its own placed *attP* site or a partially mismatched genomic pseudo *attP* site. Bxb1 integrase, another large serine recombinase, has a similar level of recombinational activity, but recognizes only its own *attB* and *attP* sites. Previously, we have used these integrases sequentially to integrate plasmid DNA into the genome. This approach relied on placing a landing pad *attP* for Bxb1 integrase in the genome by using phiC31 integrase-mediated recombination at a genomic pseudo *attP* site. In this chapter, we present a protocol for using these integrases simultaneously to facilitate cassette exchange at a predefined location. This approach permits greater control and accuracy over integration. We also present a general method for using polymerase chain reaction assays to verify that the desired cassette exchange occurred successfully.

Key words Bxb1 integrase, Cassette exchange, Homologous recombination, Phage integrase, phiC31 integrase, TALEN

1 Introduction

In genetic studies, it is beneficial to be able to integrate exogenous DNA into the genome and to generate targeted mutations. Random integration of plasmid DNA has often been used for integrating exogenous DNA, while homologous recombination has been utilized to create targeted mutations. However, random integration is undesirable because of the sacrifice of control over location and copy number, in addition to a fairly low efficiency. Spontaneous homologous recombination also has a low efficiency in mammalian cells. Because of these drawbacks, the use of transposases, retroviral and lentiviral vectors, and phage integrases has been adopted into genome engineering strategies to increase their efficiency.

Even though transposases and retroviral and lentiviral vectors are efficient at chromosomal integration, they do possess some

Shondra M. Pruett-Miller (ed.), *Chromosomal Mutagenesis*, Methods in Molecular Biology, vol. 1239,
DOI 10.1007/978-1-4939-1862-1_3, © Springer Science+Business Media New York 2015

limitations. With these methods, there is a lack of control over the number of integration events. Additionally, retroviral and lentiviral vectors have a bias for integrating near transcriptional units [1]. Use of transposases, such as *Sleeping Beauty*, results in essentially random integration [2]. Another drawback is an upper limit on the size of the DNA that can be integrated, which restricts what can be accomplished with these strategies.

To address these concerns, the site-specific resolvases Cre and FLP have been used. They do not have size limits, but do require that their recognition sites, *loxP* for Cre and *FRT* for FLP, be placed at the target genomic site [3, 4]. Placement of the recognition sites has become easier with the advent of new genome editing tools such as TALEN and CRISPR/Cas9 systems, but there is still the caveat of the bidirectionality of these resolvases. Both enzymes are capable of integration and excision without the requirement of cofactors. Because the excision reaction is favored over the integration reaction, the Cre and FLP resolvases are more useful for generating deletions than insertions.

Another group of site-specific recombinases are the phage integrases. These integrases, particularly the large serine integrases, are well-suited for integrating DNA into the genome. These integrases recombine an *attP* sequence and an *attB* sequence that are unique for each integrase, resulting in the generation of *attL* and *attR* sequences that are not substrates for the integrase in question [5]. The ΦC31 integrase from a phage of *Streptomyces* bacteria was the first of these unidirectional integrases to demonstrate the ability to integrate DNA into a mammalian genome efficiently [6–8]. ΦC31 integrase is capable of recombining a plasmid bearing its own perfect *attB*, either with a perfect *attP* site preplaced in the mammalian genome, or with endogenous sequences that partially resemble its *attP* site, known as pseudo *attP* sites [9]. There are several pseudo *attP* sites present in mammalian genomes, and their accessibility to ΦC31 integrase appears to be cell-type specific [10, 11]. Additionally, there does not appear to be an upper size limit for ΦC31 integrase. These features make ΦC31 integrase attractive for use in generating transgenic animals or cell lines with better-than-random integration. Another phage integrase, Bxb1 integrase, has been found to facilitate integration of DNA into mammalian genomes, but requires a preplaced perfect Bxb1 *attP* site, since Bxb1 integrase does not appear capable of recognizing pseudo *attP* sites [12]. We exploited this feature of Bxb1 by placing its perfect *attP* site into the genome at a ΦC31 pseudo *attP* site [13]. This placement allowed us to retarget the same locus with high efficiency by using Bxb1 integrase.

Previously, we described a method for the reproducible, precise placement of DNA at a preplaced ΦC31 *attP* landing pad in mammalian cell lines [14]. This method reduced the frequency of pseudo site integration by relying on the activation of resistance to an antibiotic that only occurred if the *attB*-containing plasmid integrated correctly at the preplaced *attP* site. However, this method results in the retention of the two antibiotic resistance genes, which may be unfavorable for therapeutic applications. Additionally, in our more recent reprogramming studies, we observed that there were a large number of different pseudo site integration locations in experiments involving mouse fibroblasts [11, 13]. Thus, we sought to develop a method to facilitate integration of exogenous DNA at a specific locus in a controlled manner, while leaving minimal exogenous genes in the genome.

Recently, we developed a method called Dual Integrase Cassette Exchange (DICE) that addresses these criteria [15]. To perform DICE, we first place a landing pad cassette into the genome (Fig. 1). This cassette contains the ΦC31 and Bxb1 *attP* sites flanking a selectable marker. In our study, we used G418 resistance as the selectable marker and GFP fluorescence as a screenable marker for this step. These markers allowed us easily to obtain correctly targeted clones. The landing pad cassette was placed into the genome

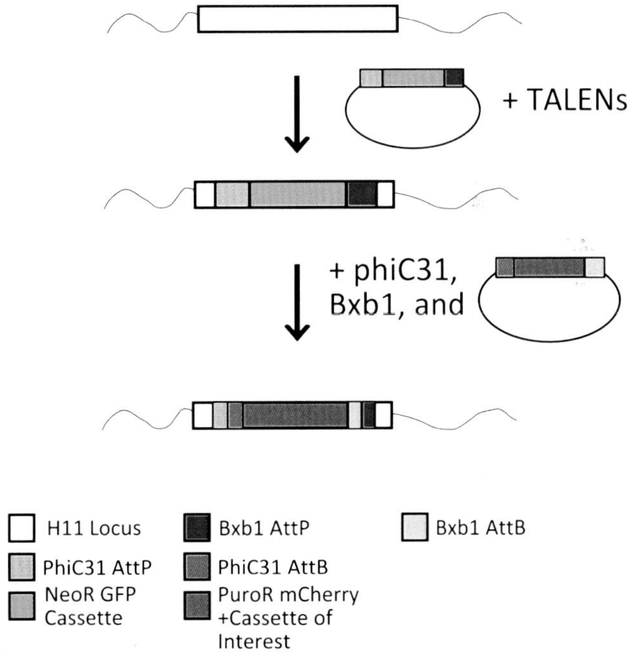

Fig. 1 DICE in an unmodified human genome. The *H11* locus is targeted with a landing pad plasmid expressing a neomycin-resistance GFP cassette flanked with ΦC31 and Bxb1 *attP* sites through homologous recombination mediated by a pair of TALENs designed against the *H11* locus. Following neomycin selection and GFP-assisted picking, DICE is mediated by use of ΦC31 and Bxb1 integrases and a donor plasmid carrying the cassette of interest flanked by ΦC31 and Bxb1 *attB* sites. Positive selection with puromycin permits growth of clones that successfully underwent DICE

through homologous recombination alone, or with the use of TAL effector nucleases to stimulate recombination frequency. It would also be feasible to use the CRISPR/Cas9 system or zinc finger nucleases to facilitate homologous recombination of the landing pad into the genome.

The location of the landing pad integration site is an important consideration. Because we desired a location that is intergenic and capable of facilitating long-term expression of transgenes, we chose the human homolog of the mouse *Hipp11* region [15]. This region, called *H11*, is located on chromosome 11 in mouse and chromosome 22 in human and had been previously shown to permit long-term expression of transgenes in a variety of murine tissues, without silencing [16, 17]. Following integration of and selection for the landing pad cassette, we perform cassette exchange by introducing three components: (1) ΦC31 integrase, (2) Bxb1 integrase, and (3) a donor plasmid containing the cassette to be exchanged, flanked by ΦC31 and Bxb1 *attB* sites (Fig. 1). Our donor cassette consisted of a puromycin resistance-mCherry fluorescent protein expression cassette under the control of the mPGK promoter and a cassette expressing various neural transcription factors under the control of the Nestin enhancer [15]. Thus, we were able to select clones that had successfully undergone DICE by choosing puromycin-resistant colonies that were mCherry-positive and GFP-negative. This method has a high efficiency and specificity and allows for precise identification of correctly integrated clones.

In this chapter, we describe our protocol for carrying out the DICE reaction. While we describe an example of targeting the *H11* locus in human pluripotent stem cells, we predict that DICE should work well at many genomic locations in a wide range of species and cell types.

2 Materials

2.1 Reagents

1. Dulbecco's modified Eagle's medium/Ham's F-12 50/50 mix (DMEM/F-12) with l-glutamine supplemented with 20 % Knockout Serum Replacement, 0.1 mM nonessential amino acids, 2 mM GlutaMAX, 0.1 mM β-mercaptoethanol, and 10 ng/mL bFGF (R&D Systems, Minneapolis, MN, USA).
2. Phosphate buffered saline (PBS).
3. Accutase (Millipore, Billerica, MA, USA).
4. Rho kinase inhibitor Y27632 (Tocris Bioscience, Bristol, UK).
5. P3 Primary Cell 4D-Nucleofector X kit L (Lonza, Walkersville, MD, USA).
6. G418.
7. Puromycin.

8. ZR Genomic DNA II kit (Zymo Research Corp., Irvine, CA, USA).
9. GoTaq Green Master Mix (Promega Biosystems, Sunnyvale, CA, USA).
10. QIAquick PCR purification kit (Qiagen, Valencia, CA, USA).
11. 0.1 % gelatin solution (Tribio, Menlo Park, CA, USA).
12. 1 mg/mL Collagenase IV solution (Stem Cell Technologies, Vancouver, Canada).

2.2 Plasmids

1. p2attNG, a plasmid containing ΦC31 and Bxb1 *attP* sites flanking a neomycin-resistance-GFP cassette, with all this flanked by two regions of homology to the *H11* locus [15] (*see* **Note 1**).
2. A pair of plasmids encoding TALENs to the *H11* locus.
3. pCS-kI, a plasmid encoding the 605-amino-acid isoform of ΦC31 integrase [18].
4. pCS-Bxb1, plasmid expressing wild-type Bxb1 integrase.
5. *attB* donor plasmid, a plasmid expressing the cassette of interest, plus a puromycin-resistance mCherry cassette under the control of the mPGK promoter, flanked by ΦC31 and Bxb1 *attB* sites.
6. PCR primers: TAL-LPF: 5′-AGTTCCAGGCTTATAGTCATTATTCCCTAA-3′, TAL-LPR: 5′-GTCTCATGAGCGGATACATATTTGAATGTA-3′, TAL-DCF: 5′-AAGCTGAGGAATCACATGGAGTGAATAGCA-3′, TAL-DCR: 5′-GGGTGGGGCAGGACAGCAAG-3′, TAL-UNF: 5′-CCAACCACCTTGACCTTTACCTCATTATCT-3′, TAL-UNR: 5′-TGCAGCTTCAACCTCCTGGGC-3′.

2.3 Cell Lines

1. H9, a female human embryonic stem cell line (WiCell, Madison, WI, USA) (*see* **Note 2**).
2. Gamma-irradiated CF1 mouse embryonic fibroblast feeder cells (Applied Stem Cell, Menlo Park, CA, USA) (*see* **Note 3**).
3. Gamma-irradiated DR4 mouse embryonic fibroblast feeder cells (Applied Stem Cell) (*see* **Note 3**).

2.4 Equipment

1. Nucleofector-4D (Lonza).
2. CO_2 incubator set at 37 °C.
3. 6-well, 35 mm, and 24-well Tissue culture plates.
4. Thermocycler.
5. Agarose gel electrophoresis equipment.
6. Microcentifuge tubes.
7. Thin-walled PCR tubes.

3 Methods

3.1 Construction of the Landing Pad Cell Line

1. Grow the H9 cells in 6-well tissue culture dishes coated with 0.1 % gelatin and 4.75×10^5 gamma-irradiated CF1 feeder cells per well in 2 mL per well of fully supplemented DMEM/F-12 media. The H9 cells should be about 70 % confluent on the day of electroporation.
2. Incubate the cells at 37 °C in a 5 % CO_2 incubator.
3. Twenty-four hours before electroporation, coat an appropriate number of 35 mm tissue culture plates with 1 mL of 0.1 % gelatin solution. Incubate these dishes for 30 min at 37 °C in a 5 % CO_2 incubator. Aspirate the solution and allow the plates to dry for 10 min at room temperature. Then plate 4.75×10^5 gamma-irradiated DR4 feeder cells per well in 1.5 mL of fully supplemented DMEM/F-12 media.
4. On the day of electroporation, aspirate the media from the H9 cells and wash with 2 mL of PBS per well. Aspirate the PBS and then add 1 mL per well of Accutase. Incubate the cells at 37 °C in a 5 % CO_2 incubator for 10 min to dissociate the cells into a single cell suspension.
5. Carefully aspirate the Accutase from each well and harvest each well using 1 mL of fully supplemented DMEM/F-12 into microcentrifuge tubes. Spin cell suspensions down at $200 \times g$ in a microcentrifuge. Resuspend cells in 1 mL of fully supplemented DMEM/F-12 with 1:1,000 Rho kinase inhibitor Y27632. Pipette each suspension back into the well each was harvested from. Incubate cells at 37 °C in a 5 % CO_2 incubator for 30 min to deplete feeders.
6. Harvest the H9 cells and electroporate $0.8–1.6 \times 10^6$ cells per cuvette using the P3 Primary Cell 4D-Nucleofector X kit L and the Nucleofector-4D with 8 μg of p2attNG and 1 μg of each TALEN-encoding plasmid according to the manufacturer's instructions. Post-electroporation, quickly reseed the contents of each cuvette onto the DR4-coated 35 mm plates previously prepared in 1.5 mL of fully supplemented DMEM/F-12 with 1:1,000 Rho kinase inhibitor Y27632. Incubate the cells at 37 °C in a 5 % CO_2 incubator for 24 h.
7. The next day, replace the media with fresh fully supplemented DMEM/F-12 with 1:100 Rho kinase inhibitor Y27632. Incubate the cells at 37 °C in a 5 % CO_2 incubator.
8. For the next 3 days, carry out selection using fully supplemented DMEM/F-12 media containing 50 μg/mL G418 with media changes each day (*see* **Note 4**). After selection, maintain cells in fully supplemented DMEM/F-12 media until well-formed colonies appear.

9. Prepare 0.1 % gelatin-coated 24-well tissue culture dishes containing 100,000 gamma-irradiated CF1 feeder cells per well. Pick single, well-isolated, GFP-positive colonies using glass picking tools and expand in each well of the prepared 24-well plate in fully supplemented DMEM/F-12 media.
10. Passage the colonies by washing once with PBS followed by incubation with 1 mL of collagenase IV solution for 10 min at 37 °C in a 5 % CO_2 incubator. Carefully aspirate the collagenase IV solution and add 1 mL of fully supplemented DMEM/F-12 media. Cut the colonies into small pieces using a pipette tip, and then transfer to a 1.5 mL microcentrifuge tube. Spin down at 300 × *g* for 5 min. Resuspend in 1 mL of fully supplemented DMEM/F-12 and divide it into two wells per clone of a 0.1 % gelatin-coated 6-well plate containing gamma irradiated CF1 feeder cells in 2 mL of fully supplemented DMEM/F-12 media.

3.2 PCR-Based Assay for Identifying Correctly Targeted Landing Pad Clones

1. For each clone from Subheading 3.1, wash one of the two wells with 2 mL of PBS and then incubate with 1 mL Accutase for 10 min at 37 °C in a 5 % CO_2 incubator. Carefully aspirate the Accutase, and harvest cells in 1 mL of fully supplemented DMEM/F-12. Spin down at 300 × *g* for 5 min. Deplete feeders from the harvested cells by replating in 1 mL fully supplemented DMEM/F-12 and incubating at 37 °C in a 5 % CO_2 incubator for 30 min. After depletion, transfer the media containing the depleted dissociated cells to a microcentrifuge tube and spin at 300 × *g* for 5 min. Aspirate media.
2. Prepare genomic DNA from each cell pellet using the ZR Genomic DNA II kit according to the manufacturer's directions.
3. Perform the PCR analysis using the primers TAL-LPF and TAL-LPR to determine if the landing pad has been homologously recombined correctly. Additionally, use the primers TAL-UNF and TAL-UNR to verify the number of alleles targeted.
4. Carry out the PCRs using GoTaq Green Master Mix and the following protocol: 95 °C for 5 min; and 35 cycles of 95 °C for 30 s; 52 °C for TAL-LPF + TAL-LPR or 56 °C for TAL-UNF + TAL-UNR for 30 s; 72 °C for 1 min, followed by 72 °C for 7 min.
5. Visualize the amplified PCR products via agarose gel electrophoresis. For the TAL-LPF + TAL-LPR PCR, the presence of a band indicates that homologous recombination successfully took place. For the TAL-UNF + TAL-UNR PCR, the presence of a band indicates that only one allele successfully underwent homologous recombination, whereas the absence of a band indicates that both alleles were successfully targeted.

3.3 DICE at the Modified H11 Locus

1. Choose a clone from Subheading 3.2 that contains only one correctly targeted allele. Prepare for DR4 feeder plates 24 h in advance of electroporation following **step 3** of Subheading 3.1.
2. On the day of electroporation, treat the clone as described in **steps 4** and **5** of Subheading 3.2.
3. Harvest the cells and electroporate $0.8–1.6 \times 10^6$ cells per cuvette using the P3 Primary Cell 4D-Nucleofector X kit L and the Nucleofector-4D with 4 μg each of pCS-kI, pCS-Bxb1, and the *attB* donor plasmid. After electroporation, quickly reseed the contents of each cuvette onto the DR4-coated 35 mm plates previously prepared in 1.5 mL of fully supplemented DMEM/F-12 with 1:1,000 Rho kinase inhibitor Y27632. Incubate the cells at 37 °C in a 5 % CO_2 incubator for 24 h.
4. The next day, follow **step 7** from Subheading 3.1.
5. For the next 2–3 days, carry out negative selection using fully supplemented DMEM/F-12 containing 500 ng/mL puromycin, replacing with fresh media each day (*see* **Note 5**). After selection, continue growing the cells in fully supplemented DMEM/F-12 until well-formed colonies appear, which is usually about 2 weeks.
6. Pick mCherry-positive, GFP-negative colonies following the procedure described in **step 9** under Subheading 3.1.
7. Passage colonies onto 6-well plates as described in **step 10** under Subheading 3.1.

3.4 PCR Assay to Verify Successful DICE

1. From each clone from Subheading 3.3, prepare genomic DNA following **steps 1** and **2** from Subheading 3.2.
2. Perform the PCR analysis using the primers TAL-DCF and TAL-DCR.
3. Carry out the PCR using GoTaq Green Master Mix and the following protocol: 95 °C for 5 min; and 35 cycles of 95 °C for 30 s; 57 °C for 30 s; 72 °C for 1 min, followed by 72 °C for 7 min.
4. Visualize the PCR products via agarose gel electrophoresis. The presence of a band indicates that DICE occurred successfully.

4 Notes

1. This protocol uses the *H11* locus in the human genome as an example, but it can be adapted to work with nearly any site. The major changes would be the method of cell culture (if not using human embryonic stem cells), the homology arms flanking the landing pad, the PCR primers, and the thermocycler protocols.

2. This protocol describes cell culture techniques and handling for human embryonic stem cells. If different cells are used, use appropriate media and techniques for those cells.
3. While gamma-irradiated feeder cells are used in this protocol, it may be difficult to obtain such cells. If that is the case, mitomycin-C-inactivated feeders can be substituted.
4. In our hands, 50 μg/mL of G418 was optimal for selection. We recommend that a kill curve is carried out to account for lot-, locus-, and cell-specific differences.
5. In our hands, 500 ng/mL of puromycin works well for positive selection. As stated in **Note 4**, we recommend performing a kill curve to determine optimal concentration for locus and cell type.

Acknowledgements

This work was supported by grants from the California Institute for Regenerative Medicine.

References

1. Bushman FD (2003) Targeting survival: integration site selection by retroviruses and LTR-retrotransposons. Cell 115:135–138
2. Yant SR, Wu X, Huang Y, Garrison B, Burgess SM, Kay MA (2005) High-resolution genome-wide mapping of transposon integration in mammals. Mol Cell Biol 25:2085–2094
3. Sauer B, Henderson N (1988) Site-specific DNA recombination in mammalian cells by the Cre recombinase of bacteriophage P1. Proc Natl Acad Sci U S A 85:5166–5170
4. O'Gorman S, Fox DT, Wahl GM (1991) Recombinase-mediated gene activation and site-specific integration in mammalian cells. Science 251:1351–1355
5. Thorpe HM, Smith MCM (1998) *In vitro* site-specific integration of bacteriophage DNA catalyzed by a recombinase of the resolvase/invertase family. Proc Natl Acad Sci U S A 95:5505–5510
6. Groth AC, Olivares EC, Thyagarajan B, Calos MP (2000) A phage integrase directs efficient site-specific integration in human cells. Proc Natl Acad Sci U S A 97:5995–6000
7. Thyagarajan B, Olivares EC, Hollis RP, Ginsburg DS, Calos MP (2001) Site-specific genomic integration in mammalian cells mediated by phage ΦC31 integrase. Mol Cell Biol 21:3926–3934
8. Groth AC, Calos MP (2004) Phage integrases: biology and applications. J Mol Biol 335: 667–678
9. Calos MP (2006) The ΦC31 integrase system for gene therapy. Curr Gene Ther 6: 633–645
10. Chalberg TC et al (2007) Integration specificity of phage ΦC31 integrase in the human genome. J Mol Biol 357:28–48
11. Karow M et al (2011) Site-specific recombinase strategy to create induced pluripotent stem cells efficiently with plasmid DNA. Stem Cells 29:1696–1704
12. Russell JP, Chang DW, Tretiakova A, Padidam M (2006) Phage Bxb1 integrase mediates highly-efficient site-specific recombination in mammalian cells. Biotechniques 40:460–464
13. Zhao C et al (2014) Recombinase-mediated reprogramming and dystrophin gene addition in mdx mouse induced pluripotent stem cells. PLoS One 9(4):e96279
14. Keravala A, Calos MP (2007) Site-specific chromosomal integration mediated by ΦC31 integrase. In: Davis G, Kayser KJ (eds) Methods in molecular biology, vol 435, Chromosomal mutagenesis. Humana Press, Totowa, NJ, pp 165–173

15. Zhu F et al (2013) DICE, an efficient system for iterative genomic editing in human pluripotent stem cells. Nucleic Acids Res 42:e34. doi:10.1093/nar/gkt1290
16. Hippenmeyer S et al (2010) Genetic mosaic dissection of Lis1 and Ndel1 in neuronal migration. Neuron 68:695–709
17. Tasic B et al (2011) Site-specific integrase-mediated transgenesis in mice via pronuclear injection. Proc Natl Acad Sci U S A 108:7902–7907
18. Farruggio AP, Chavez CL, Mikell CL, Calos MP (2012) Efficient reversal of phiC31 integrase recombination in mammalian cells. Biotechnol J 7:1332–1336

Chapter 4

Therapeutic Genome Mutagenesis Using Synthetic Donor DNA and Triplex-Forming Molecules

Faisal Reza and Peter M. Glazer

Abstract

Genome mutagenesis can be achieved in a variety of ways, though a select few are suitable for therapeutic settings. Among them, the harnessing of intracellular homologous recombination affords the safety and efficacy profile suitable for such settings. Recombinagenic donor DNA and mutagenic triplex-forming molecules co-opt this natural recombination phenomenon to enable the specific, heritable editing and targeting of the genome. Editing the genome is achieved by designing the sequence-specific recombinagenic donor DNA to have base mismatches, insertions, and deletions that will be incorporated into the genome when it is used as a template for recombination. Targeting the genome is similarly achieved by designing the sequence-specific mutagenic triplex-forming molecules to further recruit the recombination machinery thereby upregulating its activity with the recombinagenic donor DNA. This combination of extracellularly introduced, designed synthetic molecules and intercellularly ubiquitous, evolved natural machinery enables the mutagenesis of chromosomes and engineering of whole genomes with great fidelity while limiting nonspecific interactions.

Herein, we demonstrate the harnessing of recombinagenic donor DNA and mutagenic triplex-forming molecular technology for potential therapeutic applications. These demonstrations involve, among others, utilizing this technology to correct genes so that they become physiologically functional, to induce dormant yet functional genes in place of non-functional counterparts, to place induced genes under regulatory elements, and to disrupt genes to abrogate a cellular vulnerability. Ancillary demonstrations of the design and synthesis of this recombinagenic and mutagenic molecular technology as well as their delivery and assayed interaction with duplex DNA reveal a potent technological platform for engineering specific changes into the living genome.

Key words Triplex-forming oligonucleotide (TFO), Triplex-forming peptide nucleic acid (PNA), Donor DNA, Recombination, Molecular delivery, Genome targeting, Genome editing, Genome mutagenesis, Genome engineering

1 Introduction

For genome mutagenesis, significant sequence-specificity metrics should be satisfied. For therapeutic genome mutagenesis these standards are raised even higher so that significant on-target and off-target safety metrics are met as well. Among the means of chro-

Shondra M. Pruett-Miller (ed.), *Chromosomal Mutagenesis*, Methods in Molecular Biology, vol. 1239,
DOI 10.1007/978-1-4939-1862-1_4, © Springer Science+Business Media New York 2015

mosome and whole genome mutagenesis methods that meet these raised standards are those that utilize the intracellular phenomena of homologous recombination using extracellularly introduced recombinagenic donor DNA and mutagenic triplex-forming molecular technology.

Mutagenesis by homologous recombination involves the artificial and deliberate exchange and integration of disparate pieces of DNA that share some sequence similarity, but with tolerance for dissimilarity. Naturally occurring homologous recombination transpires to create genetic variation and diversification on a microscopic level in order to tolerate and adapt during the course of evolution at a macroscopic level. Artificial insertion of DNA sequences into human genomic loci, such as for the chromosomal beta-globin locus, has been achieved by homologous recombination [1]. The technological applications for homologous recombination include intracellular gene targeting and editing [2] and genome and protein engineering [3]. Specifically, the nucleotide excision repair, or NER, pathway has been implicated to be involved in triplex-formation induced recombination [4] and this induced recombination has even been triggered by intracellular generation of single-stranded DNA for chromosomal triplex formation [5]. The endogenous repair machinery recruited for this NER-based recombination by the triplex-forming molecules utilizes includes xeroderma pigmentosum, complementation group A (XPA) protein [6, 7], xeroderma pigmentosum, complementation group C (XPC) [8, 9], replication protein A (RPA) [6], and Cockayne's syndrome, group B (CSB) protein [10], among others, and complexes thereof to reduce nonspecific recombination. In doing so, targeted chromosome and genome mutagenesis is feasible in mammalian cells in a safe and effective manner [11, 12].

Cells proficient in this repair machinery can be treated with recombinagenic donor DNA molecules as templates for genome editing at a recombination-potentiated editing site. These recombinagenic donor DNA molecules are single-stranded and nearly homologous to the genome editing site and neighboring bases on one strand of an allele of the chromosome, except for the mismatched nucleobases, as genome mutation, insertion, or deletion edits. This recombinagenic donor DNA preferentially localizes directly to the homologous genome editing site upon transfection. Given that there are a number of possible recombinagenic donor DNA, that vary in sequence length, sequence off-centering from the genome editing site, and sequence homology and mismatches to the strand to be edited, or the complementary strand, a variety of molecules can be designed and empirically evaluated for a genomic locus for recombination efficiency. Further efficiency, along with specificity, can be gained by coupling the activities of this recombinagenic donor DNA with the mutagenic capacity of triplex-forming molecules. In doing so for gene editing, a 51-mer recombinagenic donor DNA has designed to correct the beta-globin gene from mis-splicing (Fig. 1a), a 54-mer recombinagenic donor DNA has been designed to induce the gamma-globin gene from postnatal dormancy and place it under exquisite control of the in vivo hypoxia microenvironment (Fig. 1b), and two 60-mers recombinagenic donor DNA have been designed to disrupt the *CCR5* gene with premature stop codons (Fig. 1c).

Since recombinagenic donor DNA are foreign matter, they encounter the cellular defense machinery as well. In order to combat this, recombinagenic donor DNA can be synthetically modified to resist these defenses. For example, three nucleotides at each terminus of the recombinagenic donor DNA are modified to phosphorothioate linkages, so that phosphodiester linkage recognizing nucleophilic enzymes are unable to attack the molecule (Fig. 1d). While this modification resists nucleases from digesting the recombinagenic donor DNA, this internucleotide linkage modification does not prevent the recombination machinery proteins from using it as a template for recombination.

While modest levels of recombination and, thus, modification frequency occur when recombinagenic donor DNA is utilized by the recombination machinery, significantly greater levels are achieved through the use of triplex-forming molecules that heighten the recruitment and activity of this machinery in a dose-dependent manner. It has only been a little over six decades since the discovery of the duplex structure of deoxyribose nucleic acid (DNA) was made by Watson and Crick [13]. Four brief years after this discovery, the discovery of the triplex structure of DNA, whereby a third strand of polyuracil bound a duplex of polyadenine and polyuracil, was made by Felsenfeld, Davies, and Rich [14]. In the mid-1980s, further elucidation not just of their structure, but of some of their mechanism of action, such as sequence-specific distortion and cleavage was made by Dervan and coworkers as well as by Helene and coworkers [15, 16].

These discoveries have led to greater understanding and application of a diverse array of triplex-forming molecules that are designed for a variety of purposes. Triplex-forming molecules that are composed of oligonucleotides, such as those used in the first triplexes discovered, are termed triplex-forming oligonucleotides (TFOs). These oligonucleotides are made of nucleobases and linked by phosphodiester linkages, except for some notable synthetic modifications that are suited to their genome binding and nuclease resistance roles. The TFO molecule also has an orientation, just as other oligonucleotides, whereby it is parallel or antiparallel in polarity to the polypurine strand of genomic DNA. Furthermore, the major groove binding to duplex DNA necessitates Hoogsteen or reverse Hoogsteen base pairing, as the natural DNA Watson–Crick donor–acceptor base pairing atoms are already occupied. Thus, Hoogsteen base pairing occurs so that adenine, or A, guanine, or G, and thymine or T, base pair with adenine, guanine,

a recombinagenic donor DNA for beta-globin gene editing

51-mer IVS2 donor: 5'-GTT CAG CGT GTC CGG CGA GGG CGA GgT GAG TCT ATG GGA CCC TTG ATG TTT-3'

NOTE: donor is sense

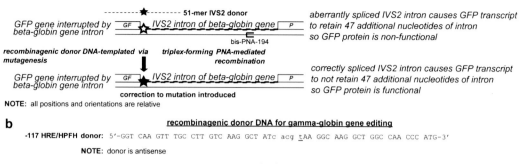

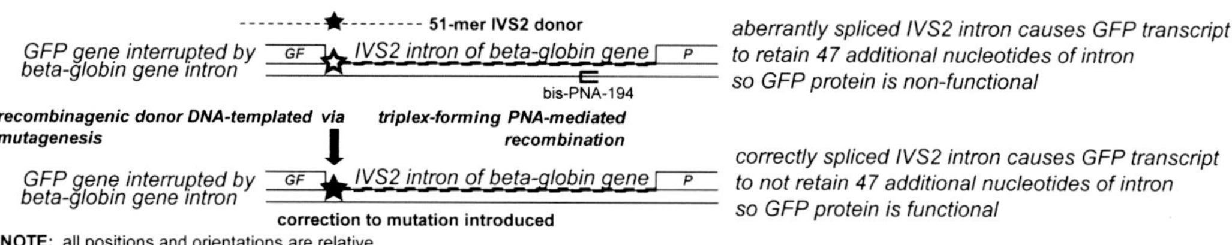

51-mer IVS2 donor
GFP gene interrupted by beta-globin gene intron
GF
IVS2 intron of beta-globin gene
P
bis-PNA-194
aberrantly spliced IVS2 intron causes GFP transcript to retain 47 additional nucleotides of intron so GFP protein is non-functional
recombinagenic donor DNA-templated via mutagenesis
triplex-forming PNA-mediated recombination
GFP gene interrupted by beta-globin gene intron
GF
IVS2 intron of beta-globin gene
P
correction to mutation introduced
correctly spliced IVS2 intron causes GFP transcript to not retain 47 additional nucleotides of intron so GFP protein is functional
NOTE: all positions and orientations are relative

b recombinagenic donor DNA for gamma-globin gene editing

-117 HRE/HPFH donor: 5'-GGT CAA GTT TGC CTT GTC AAG GCT ATc acg tAA GGC AAG GCT GGC CAA CCC ATG-3'

NOTE: donor is antisense

gamma-globin promoter γ-194-3K **-117 HPFH/HRE donor** *low/absent gamma-globin expression*

recombinagenic donor DNA-templated via mutagenesis ***triplex-forming PNA-mediated recombination***

gamma-globin promoter **HPFH mutation and HRE site introduced** *increased gamma-globin expression and responsive to hypoxia*

NOTE: all positions and orientations are relative

c recombinagenic donor DNA for CCR5 gene editing

CCR5-597 donor: 5'-TT TAG GAT TCC CGA GTA GCA GAT GAC Ccc tca gAG CAG CGG CAG GAC CAG CCC CAA GAT G-3'

CCR5-591 donor: 5'-AT TCC CGA GTA GCA GAT GAC CAT GAC Agc tta gGG CAG GAC CAG CCC CAA GAT GAC TAT C-3'

NOTE: donors are antisense

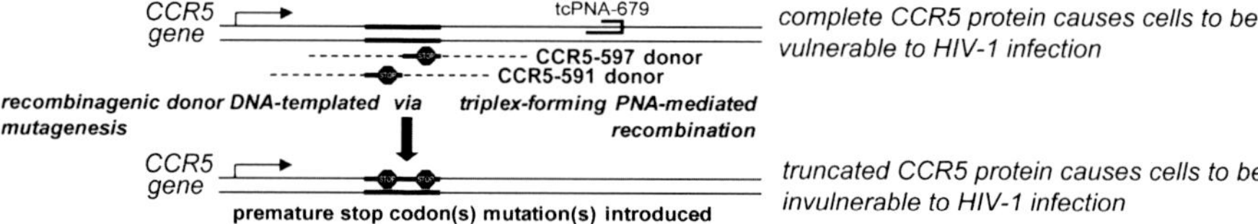

CCR5 gene
tcPNA-679
complete CCR5 protein causes cells to be vulnerable to HIV-1 infection
CCR5-597 donor
CCR5-591 donor
recombinagenic donor DNA-templated via mutagenesis
triplex-forming PNA-mediated recombination
CCR5 gene
truncated CCR5 protein causes cells to be invulnerable to HIV-1 infection
premature stop codon(s) mutation(s) introduced

NOTE: all positions and orientations are relative

d

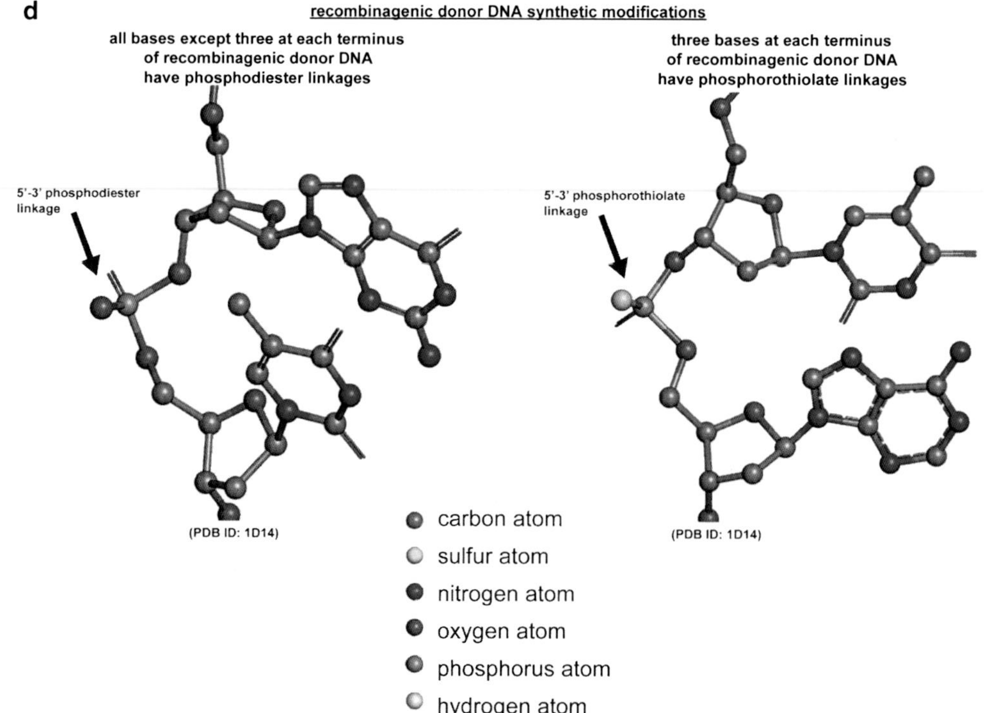

recombinagenic donor DNA synthetic modifications
all bases except three at each terminus of recombinagenic donor DNA have phosphodiester linkages
three bases at each terminus of recombinagenic donor DNA have phosphorothiolate linkages
5'-3' phosphodiester linkage
5'-3' phosphorothiolate linkage
(PDB ID: 1D14)
(PDB ID: 1D14)
carbon atom
sulfur atom
nitrogen atom
oxygen atom
phosphorus atom
hydrogen atom

and adenine, respectively, of the Watson–Crick-base-paired DNA duplex to form the triplex. A special case occurs with cytosine, or C, on this third strand as its N3 atom lacks the hydrogen atom at physiological pH to be a Hoogsteen base pairing hydrogen bond donor. Thus, while TFOs in which this N3 atom is protonated in acidic pHs are able to Hoogsteen base pair with duplex DNA to form triplex structures, an unnatural nucleobase such as pseudoisocytidine or J, a, C-nucleoside analog that mimics the N3 protonation of cytosine in triplex-forming peptide nucleic acids (PNAs), can Hoogsteen base pair with the guanine of a Watson–Crick base pair without this pH dependency.

The major groove of duplex genomic DNA provides the steric and conformational clearance for TFOs and other triplex-forming molecules to bind and form this triplex molecule. In the triplex code dictated by Hoogsteen base pairing, the third strand triplex-forming molecule can bind duplex genomic DNA in a sequence-specific manner in either a parallel or antiparallel motif. The antiparallel purine motif involves a polypurine TFO that binds to the purine strand of the duplex genomic DNA through reverse Hoogsteen base pairing and in an orientation that has antiparallel polarity [17]; whereas the parallel pyrimidine motif involves a polypyrimidine TFO that binds to the purine strand of the duplex genomic DNA through regular, not reverse, Hoogsteen base pairing and in an orientation that has parallel polarity [18, 19].

The chemistry of triplex-forming molecules can be further modified from TFOs in triplex-forming PNAs. These triplex-forming PNAs have been designed to upregulate the mutagenic potential of a recombination machinery interacting with recombinagenic donor DNA for the beta-globin gene (Fig. 2a), the gamma-globin gene (Fig. 2b), and the *CCR5* gene. In these triplex-forming PNAs, not only are the termini internucleoside linkages, but rather the all of the internucleoside linkages are modified so that the negatively charged phosphodiester linkages are replaced by uncharged polyamide linkages [20] (Fig. 2d). This neither wholly nucleic nor amino acid biopolymer is not recognized by cellular nucleases and proteases and, in addition, is better electrostati-

Fig. 1 (continued) Synthetic recombinagenic donor DNA molecules and their genomic editing sites. Recombinagenic donor DNA molecules have been designed to edit genomic sites for (**a**) the correction of a splice site mutation in the beta-globin gene to reestablish functional beta-globin expression, (**b**) the induction of expression, and regulation by hypoxia, of a silenced, but functional, fetal gamma-globin gene to supplant non-functional beta-globin expression in adults, (**c**) and the disruption of the *CCR5* gene to eliminate a receptor-mediated entry of HIV-1. (**d**) For these exogenously introduced recombinagenic donor DNA molecules, positive design using Watson–Crick oriented base pairing with genomic DNA is implemented, as well as negative design using synthetic modifications to the linkages at each terminus to prevent endogenous nuclease-mediated degradation. Linkages for recombinagenic donor DNA molecular modeled from [97]

a

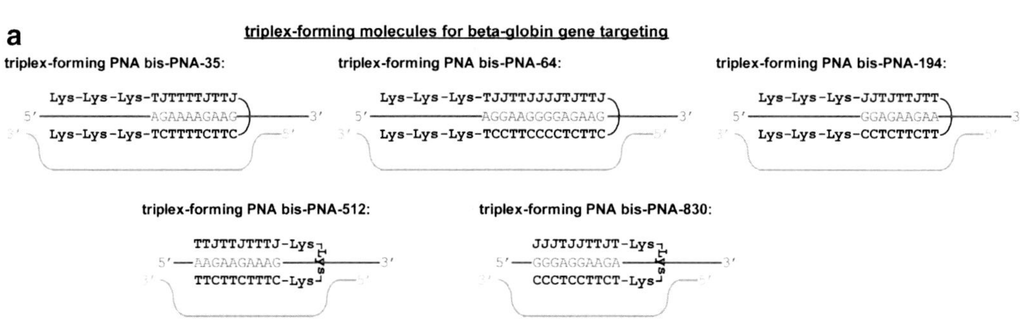

NOTE: all positions and orientations are relative

b

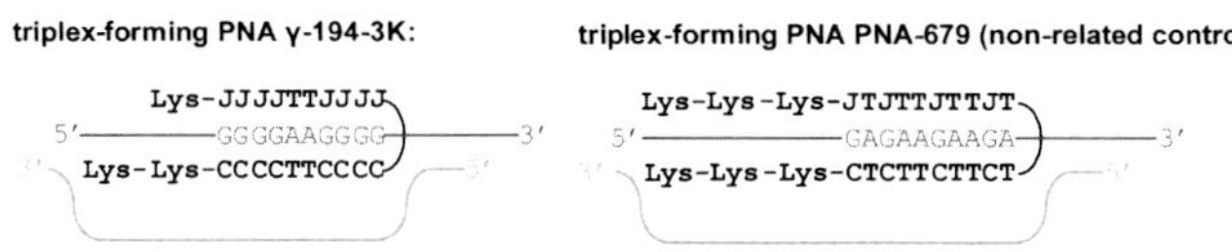

NOTE: all positions and orientations are relative

c

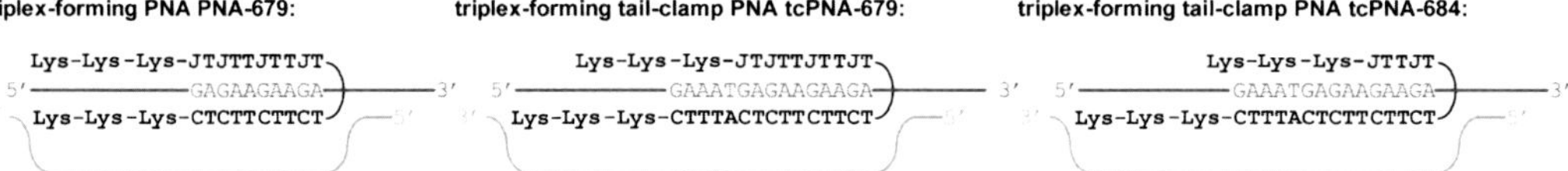

NOTE: all positions and orientations are relative

d

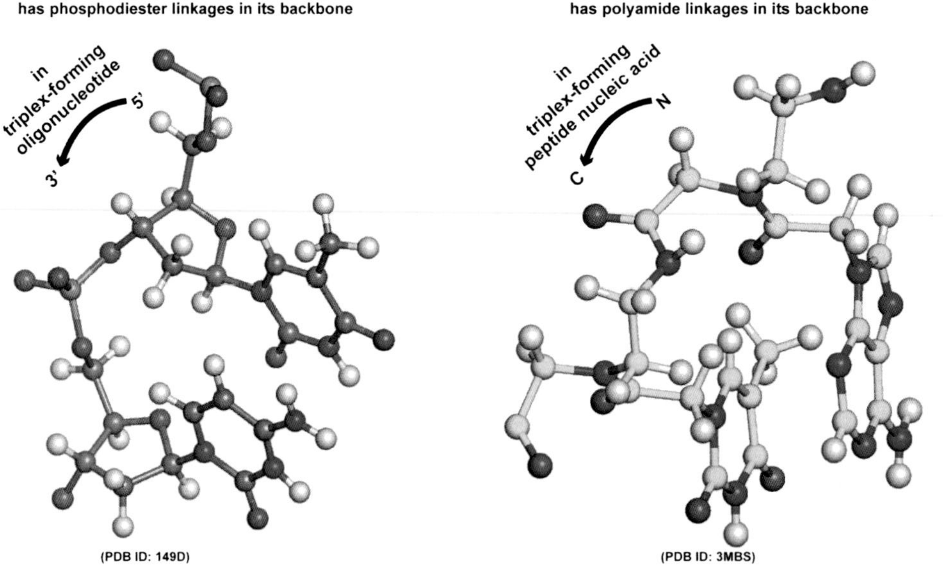

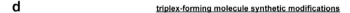

- carbon atom (in triplex-forming oligonucleotide)
- carbon atom (in triplex-forming peptide nucleic acid)
- nitrogen atom
- oxygen atom
- phosphorus atom
- hydrogen atom

cally complemented by the negatively charged genomic DNA phosphodiester linkages [21]. Furthermore, triplex-forming PNAs are resistant to polymerase recognition and thus will not contribute to the crosstalk noise by itself providing a template for replication or transcription [22].

Antiparallel and parallel triplex-forming DNA stimulates homologous recombination in human cells [19, 23]. TFOs and triplex-forming PNAs retain the ability to bind sites on genomic and episomal DNA in a recombination-like motif with high thermodynamic stability to form triplexes [24]. Unlike the recombinagenic donor DNA molecule, which localizes directly to the genome editing site, the mutagenic triplex-forming molecule can potentiate the recombination machinery upon binding to a genome targeting site that is immediately overlapping the genome editing site, or as distant as almost as a kilobase away. Under acidic conditions, the natural base cytosine in a TFO can be protonated so that it can Hoogsteen bind to guanine bases in the genomic target. Similar to TFOs, triplex-forming PNAs can form triplexes in the major groove of DNA [25] under normal physiological conditions using the unnatural base pseudoisocytidine. By linking a triplex-forming PNA with a strand-invading PNA through a flexible linker, a triplex-forming bis-PNA can simultaneously Hoogsteen and Watson–Crick hydrogen base pair to the same base in duplex genomic DNA in a sequence-specific manner at physiologically high melting temperatures [26]. Further extensions of the strand-invading PNA "tail" Watson–Crick base pairing domain beyond that of the triplex PNA "clamp" Hoogsteen base pairing domain creates a triplex-forming tail-clamp PNA that garners greater affinity as well as specificity for the DNA targeting site [27]. These binding affinities of TFOs and triplex-forming PNAs have been demonstrated to be considerable enough in vitro to achieve a variety of perturbations to cellular processes, including the obstructing DNA polymerization [28], inhibiting transcription initiation and elongation [29], and preventing sequence-specific protein binding [30].

Fig. 2 (continued) Synthetic mutagenic triplex-forming molecules and their genomic targeting sites. Triplex-forming molecules have been designed to target genomic sites, thereby upregulating homologous recombination with the recombinagenic donor DNA molecules, or as non-related controls, to examine targeting specificity, for (**a**) the beta-globin gene, where triplex-forming bis-PNAs are used, (**b**) the gamma-globin gene, where triplex-forming PNAs were used, (**c**) and the *CCR5* gene, where triplex-forming PNAs and triplex-forming tail-clamp PNAs were used. (**d**) For these exogenously introduced triplex-forming molecules, positive design is implemented based on major groove binding Hoogsteen oriented base pairing and strand-invading Watson–Crick oriented base pairing with genomic DNA, as well as negative design based on synthetic modifications to the linkages in the backbone to prevent endogenous nuclease and protease-mediated degradation and enhance electrostatic complementarity to negatively charged genomic DNA. Linkages for triplex-forming oligonucleotide and triplex-forming peptide nucleic acid molecular modeled from [98] and [99], respectively

In addition, the conformational capacity of triplex-forming PNAs to potentially form D-loop open DNA conformations have been demonstrated to act as artificial transcriptional promoters [31–35] and diethylene glycol (miniPEG)-based triplex-forming gamma PNAs have been demonstrated to act as an antisense gene targeting agent to silence the *CCR5* gene [36]. This triplex-forming molecular activity can further alter genomic conditions by inducing [37] or inhibiting [38] DNA transcription, DNA replication [39–41], and protein–DNA interactions [42–45], and promoting site-specific DNA damage [10, 46–48], mutagenesis [49, 50], or recombination on chromosomal and episomal DNA [7, 51, 52].

There still exist many challenges in efficiently delivering recombinagenic donor DNA and mutagenic triplex-forming technology in vitro and in vivo, though a number of strategies have advanced the state-of-the-art (*see* Subheading 3). Interactions of triplex-forming PNAs with cellular and nuclear membranes and negatively-charged genomic DNA have been improved through the decoration with positively charged lysine residues [37] and conjugation to cell-penetrating peptides such as Antennapedia and transactivator of transcription (TAT) [53] as well as polymeric nanoparticles [54]. This is significant, since upon arrival into the cytoplasm, the functionality of triplex-forming molecules can be inhibited by cellular conditions, such as concentrations of cytosolic potassium and magnesium. Within the nucleus, triplex-forming molecular interaction may also be limited by the accessibility of its target genomic site as a result of its chromosomal location and interaction with other genomic components, such as chromatin, regulatory proteins, and other molecules competing for access to the same genomic targeting site. A nuclear localization signal when conjugated to TFO and triplex-forming PNA molecules has been shown to increase directed delivery to the nucleus [55]. The aforementioned modifications, among others, have been used to increase binding affinity as well. For example, binding affinity of TFOs in vitro has been increased by modifying the linkages between bases, such as by replacing the phosphodiester linkages with cationic phosphoramidate linkages, *N*, *N*-diethylethylene-diamine or *N*, *N*-dimethyl-aminopropylamine [56]. In G-rich oligonucleotides, the use of *N*, *N*-diethylethylene-diamine-modified bases may also mitigate G-quartet formation. The use of pseudoisocytosine, J, rather than cytosine, C, in triplex-forming molecules in order to Hoogsteen pair with guanine, G, in a DNA targeting site at physiological pH has already been mentioned. Thus, in addition to the linkages between bases, the bases themselves can be modified to increase the binding affinity of TFOs in the pyrimidine motif, for example by using 5-methyl-2-deoxyuridine and 5-methyl-2-deoxycytidine, as too can the sugar moieties to which they are covalently linked, for example to a 2-*O*-(2-aminoethyl)-ribose [57, 58].

Upon delivery, triplex formation at a specific targeting site leads to gene targeting by provoking the cell's own DNA repair pathways, primarily the NER and recombination machinery, to localize near that site in a similarly safe manner to that which has evolved already to repair naturally occurring triplex adducts [59]. This a versatile mechanism of mutagenesis, as the inherent sequence specificity of a triplex-forming molecule can be coupled to a DNA damage-causing mutagen that is not sequence or structure specific, such as psoralen (pso), to a choromosomal locus of interest [60, 61]. These pso-TFOs have induced mutations to a target site on plasmids in vitro, when transfected into mammalian cells, and on chromosomes [46, 49, 62]. Furthermore, the influence of the cell cycle on genome accessibility and recombination machinery availability has been elucidated, showing that by synchronizing and treating cells in late S-phase this targeted mutagenesis could be increased 5.5-fold over those in G1, and 2.5-fold over cells in early S-phase [63]. By modulating the cell cycle or transcriptional state, it may be possible to increase the efficiency of triplex-forming molecule induced mutagenesis.

The detection and quantification of these mutations induced by triplex-forming molecules, such as TFOs, on plasmids in vitro, has been developed in a blue and white cell reporter system. This reporter system is composed of the *supF* reporter gene, which encodes an amber suppressor tyrosine tRNA having a TFO targeting site, cloned into an SV40 plasmid vector. When cells carrying this cloned plasmid vector are plated in the presence of 4-chloro-5-bromo-3-indolyl-β-D-galactopyranoside and isopropyl-β-D-1-thiogalactopyranoside, bacteria having an amber mutation in the *lacZ* gene but a functional copy of the *supF* gene produce blue colonies. However, a nonfunctional, or mutated, copy of the *supF* gene due to the mutagenic triplex-forming molecule produces white colonies. The number and ratio of white colonies to blue colonies can be used to calculate the mutagenesis frequency [46, 64]. Using this reporter system, mutagenesis frequencies due to pso-TFOs have been observed and quantified in murine cells containing a chromosomally integrated copy of the *supF* reporter gene [65]. Whereas psoralen has been shown to induce mutations, formation of a triplex structure alone using either TFOs or triplex-forming PNAs has also been shown to be sufficient to stimulate mutagenesis and treated by the cellular repair and recombination machinery accordingly [25, 65]. Extending these studies even further, in transgenic mice containing the *supF* reporter gene integrated in the chromosome, intraperitoneal injection of a TFO, AG30, also led to site-specific mutations [49]. Note that when this mutagenesis occurs at a low frequency tolerated by the cell studies have indicated that the cell fate is one of repairing the target site and environs by upregulating the frequency of homologous recombination [66]. To exploit this idea, pso-TFOs have been used to create site-specific damage in order to sensitize a target site for subsequent homologous recombination [67, 68]. It was found that not only could pso-TFO-associated DNA damage increase intrachromosomal recombination [4], but also a TFO without a DNA-damaging agent was sufficient to induce homologous recombination [7, 52, 69]. Triplex formation increased the level of recombination at a targeting site and led to gene correction of a specific mutation [7]. Using a plasmid with two tandem *supF* genes, each containing different point mutations, increased intramolecular recombination upon binding of a TFO was demonstrated and resulted in gene correction of one copy of the gene [67]. Furthermore, homologous recombination was stimulated by TFOs with or without pso at a frequency of 1 %, which is 2,000-fold over with the following, and with italics on the words thymidine kinase: background in the nuclei of murine cells containing two mutant copies of the herpes simplex virus *thymidine kinase* gene [52]. Given the potential for a TFO to be mutagenic, and given the potential for the recombination machinery to be recruited by this mutagenic induction, it then followed that linking a TFO to a short donor fragment, would act to upregulate recombination near a co-localized editing site and recombinagenic donor DNA [51]. That recombinagenic DNA donor fragment is almost homologous to the targeting site, except for the edited bases [70]. For certain editing sites, the more effective designed recombinagenic donor DNA have been antisense (homologous to the transcribed strand) [69] but this is not universally the case for other sites and sense counterparts [26]. A designed recombinagenic donor also need not be for an editing site near the targeting site to be effective since recombination has been detected at sites up to three-fourths of a kbp away from the targeting site [71]. Site-specific intermolecular homologous recombination in a singe genomic targeting site up to a frequency of 0.11 % has been achieved in mammalian cells by TFOs designed employing either motifs [69].

Recombinagenic donor DNA and mutagenic triplex forming molecules, when used for gene editing and targeting of chromosomes and genomes, has numerous potential therapeutic applications. In particular, the editing and targeting of a gene or genes in living cells, and the subsequent propagation of the results of those targetings and editings in progeny provides a means of permanently transforming cells from disease-prone to disease-free. In order to implement recombinagenic donor DNA and mutagenic triplex-forming molecular technology, a series of methods are recommended, such as for identifying genomic editing and targeting sites, designing the synthetic molecules to interact with these sites, and then assessing the binding affinities and modification frequencies (*see* Subheading 3).

There are a variety of therapeutic gene targeting and editing capabilities of triplex-forming oligonucleotides and recombinagenic donor DNA molecules. Among these capabilities are gene correction, induction with regulation, and disruption, as discussed below.

For gene correction, donor DNA carry one or more corrected bases that will be utilized by a homologous recombination machinery, which is upregulated due to a triplex-forming molecule, to correct a particular disease-causing gene. In the case of a hematologic disorder, beta-thalassemia a common mutation, that of a guanine to an adenine mutation (G→A), on the second intron at position 1 (IVS2-1) of the beta-globin gene residing on the short arm of human chromosome 11 has been successfully modified to a healthy state using recombinagenic donor DNA and mutagenic triplex-forming technology in living cells [26, 72]. In a dual-pronged approach, using Chinese hamster ovary (CHO) GFP reporter cells the correction of this particular splicing mutation was demonstrated, as was the creation of this beta-thalassemia mutation in healthy human CD34+ (hCD34+) cells.

To elaborate, the CHO GFP reporter cell system integrated a genomic green fluorescent protein (GFP) gene interrupted by the IVS2 intron of the beta-globin gene with a G→A mutation at its position 1 [26]. When this mutation went uncorrected, then the expressed transcript from this locus was not correctly spliced and retained an 47 additional nucleotides arising from the beta-globin intron causing the GFP protein to be improperly translated and the CHO GFP reporter cell carrying it to not be fluorescent. However, if this mutation is corrected using a triplex-forming bis-PNA at position 194 in conjunction with a recombinagenic donor DNA carrying the A→G correction at position 1, then proper splicing and translation occurs so that the CHO GFP reporter cell carrying it does become fluorescent. Each fluorescent or non-fluorescent cell can be accurately measured by fluorescence activated cell sorting (FACS) of these CHO model system cells and the recombinagenic and mutagenic potential of donor DNA and triplex-forming bis-PNA to cause gene correction can be quantified. In addition to the doses of triplex-forming bis-PNAs and donor DNAs to further increase their availability from cellular sequestration by lysosomes chloroquine can be added to the treatment regimen. The correction percentage can also be modulated by synchronizing the CHO cells at various cell cycle phases just before the treatment regimen as well as increase accessibility of the genomic DNA (Table 1).

These modifications can be detected at the phenotypic level, and also at the genotypic level. A phenotypic assay can be such as the aforementioned fluorescent protein reporter system can be used to detect the consequence of a modifiction or at the genomic level, allele-specific polymerase chain reaction (PCR) can be used to detect a sequence modification. Extending this further, allele specific quantitative PCR (qPCR) can detect modification as well as quantify modification frequencies in treated cells relative to cells

Table 1
Recombinagenic and mutagenic frequencies of therapeutic targeting and editing of genomic loci to ameliorate disease etiologies

Disease etiology	Genomic locus	Recombinagenic and mutagenic molecules	Cell type	Molecular delivery modality	Recombinagenic and mutagenic frequency (%)
Beta-thalassemia	Beta-globin	Donor DNA	CHO	electroporation	0.05
Beta-thalassemia	Beta-globin	Donor DNA+ triplex-forming PNA	CHO	electroporation	0.20
Hemoglobinopathies, such as sickle cell disease	Gamma-globin	Donor DNA	K562	electroporation	0.29
Hemoglobinopathies, such as sickle cell disease	Gamma-globin	Donor DNA+ triplex-forming PNA	K562	electroporation	1.63
HIV-1	CCR5	Donor DNA+ triplex-forming PNA	hCD34+	nucleofection	0.10
HIV-1	CCR5	Donor DNA+ triplex-forming PNA	hCD34+	nanoparticles	0.91
HIV-1	CCR5	Donor DNA+ triplex-forming PNA	THP-1	electroporation	2.46

Recombinagenic and mutagenic frequencies on the chromosome or episome are influenced by a number of factors, including the particular endogenous genomic locus, exogenously introduced molecules, cell type, and molecular delivery modality

with genomes having known modification frequencies. Furthermore, at the transcriptional mRNA level, combined quantitative reverse transcriptase real time PCR (qRT-PR) can show the presence and amount of gene expression from the modification of the genome.

For gene induction and regulation disruption, donor DNA and triplex-forming molecule modify the silenced promoters of functional but dormant genes and perturb them by editing in a regulatory element. These induced gene products can mitigate hematologic disease [37]. For example, a cause of thalassemias, sickle cell disease, and other hemoglobinopathies is a non-functional beta-globin gene and subunit in adult hemoglobin. The hereditary persistence of fetal hemoglobin (HPFH) is a benign condition in which a functional, fetal gamma-globin subunit is expressed in adults and can outcompete the polymerization of non-functional beta-globin subunits with alpha-globin subunits, can alleviate hemoglobinopathies, such as thalassemias, arising from the non-functional beta-globin [73, 74]. Chin, Reza, and Glazer demonstrated that the targeted induction of the functional fetal gamma-globin gene also on the short arm of human chromosome 11 in cells carrying the non-functional adult beta-globin gene can mimic HPFH and restore the working state of hemoglobin, thus lessening the severity of these diseases. A triplex-forming bis-PNA targeting the gamma-globin promoter along with a HFPH recombinagenic donor DNA at the -117 position that also edited in a hypoxia responsive element (HRE) to regulate the HPFH state was used in the treatment. An unrelated control triplex-forming bis-PNA was also evaluated to examine whether the triplex-forming PNAs were acting as artificial promoters of gene expression. A triplex-forming PNA and either -117 HRE/HPFH donor DNA or control donor that introduced the HPFH mutation but not HRE were used to treat hCD34+ cells, that were subsequently subjected to a hypoxic environment for two days. The harvested RNA of treated cells, as well as those of control cells from a normal oxygen tension environment, were subjected to qRT-PCR with TaqMan® probes indicated that gamma-globin was being expressed and regulated by hypoxia. In contrast, cell samples treated with a gamma-globin triplex-forming PNA and an unrelated control triplex-forming PNA both showed minimal induction of expression of gamma-globin, thus suggesting that the donor DNA editings were the cause for these changes. Allele-specific qPCR relative amplification values were correlated with modification frequencies to demonstrate that greater recombinagenic and mutagenic frequencies were achieved when using donor DNA and triplex-forming PNA concurrently (Table 1).

For gene disruption, donor DNA and triplex-forming tail-clamp PNA create premature stop codons and, in turn, create non-functional gene products. Premature stop codons can thus stop the production of a cellular component that makes the cell vulnerable to disease or infection. A naturally occurring mutation, called *CCR5-delta32* to the chemokine receptor gene, prevent proper CCR5 chemokine receptor trafficking and ultimately an inability of human immunodeficiency virus, type 1 (HIV-1) to infect human cells [75]. This suggests various strategies of recapitulating this mutation to prevent HIV-1 propagation, such as by intracellular immunization by a CCR5 blocking single-chain antibody therapy [76], or by non-autologous transplantation of stem cells homozygous for this *CCR5*-delta32 mutation [77], or by knockdown siRNAs of CCR5 delivered by lentiviral vectors [78], but, given their transient nature and infectious delivery mechanism respectively, their translational potential as therapeutics is uncertain [79]. Disruption of the *CCR5* gene by using donor DNA and tail-clamp PNA molecules delivered to hCD34+ hematopoietic stem cells

(HSCs) by nanoparticles successfully resulted in permanent, heritable, non-immunogenic, and non-infectious disruption of HIV-1 [27, 80]. the self-renewal and heritable capabilities of (HSCs) are advantageous for gene targeting and editing therapeutic intervention that persists and propagates further. For HIV-1 infections, hCD34+ HSCs are also advantageous since they have been shown to remain unaffected [81].

Co-transfection of a triplex-forming tail-clamp PNA and donor DNA led to the greatest amount of gene disruption, enough to nullify CCR5 expression over single-cell clones and their progeny. Specific on-target effects were 43-fold greater than the negligible off-target effects, as determined by DNA sequencing from single-cell cloning assays. In doing so, the differentiation capacity of THP-1 human acute monocytic leukemia cells into a macrophage-like state was maintained, as was high levels of CCR5 cell surface expression. The triplex-forming PNAs and donor DNAs introduced to create both 591 and 597 gene disruptions double mutant resulted in differentiated parental THP-1 cells with negligible cell surface staining for CCR5.

This disruption of the CCR5 protein now led to resistance to HIV-1 infection. Cloned THP-1 cells treated with these molecules were isolated, expanded, and induced to express *CCR5* and then challenged with live HIV-1. The treated cells, as opposed to the parental cells, had significantly lower core protein p24 antigen levels, which indicate the amount of HIV-1 infection in cells.

Molecular delivery of these mutagenic and recombinagenic molecules using electroporation, nucleofection, cell-penetrating peptides, and nanoparticles, have been applied to in vitro and in vivo targeting and site-specific editing [54]. Human HSCs engrafted in NOD-*scid* $IL2r\gamma^{null}$ mice were treated with biodegradable nanoparticles carrying recombinagenic triplex-forming PNAs, recombinagenic donor DNA molecules, or both. An in vitro screen indicated that nanoparticles encapsulating both more effectively achieved genome targeting and editing of human hematopoietic cells [82]. Further direct deep sequencing of the *CCR5* gene had modification on-target at frequencies of 0.43 % in these stem cells found in the spleen, and at 0.05 % in the bone marrow, and off-target at two order of magnitudes lower off-target to related genes.

Thus, recombinagenic donor DNA and mutagenic triplex-forming molecular technology have the demonstrated potential and performance record to target and edit chromosomes and engineer genomes in a safe and effective heritable fashion. This technology may play a pivotal therapeutic role in transforming and then transferring a patient's own once disease-causing cells to those that are disease-free by genomes and chromosomes that have had their genes corrected, induced, regulated, or disrupted by recombinagenic donor DNA and mutagenic triplex-forming molecules.

2 Materials

Synthetic molecules, such as the single-stranded recombinagenic donor DNA molecules and the mutagenic triplex-forming molecules, TFO, triplex-forming peptide nucleic acid, mentioned herein, can be procured from commercial vendors worldwide, an institutional oligonucleotides facility, or synthesized in the laboratory. Upon synthesis, it is highly recommended that these oligonucleotides be purified using high-pressure liquid chromatography or gel purification to remove synthesis reagents or impurities prior their use in the treatment of cells. Further best practices are highly recommended for industrial formulation, scale-up, packaging, and quality control.

2.1 Recombinagenic DNA Donors

1. Recombinagenic donor DNA can be synthesized in the laboratory using sequentially coupled solid-phase phosphoramidite chemistry, consisting of synthesis cycles of deblocking (detritylation) of a acid-labile DMT (4,4′-dimethoxytrityl) group on the 5′-hydroxyl group, coupling with a nucleoside phosphoramidite in acetonitrile that is catalyzed by an acidic azole catalyst, capping with a mixture of acetic anhydride and 1-methylimidazole, and lastly oxidation with iodine and water having a weak base (*see* **Notes 1** and **2**).

   Alternatively, long double-stranded recombinagenic donor DNA can be synthesized by PCR amplification from a plasmid carrying the donor sequence of interest.
2. Recombinagenic donor DNA can be procured from an institutional oligonucleotide synthesis facility.
3. Recombinagenic donor DNA can be procured from a commercial supplier, such as The Midland Certified Reagent Company (Midland, TX, USA) or Panagene (South Korea) (*see* **Note 3**).

2.2 Triplex-Forming Molecules (TFOs and Triplex-Forming PNAs)

1. TFOs can be synthesized can be synthesized in the laboratory using, again, sequentially coupled phosphoramidite chemistry.
2. TFOs can be procured from an institutional oligonucleotide synthesis facility.
3. TFOs can be procured from a commercial supplier, such as The Midland Certified Reagent Company (*see* **Notes 4** and **5**).
4. Triplex-forming PNAs can be synthesized in the laboratory using an α-amino protecting group based on 9-fluorenylmethyloxycarbonyl, or Fmoc, or based on tert-butyloxycarbonyl, or Boc, solid-phase peptide synthesis chemistry. Fmoc and Boc are protecting groups that permit proper peptide linked moiety synthesis by blocking nonspecific reactions. The chemistry for either consists of an initial activation step, followed by synthesis

cycles of coupling and deblocking of the Fmoc or Boc protecting group, and then a final deblocking step (*see* **Note 6**).

5. Triplex-forming PNAs can be procured from an institutional peptide synthesis facility (*see* **Note 7**).
6. Triplex-forming PNAs can be procured from a commercial supplier, such as Bio-Synthesis (Lewisville, TX, USA), PNA BIO (Thousand Oaks, CA, USA), or Panagene (*see* **Notes 8–10**).

2.3 Cells and Vectors

1. Various mammalian cell types have been amenable to treatment with recombinagenic and mutagenic molecules including: Chinese hamster ovary (CHO) cells, human K562 myelogenous leukemia cells, human THP-1 acute monocytic leukemia cells, and human CD34+ primary hematopoietic progenitor cells.
2. Various plasmid vectors have been suitable for gel shift assays due to their supercoiled conformations reminiscent of genomic DNA (Fig. 3a) and allele-specific qPCR assays due to their relatively smaller size as compared to the human genome (Fig. 5a), such as pBlueScript II-SK (Agilent Technologies, Santa Clara, CA, USA) and *FLuc+* from pGL3-Basic Vector (Promega, Madison, WI, USA).

2.3.1 Chinese Hamster Ovary (CHO) Cell Cultures

1. Chinese hamster ovary cell medium: Ham's F12 medium, 10 % supplementation of fetal bovine serum, and 2 mM of L-glutamine. Vacuum-filter.
2. If genomic cassettes are integrated in the genome, such as those in the cells of the beta-globin fluorescent assay (Fig. 4a), then in order to maintain the integrations further supplementation of this aforementioned culture medium with a selection marker, such as hygromycin, that is also filtered, is highly suggested (*see* **Note 11**).

2.3.2 Human K562 Myelogenous Leukemia Cell Cultures

1. K562 myelogenous leukemia cell medium: RPMI medium, 10 % supplementation of fetal bovine serum. Vacuum-filter.
2. If viability of these cell cultures is compromised due to contamination, antibiotics may be considered, such as penicillin and streptomycin at the requisite concentrations (*see* **Note 12**).

2.3.3 Human Primary CD34+ Hematopoietic Progenitor Cell Cultures

1. The StemSpan Serum-Free Expansion Medium (StemCell Technologies Inc., Vancouver, Canada): bovine serum albumin, recombinant human insulin, human transferrin (iron-saturated), 2-mercaptoethanol, Iscove's Modified Dulbecco's Medium, and other supplements.
2. To permit proper maintenance of human primary CD34+ hematopoietic progenitor cells in this media, a supplement of StemSpan CC110 cytokine mixture (StemCell Technologies Inc.) should also be used. The StemSpan CC110 cytokine mixture: 100 ng/mL rh Flt-3 Ligand, 20 ng/mL rh IL-3, 20 ng/mL rh IL-6, and 100 ng/mL rh Stem Cell Factor (*see* **Note 13**).

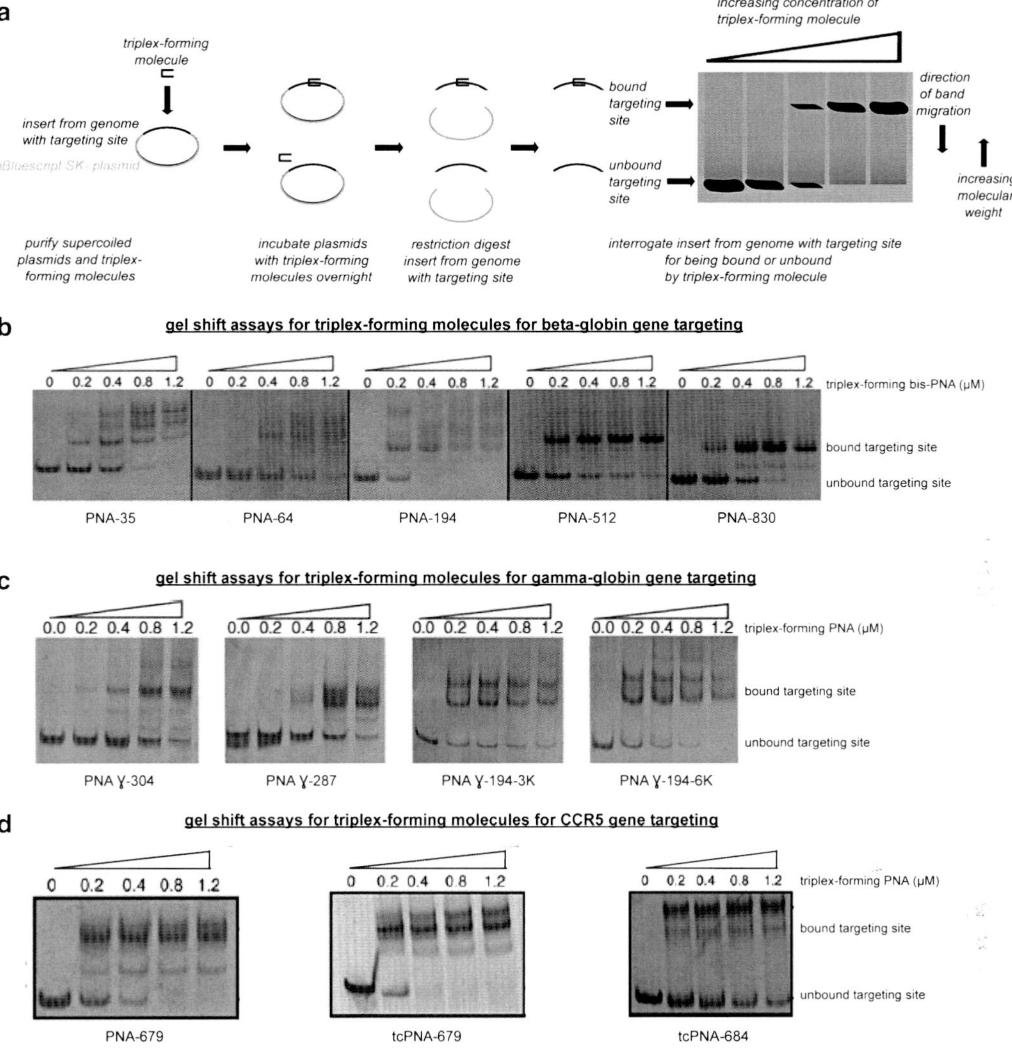

Fig. 3 Assessing triplex-forming molecule binding to targeting sites with gel shift assays. Gel shift assays are used to (**a**) assess the binding potential of triplex-forming molecules to genomic DNA, using supercoiled plasmid DNA with an insert from the genome with the targeting site as surrogates, and a shift in position of the visualized band in an electrophoresed gel to indicate that the triplex-forming molecule has bound this targeting site in (**b**) the beta-globin gene, (**c**) the gamma-globin gene, (**d**) and the CCR5 gene

2.4 Solutions, Buffers, and Other Reagents

1. The triplex-binding buffer in the TFO binding assay: 10 mM of Tris–HCl (pH 7.6), 0.1 mM of $MgCl_2$, 1 mM of spermine, 10 % of glycerol (with or without 140 mM potassium) (*see* **Note 14**).
2. The silver stain solution in the triplex-forming PNA binding assay: sodium borohydrate in 0.1 % silver nitrate (Sigma Aldrich, St. Louis, MO, USA), e.g., 1.0 g of $AgNO_3$ in 1.0 L of dH_2O (*see* **Note 15**).

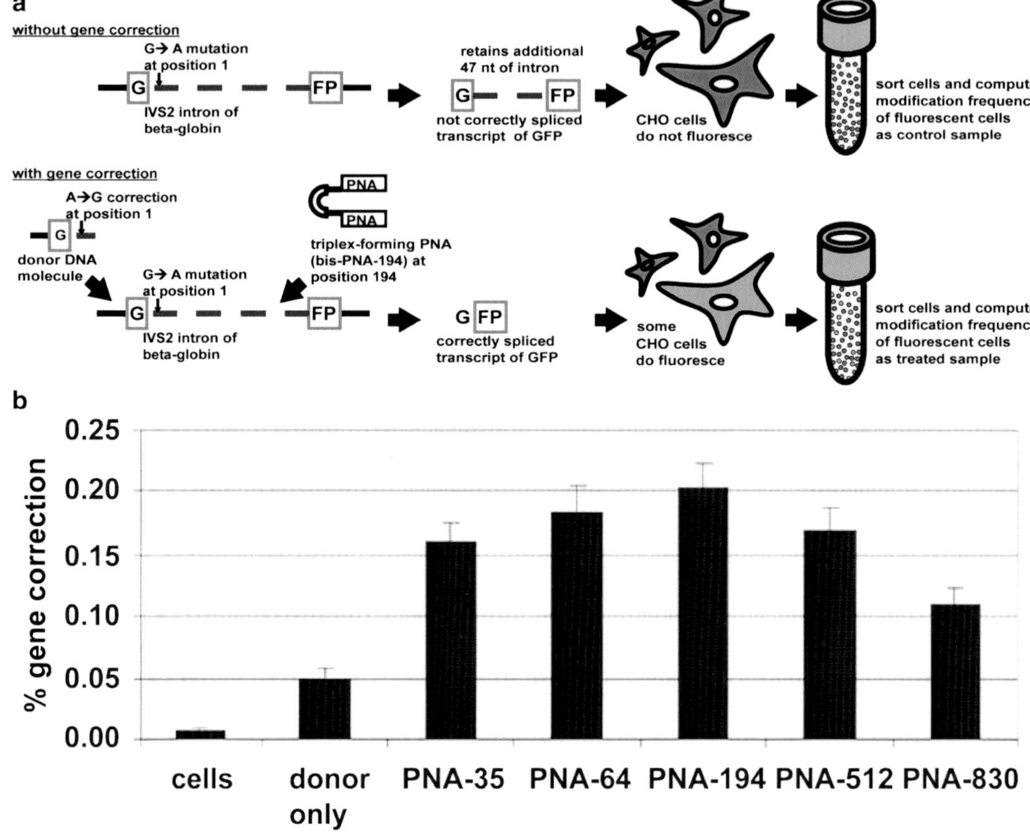

Fig. 4 Quantifying recombinagenic and mutagenic frequencies of gene correction with fluorescent assays of the beta-globin genomic locus. Fluorescent assays are used to (**a**) quantify the recombinagenic and mutagenic frequency of recombinagenic donor DNA or recombinagenic donor DNA with triplex-forming molecules, in the gene correction of the beta-globin genomic locus to permit correct splicing of transcript of GFP fluorescent reporter, (**b**) with varying frequencies of gene correction attributed to the particular molecules used in the treatment

3. The developer solution in the triplex-forming PNA binding assay: 15.0 g of NaOH, 0.1 g of $NaBH_4$, 5.0 mL of formaldehyde (*see* **Note 16**).

3 Methods

3.1 Recombinagenic donor DNA and Their Editing Sites

A recombinagenic donor DNA is an exogenously introduced molecule, which is nearly homologous to the genomic DNA editing site of interest sans the mismatched, inserted, or deleted bases that it is designed to harbor, that the endogenous cellular homologous recombination machinery utilizes as a template to introduce recombinagenic edits to the genome with great specificity.

3.1.1 Design of Recombinagenic donor DNA

The sequence length of single-stranded recombinagenic donor DNA molecules ranges from 30 to 2,000 bases in length, with sequence content that is homologous to the editing site, sans the mismatching base edits to be made flanked by a sufficient number of matching bases, and with sequence binding site proximity to within 750 bp of the triplex-forming molecule's binding site [69], and sequence orientation either as antisense (binding to the sense strand of the DNA) or sense (binding to the antisense strand of the DNA) (*see* **Note 17**). It is highly suggested that recombinagenic donor DNA molecules of both orientations are synthesized and used to treat cells, as overall mutagenic recombination efficiencies with genomic DNA editing site can vary due to a number of features of the DNA donors as well of the genomic DNA editing site by themselves, such as the intramolecular self-assembly or folding potential of either, and their interactions together, such as the intermolecular self-assembly or hybridization potential of both in the cellular milieu. Further considerations include additional cellular and genomic components, such as other potentially hybridizable nucleic acid molecules and chromatin structure, respectively, that can adversely affect the availability of recombinagenic donor DNA and the accessibility of the genomic editing site.

3.1.2 Synthesis of Recombinagenic donor DNA

Recombinagenic donor DNA molecules can be synthesized internally in the laboratory through sequentially coupled solid-phase phosphoramidite chemistry or long double-stranded donors can be synthesized by PCR amplification of a plasmid, or procured externally from an institutional oligonucleotide facility or commercial supplier. To inhibit nuclease degradation, additional synthesis modifications of the first and last three DNA donor bases is highly suggested, so that they are covalently connected to each other and to the rest of the DNA donor bases by phosphorothioate, rather than phosphodiester, linkages.

3.2 Mutagenic Triplex-Forming Molecules and Their Targeting Sites

A triplex-forming molecule is also an exogenously introduced molecule, which binds the major groove of duplex DNA targeting site to form a triplex, that, in turn, upregulates the recruitment of the endogenous cellular recombination machinery for mutagenic purposes upon the genome with the recombinagenic donor DNA with great specificity.

The concurrent treatment of the cellular genome with a highly specific triplex-forming molecule and a highly specific recombinagenic donor DNA ensures conditionally co-localized mutagenesis and recombination at targeting and editing sites in a safe and efficacious manner.

3.2.1 Design of Triplex-Forming Oligonucleotides (TFOs)

TFOs bind with high affinity to the genomic DNA duplex, either in the antiparallel purine motif, by being polypurine in TFO sequence content and binding antiparallel to the polypurine strand of the genomic DNA duplex, or in the parallel pyrimidine motif, by being polypyrimidine in TFO sequence content and binding parallel to the polypurine strand of the genomic DNA duplex [83] in both antiparallel and parallel motifs a genomic targeting site of successive homopurine, or polypurine, bases, at a length of 14–30 bases, should be chosen.

If this chosen genomic targeting site of polypurines is A-rich, then the pyrimidine motif is preferred and requires the TFO sequence content to be C and T or their analogs (C+ will form two Hoogsteen bonds with a G in a G:C base pair, and T will form two Hoogsteen bonds with the A in an A:T base pair) [83].

Alternatively, if this chosen genomic targeting site of polypurines is G-rich, then the purine motif is preferred and requires the TFO sequence content to be A (or T) and G or their analogs (G forms two reverse Hoogsteen bonds with the G in a G:C base pair, and A forms two reverser Hoogsteen bonds with the A in an A:T base pair) [83].

Like recombinagenic donor DNA, additional cellular and genomic components, such as other potentially hybridizable nucleic acid molecules and chromatin structure, respectively, can adversely affect the availability of TFOs and the accessibility of the genomic editing sites, and thus several genome targeting sites should be identified, for which TFOs should be designed and empirically evaluated accordingly.

3.2.2 Synthesis of Triplex-Forming Oligonucleotides (TFOs)

TFOs can also be can be synthesized internally in the laboratory through sequentially coupled solid-phase phosphoramidite chemistry, or procured externally from an institutional oligonucleotide facility or commercial supplier to create a molecule with a high binding affinity to the genomic DNA duplex. To inhibit nuclease degradation, additional synthesis modifications of the bases is highly suggested, so that they are covalently connected to each other by phosphoramidite, *N,N*-diethylethylenediamine (DEED), or phosphorothioate linkages, rather than phosphodiester, linkages (*see* **Note 18**).

3.2.3 Design of Triplex-Forming Peptide Nucleic Acids (PNAs)

Triplex-forming PNAs also bind with high affinity to the genomic DNA duplex, either in aforementioned the antiparallel purine motif or parallel pyrimidine motif. Note again that in both antiparallel and parallel motifs a genomic targeting site of successive homopurine, or polypurine, bases, but, as opposed to TFOs, only at a lesser length of 8–10 bases due to PNA's greater electrostatic complementarity, should be chosen. A preference for the pyrimidine motif for A-rich genomic targeting sites and for the purine motif for the G-rich genomic targeting sites also remains. Furthermore, as opposed to TFOs, triplex-forming PNAs in the pyrimidine motif can forgo the N3 atom protonation requirement of the C+ base in order to form Hoogsteen bonds with a G in a G:C base pair, a requirement that cannot usually be met in a physiological or therapeutic setting, by instead using a cytosine analog, pseudoisocytosine, J, to create two Hoogsteen bonds with the G in a G:C base pair on this triplex-forming strand [84].

In addition to the triplex-forming PNA, a second strand-invading PNA can be designed so that they form a triplex-forming bis-PNA. Usually this triplex-forming bis-PNA contains two pyrimidine PNAs, one that binds as a triplex to the duplex genomic DNA via Hoogsteen bonds and another that displaces one of the two duplex genomic DNA strands and binds the other genomic DNA strand via Watson–Crick bonds. These two pyrimidine PNAs are covalently bonded through a conformationally flexible linker, typically composed of (8-amino-3,6-dioxaoctanoic acid). Furthermore, the orientation of the N-to-C termini polyamide linkages in this bis-PNA peptide aligns in the same direction as the 5′–3′ phosphodiester linkages in the bound genomic DNA strand.

Furthermore, this triplex-forming bis-PNA design can be developed so that the stranding-invading PNA Watson–Crick bonding portion is lengthened into a "tail" portion beyond those genomic DNA bases bound by the triplex-forming Hoogsteen bonding, or "clamp," portion. This triplex-forming tail-clamp PNA facilitates greater specificity by expanding the genomic DNA targeting site beyond the polypurines, and also lends itself to greater affinity created by the additional Watson–Crick bonding [27].

Additional affinity for the triplex-forming bis-PNA or tail-clamp PNA to the genomic DNA can be implemented by design through the conjugation of lysine residues, which are positively charged at physiological pH. This covalent addition of several lysine residues on the Watson–Crick bonding strand or as part of the linker confers electrostatic complementarity to the negatively charged genomic DNA and thus facilitates strand invasion. As an additional benefit, these positively charged lysine residues promote molecular transport into cells [85]. This transcellular uptake and nuclear affinity phenomena are not limited to lysine residues, but possible through other cell-penetrating peptide conjugations [86], including Antennapedia, which is composed of the third helix of the *Drosophila* homeodomain transcription factor [87, 88], a transduction domain from the trans-activator of transcription (Tat) protein, which is translated from the HIV genome as YGRKKRRQRRR [89], and a nuclear localization signal also of Tat, which is translated from the HIV genome as GRKKR [90, 91].

3.2.4 Synthesis of Triplex-Forming Peptide Nucleic Acids (PNAs)

Like TFOs, triplex-forming PNAs can be synthesized internally in the laboratory, or procured externally from an institutional facility or commercial supplier, in this case for peptide synthesis, to create a molecule with a high binding affinity to the genomic DNA duplex. Unlike TFOs, which can be synthesized using sequentially coupled solid-phase phosphoramidite chemistry, triplex-forming PNAs can be synthesized using Fmoc or Boc solid-phase peptide synthesis chemistry. Given the hybrid nature of triplex-forming PNAs, i.e., that they have nucleic acid bases but polyamide linkages and, thus, are neither wholly nucleic acids nor peptides, they are naturally resistant to cellular nucleases and proteases. Furthermore, as opposed to the negatively charged phosphodiester linkages in TFOs, the uncharged polyamide linkages in triplex-forming peptide-nucleic acids experience lesser electrostatic repulsion from the negatively charged phosphodiester linkages of the genomic DNA.

Like recombinagenic donor DNA and TFOs, additional cellular and genomic components, such as other potentially hybridizable nucleic acid molecules and chromatin structure, respectively, can adversely affect the availability of triplex-forming PNAs and the accessibility of the genomic editing sites, and thus several genome targeting sites should be identified, for which triplex-forming PNAs should be designed and empirically evaluated accordingly (*see* **Note 19**).

3.3 Molecular Delivery of Mutagenic and Recombinagenic Molecules

A number of delivery modalities have been developed to transport mutagenic and recombinagenic molecules into cells, including electroporation, nucleofection, cationic lipids, cell-penetrating peptides, and nanoparticles.

3.3.1 Electroporation

Electroporation utilizes an electric current between two conductive plates to temporarily destabilize the cellular membrane and thus permit the exogenously located recombinagenic and mutagenic molecules to enter the cells. These conductive plates are embedded in sterile cuvettes that hold the cells, the treatment recombinagenic molecules, and the mutagenic molecules, suspended in their native media or phosphate buffered solution (PBS). Upon electroporation, it is not unusual for a fraction of the cells to undergo apoptosis immediately, and thus appearing as cellular debris floating or falling out of suspension, or shortly thereafter when the electroporated sample is rescued onto fresh media plates for expansion. When electroporating a new cell type, it is suggested that, rather than recombinagenic or mutagenic molecules, a fluorescent reporter plasmid such as GFP be used as a surrogate to optimize the conditions for electroporation of the exogenously introduced molecules. After optimizing electroporation conditions, it is suggested that a dose response standard curve be performed to optimize the subsequent cellular utilization conditions for the exogenously introduced molecules.

3.3.2 Nucleofection

Nucleofection as a delivery modality for recombinagenic and mutagenic molecules can be applied to cells that are particularly not amenable to the aforementioned electroporation protocol. The nucleofection cuvettes and solutions are adapted to particular cell types to permit greater molecular uptake. However, as a consequence of its greater disturbance of the cellular membrane, only approximately a tenth of the cells that undergo nucleofection remain viable after a day [54]. Thus, while nucleofection as a delivery modality may be effective when prototyping the recombinagenic and mutagenic molecules in a laboratory setting, other modalities can be considered when translating the fully developed molecules in a therapeutic setting.

3.3.3 Cationic Lipids

The recombinagenic and mutagenic molecules can also be encapsulated in lipids that are cytosed by the cell membrane. However, there are electrostatic considerations to be met in the encapsulation process. Cationic lipids readily encapsulate the anionic recombinagenic donor DNA or TFOs without lysine residue conjugations; however, they are unable to encapsulate the net charge neutral, or if having lysine residues conjugations then net charge positive, triplex-forming PNAs. Thus cationic lipids are often useful when delivering recombinagenic donor DNA to the genomic DNA editing site, but this approach does not benefit from the concurrent delivery of the triplex-forming PNA to the corresponding targeting site that ultimately yields greater modification frequencies.

3.3.4 Cell-Penetrating Peptides

The recombinagenic and mutagenic molecules can also be conjugated with peptides that signal the cell for cellular transmembrane transport. These cell-penetrating peptides can be developed either by-design or empirically, or biologically inspired from nature, such as from the Antennapedia or Tat protein domains of the *Drosophila* homeodomain transcription factor and HIV Tat proteins, respectively. while these cell-penetrating peptides facilitate uptake without the harshness of electroporation and nucleofection, they may or may not inadvertently affect the interaction of the molecules to which they are conjugated with the genomic DNA. Once again, it is suggested that various designs of these molecules with various conjugations be designed and empirically evaluated.

3.3.5 Nanoparticles

The recombinagenic and mutagenic molecules and also be encapsulated in nanoparticles that are cytosed by the cell membrane. The nanoparticle formulations vary and thus are often more versatile than the cationic lipids to ferry a host of neutral, net positively, or net negatively charged molecules, either alone or encapsulated together across the cell membrane. For example, nanoparticles with the requisite molecular payload can be constructed using double emulsion solvent encapsulation of poly-lactic-co-glycolic acid (PLGA) alone or with poly-ethylene glycol (PEG), both suitable for a therapeutic setting across the world as they are well-characterized and regulatory agency-approved biopolymers [92]. Furthermore, upon molecular uptake, these nanoparticles often disintegrate into metabolites that are not cytotoxic and do not further interfere with the interactions of the recombinagenic and mutagenic molecules with the genomic DNA [93] (*see* **Note 20**).

3.4 Assessing Triplex-Forming Molecule Binding to Targeting Sites with Gel Shift Assays

3.4.1 TFO Binding Gel Shift Assay

A gel shift assay can be performed to assess the binding of TFOs. In preparation, a duplex DNA fragment with the TFO targeting site can be created by synthesizing and then annealing complementary oligomers, where one of the oligomers contains the site. To isolate successfully duplexed DNA from the non-duplexed oligomers, 5′ end-labeling with T4 Polynucleotide kinase and (γ-32P) dATP can be done with 10^{-6} M of duplexed DNA in a total volume of 20 μL. To purify this labeled, duplexed DNA, a 15 % polyacrylamide electrophoretic gel can be prepared and used for electroelution of the duplex DNA. To concentration the purified duplex, a centrifugal column concentrator, such as a Centricon-3 column from Millipore can be applied (*see* **Note 21**).

To assess the TFO binding, incubate a series of reactions of fixed concentration of duplex DNA and increasing concentrations of TFO in TFO triplex-binding buffer at 37 °C for 12–24 h. A typical reaction is 20 μL, with 2 μL (10^{-6} M) of labeled duplex and 2 μL of tenfold dilutions of TFO (10^{-12}–10^{-7} M). This incubation will permit the steady-state binding of the TFO to its targeting site on the duplex DNA. To resolve the state of binding, the reactions should be electrophoresed on a 12 % native (19:1 acrylamide–bis-acrylamide) gel at 60–70 V in gel running buffer to discriminate the bound triplex from the unbound duplex DNA. To image the run gel, a PhosphorImager can be used and each band shift can be assessed. Image processing software can be utilized to estimate relative kinetic dissociation constants and, more generally, ranges of TFO concentrations for subsequent empirical studies.

3.4.2 Triplex-Forming PNA Binding Gel Shift Assay

A gel shift assay can also be performed to assess the binding of triplex-forming PNAs (Fig. 3a).

In preparation, plasmid DNA with the target site, from a genomic insert or mutagenized plasmid segment or annealed duplex DNA insert, flanked by restriction sites should be prepared. This plasmid DNA will afford the supercoiled DNA duplex that is reminiscent of genomic DNA.

To assess the triplex-forming PNA binding, incubate a series of reactions a fixed concentration of this plasmid DNA, such as 2 μg in a final volume of 10 μL TE, with the target site and increasing concentrations of triplex-forming PNAs ranging from 0 to 1 μM, 10 μM of KCl at 37 °C for 24 h. Post-incubation, each reaction should be restriction digested in a final volume of 20 μL at a temperature at which all restriction enzymes remain active for 1–2 h. Then, each digestion should be stopped by heat inactivation and electrophoresed to yield the targeting site without the rest of the plasmid and with the bound triplex-forming PNA. The electrophoresis components include DNA loading dye for each sample and migration on a 10 % polyacrylamide gel or 8 % native gel (19:1 in TBE) until bands are well separated. In other words, a typical electrophoresis involves a 8 % native gel that can be made with 10 mL of 40 % 19:1 bis-acrylamide–acrylamide, 10 mL of 5× TB, 200 μL of 0.5 M EDTA, 29.25 mL of dH_2O, 500 μL of 10 % ammonium persulfate, and 50 μL of TEMED, for a final volume of 50 mL. To resolve the resulting gel shift, initially stain with silver stain solution for 10 min, and then rinse the gel with dH_2O for 1 min, and then develop with developer solution for 1 min, and again rinse the gel with dH_2O for 1 min. With increasing concentrations of triplex-forming PNA, a single band representing the duplex DNA fragment containing the targeting site should transition to multiple bands, that represent the duplex DNA fragment with different triplex-binding modes [94], to shift higher, i.e., migrate less in the gel (*see* **Note 22**). To image the run gel, a standard computer scanner and scanning software can be used. In this manner, the triplex-forming PNA binding potentials for various designed synthetic molecules have been assessed for beta-globin targeting sites (Fig. 3b), gamma-globin targeting sites (Fig. 3c), and CCR5 targeting sites (Fig. 3d).

3.5 Quantifying Recombinagenic and Mutagenic Frequencies of Gene Correction with Fluorescent Assays of the Beta-Globin Genomic Locus

Fluorescent assays permit a facile means of sorting, counting, and thus quantifying cells that have undergone treatment with recombinagenic and mutagenic molecules. CHO reporter cells having a genome that harbors a split GFP transcript interrupted by the beta-globin IVS2 intron splicing mutation that causes beta-thalassemia provides a platform for such quantification of gene correction. Treatment of approximately one million of these CHO reporter cells with the recombinagenic donor carrying the A to G correction at position 1 of this IVS2 intron leads to a transcript of GFP that is also correctly spliced, thus yielding cells that fluoresce green (Fig. 4a). Further concurrent treatment of the CHO reporter cells with recombinagenic donor DNA and the triplex-forming PNA yields more cells that fluoresce green, and thus a greater gene correction frequency (Fig. 4b). Sort cells by fluorescent activated cell sorting (FACS) and analyze using flow cytometry software, such as FlowJo (*see* **Note 23**).

3.6 Quantifying Recombinagenic and Mutagenic Frequencies of Gene Induction and Regulation with Allele-Specific qPCR Assays of the Gamma-Globin Genomic Locus

Allele-specific PCR is a technique that is sensitive to genomic changes, and one that can combined with a qPCR technique that provides quantification by amplification, to produce an allele-specific qPCR assay. Different applications of this powerful technique can provide both qualitative and quantitative assessments of genome modification frequencies after treatment with recombinagenic donor DNA and triplex-forming molecules.

3.6.1 Genotyping Analysis by Allele-Specific PCR

After treating cells containing the target gene with TFO, triplex-forming PNA, and/or donor DNA, gene modification can be detected by allele-specific PCR [95, 96]. Harvest genomic DNA from the treated cells and dilute it to approximately 50 μg/μL. A forward primer can be designed with its 3′ end containing the desired mutation, and the reverse primer should be similar in length and T_m. Add 50 ng of the genomic DNA to a 25 μL PCR reaction and a gradient run to determine the optimal annealing temperature of the primers. Plasmids containing the wild-type and mutant gene can be used as control templates (*see* **Note 24**). The expected results are that a band should be present on the gel in lanes with mutant template and no bands for wild-type DNA (*see* **Note 25**). When combined with qPCR, allele-specific qPCR can use relative known frequencies of surrogates to extrapolate the gene induction and regulation frequency of, for example the usually dormant gamma-globin gene (Fig. 5a). After treatment with the recombinagenic and mutagenic molecules, the treated genome can be harvested and its unknown modification frequency can be extrapolated by correlating its allele-specific qPCR amplification values alongside those of known concentrations and thus modification frequencies of plasmids carrying the same uninduced gene sequence, i.e., the endogenous plasmid, and induced gene sequence, i.e., the codon-modified plasmid, in a milieu of genomic DNA, as well as those of a known concentration the untreated genome (*see* **Note 26**). In the case of gamma-globin, the co-treatment with the designed recombinagenic donor DNA and mutagenic triplex-forming PNA molecules yield a sequence modification frequency that is more than 560 % greater than that of treatment with the designed recombinagenic donor DNA alone (Fig. 5b).

3.7 Quantifying Recombinagenic and Mutagenic Frequencies of Gene Disruption with Single-Cell Cloning Assays of the CCR5 Genomic Locus

To quantify the amount of gene modification in cells treated with TFO, triplex-forming PNA, and/or donor DNA, the genomes of single-cell clones can be directly expanded and counted or deep sequenced [27, 82].

For counting and frequency quantification, such as those for the CCR5 gene, the treated cells can be replated at lower dilution. Cells can be treated in bulk and then diluted into multi-well dishes so that approximately a single cell on average occupies a single well. Each single-cell clone can proliferate within its dish to a density sufficient to extract the genomic DNA. a population of cell clones, rather than a single cell, is used to provide sufficient identical genomes for allele-specific PCR or direct sequencing (Fig. 6a). Based on the presence or absence or absence of the genome modification in each population, the cell that gave rise to that population can be assigned the same state, and all cells can thus be categorized to compute the modification frequency.

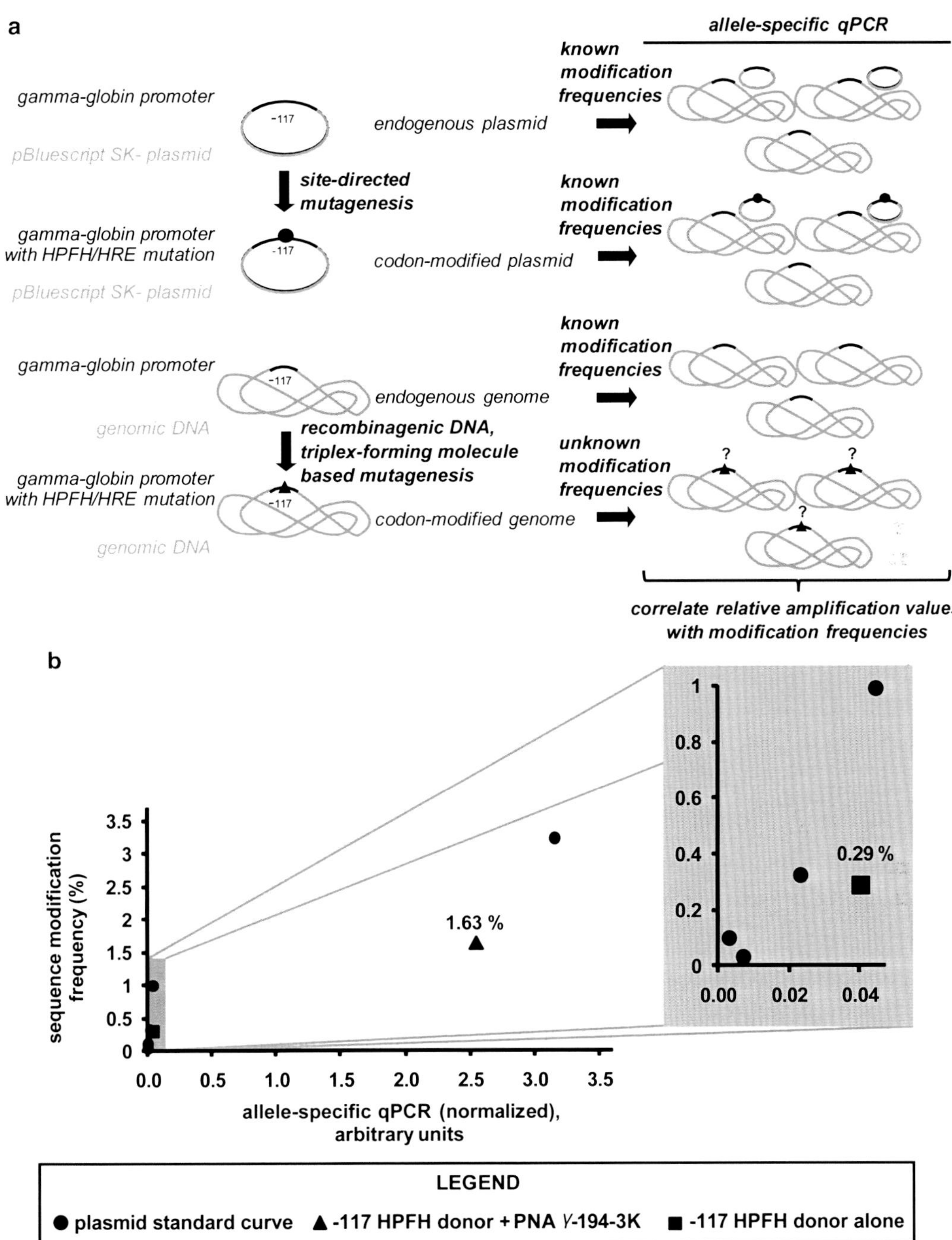

Fig. 5 Quantifying recombinagenic and mutagenic frequencies of gene induction and regulation with allele-specific qPCR assays of the gamma-globin genomic locus. Allele-specific qPCR assays are used to (**a**) quantify the recombinagenic and mutagenic frequency of recombinagenic donor DNA or recombinagenic donor DNA co-transfected with mutagenic triplex-forming molecules, in the gene induction and regulation of the gamma-globin genomic locus by correlating the relative qPCR amplification values with modification frequencies, (**b**) with varying frequencies of gene induction and regulation attributed to the particular molecules used in the treatment

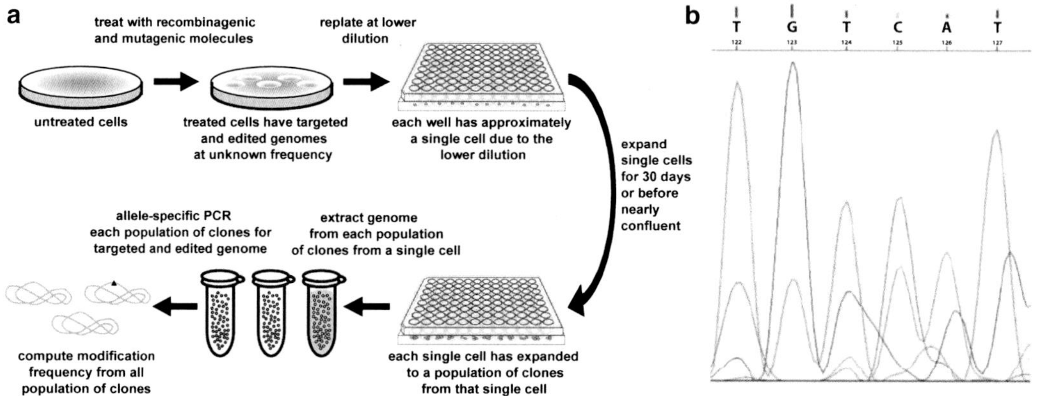

Fig. 6 Quantifying recombinagenic and mutagenic frequencies of gene disruption with single-cell cloning assays of the CCR5 genomic locus. Single-cell cloning assays are used to (**a**) quantify the recombinagenic and mutagenic frequency of recombinagenic donor DNA alone or recombinagenic donor DNA co-transfected with mutagenic triplex-forming molecules, in the gene disruption of the CCR5 genomic locus by expanding each single cell to a population of clones sufficient in number to enable allele-specific PCR-based verification and calculation of modification frequency, (**b**) with sequencing chromatographic mixed peaks indicating a heterozygous clone (wild type sequence being TGTCAT; modified sequence being CTGAGG)

For direct deep sequencing, forward and reverse primers can be designed to insert barcode-tagging and to encapsulate the desired locus with the location of the mutation position centered, and be similar in length and T_m. Determine the optimal anneal temperature of the primers using a gradient run and 50 ng of the genomic DNA to a 25 μL PCR reaction. Gel-electrophorese the PCR products and purify the desired bands using a gel purification kit. Ligate the direct deep sequencing PCR samples to adapters and sequence with 75 base-pair paired-end reads on an Illumina HiSeq platform. The expected results are that a band should be present on the gel representing the amplified locus that was flanked by the primers. If one of the two alleles undergoes modification, a heterozygous clone with be created, as evidenced by mixed peaks in sequencing chromatograms (Fig. 5b). The ratio of the number of clones harboring the targeted modification to the total number of single-cell clones assayed represents the targeting frequency.

4 Notes

1. To increase resistance to intracellular nucleases, and thus bioavailability, the synthesis of the recombinagenic donor DNA should proceed with phosphorothioate, rather than phosphodiester, termini linkages on the first three bases on 5′ end and last three bases on 3′ end, and then these three subsequences,

i.e., the two phosphorothiolated termini and the intervening phosphodiester sequence, can be linked.

2. If recombinagenic donor DNA is synthesized on a solid support, then a final deblocking followed by cleavage and deprotection step is necessary to free the final product from the support surface.
3. In order to maintain the integrity of recombinagenic donor DNAs when not in use, they should be aliquoted into smaller portions and stored at −20 °C. Also when in use for experiments, aliquots of recombinagenic donor DNA can be thawed and kept on ice.
4. To increase resistance to intracellular nucleases, and thus bioavailability, the synthesis of TFOs can progress with phosphoramidate, *N*, *N*-diethylethylenediamine (DEED), or phosphorothioate linkages.
5. The TFOs should be modified with one or more amine groups on the 3′ end to promote resistance to nucleases.
6. If triplex-forming peptide nucleic acid is synthesized on a solid support, then a final deblocking followed by cleavage and deprotection step is necessary to free the final product from the support surface.
7. Unnatural bases such as pseudoisocytosine should be procured and supplied to the institutional peptide synthesis facility.
8. The triplex-forming PNA should be modified with one or more terminal lysine residues to increase electrostatic complementarity, and thus affinity, to negatively charged genomic DNA.
9. While synthetically modified TFOs and triplex-forming PNAs are less prone to intracellular nuclease and nuclease and protease degradation, respectively, the latter also affords better electrostatic complementarity with the negatively charged genomic DNA. As such, triplex-forming PNAs are recommended over TFOs for use when possible.
10. Triplex-forming PNAs and TFOs should be aliquoted and keep in the −20 °C freezer for storage. Also during usage for experiments, an aliquot of triplex-forming PNAs can be thawed but kept cold on ice.
11. An antibiotic like hygromycin is light-sensitive. Thus, exercise caution when supplementing medium with hygromycin by shielding the medium flask with aluminum foil wrapping and/or use amber glass containers.
12. Antibiotics like penicillin and streptomycin are not light-sensitive. Shielding with aluminum foil wrapping or amber glass containers is not necessary.

13. Human primary CD34+ hematopoietic progenitor cells cannot be maintained indefinitely in their pluripotent state without differentiating. It is highly suggested that fresh or pre-frozen aliquots of these cells be procured immediately prior to use and then used within a few days.
14. This buffer has been composed so that it is appropriate for evaluating TFO and triplex-forming PNA binding under physiological conditions.
15. For ideal gel staining, silver stain solution should be fresh and, if possible, during the time the loaded gel is running through the electrophoresis apparatus.
16. For ideal gel developing, developer solution should be fresh and, if possible, during the time the loaded gel is being stained.
17. The optimal donor length and offsets from the recombination center will vary from editing site to site. This often makes it necessary to try various donor candidates and at various concentrations.
18. To prevent degradation by 3′ exonucleases it is recommended that TFOs have a 3′ end cap of an amine group chemical modification.
19. Triplex-forming PNAs do not require this chemical modification, since their peptide backbones make them resistant to degradation.
20. Use transfection methods and parameters that are suitable for the cell type in use. Some of these transfection methods include digitonin permeabilization, electroporation, cationic lipids, or nanoparticles. It is advisable to optimize the transfection method and parameters that provide high delivery efficiency and low cell damage and/or death.
21. The appropriate size Centricon column should be used based on the molecular weight of the duplex and/or triplex molecule to be filtered.
22. Rather than a single band, multiple bands may appear during the gel shift. This is not unexpected, since many triplex-forming PNAs have the ability to form multiple complex triplex structures with their target DNA.
23. Immediately after harvesting CHO reporter cells by trypsining the focal adhesions and washing with PBS, but prior to cell sorting, it is recommended that they be sterile-filtered through a 40 μm filter to further separate clusters of cells into individuals for sorting.
24. First the wild-type gene can be cloned, to which mutations can be introduced with site-directed mutagenesis via the QuikChange Site-Directed Mutagenesis Kit (Stratagene, LaJolla, CA, USA).

25. Initially, 40 cycles of PCR can be applied during allele-specific PCR using genomic DNA. In general, this provides sufficient amplification over background. For allele-specific PCR using plasmid DNA, since it is much easier to PCR-amplify from the latter, it is preferable to reduce the concentration rather than modify the allele-specific PCR cycling program, if possible.
26. Codon modifications, are silent modifications, and thus permit the wild-type protein to be transcribed and translated from a modified gene sequence of usually 6–8 base pairs, necessary for the sensitivity threshold of allele-specific PCR.

Acknowledgement

We gratefully acknowledge members of the Glazer Laboratory for helpful discussions. This work was supported by a National Institutes of Health (NIH) grant R01HL082655 and by a Doris Duke Innovations in Clinical Research Award (to P.M.G.). A National Institute of Diabetes and Digestive and Kidney Diseases Experimental and Human Pathobiology Postdoctoral Fellowship from NIH grant T32DK007556 also provided support (to F.R.). The authors declared no conflict of interest.

References

1. Smithies O, Gregg RG, Boggs SS, Koralewski MA, Kucherlapati RS (1985) Insertion of DNA sequences into the human chromosomal beta-globin locus by homologous recombination. Nature 317:230–234
2. Thomas KR, Capecchi MR (1987) Site-directed mutagenesis by gene targeting in mouse embryo-derived stem cells. Cell 51:503–512
3. Reza F, Zuo P, Tian J (2007) Protein interfacial pocket engineering via coupled computational filtering and biological focusing criterion. Ann Biomed Eng 35:1026–1036
4. Faruqi AF, Datta HJ, Carroll D, Seidman MM, Glazer PM (2000) Triple-helix formation induces recombination in mammalian cells via a nucleotide excision repair-dependent pathway. Mol Cell Biol 20:990–1000
5. Datta HJ, Glazer PM (2001) Intracellular generation of single-stranded DNA for chromosomal triplex formation and induced recombination. Nucleic Acids Res 29:5140–5147
6. Vasquez KM, Christensen J, Li L, Finch RA, Glazer PM (2002) Human XPA and RPA DNA repair proteins participate in specific recognition of triplex-induced helical distortions. Proc Natl Acad Sci U S A 99:5848–5853
7. Datta HJ, Chan PP, Vasquez KM, Gupta RC, Glazer PM (2001) Triplex-induced recombination in human cell-free extracts - dependence on XPA and HsRad51. J Biol Chem 276: 18018–18023
8. Thoma BS, Vasquez KM (2003) Critical DNA damage recognition functions of XPC-hHR23B and XPA-RPA in nucleotide excision repair. Mol Carcinog 38:1–13
9. Thoma BS, Wakasugi M, Christensen J, Reddy MC, Vasquez KM (2005) Human XPC-hHR23B interacts with XPA-RPA in the recognition of triplex-directed psoralen DNA interstrand crosslinks. Nucleic Acids Res 33: 2993–3001
10. Wang G, Seidman MM, Glazer PM (1996) Mutagenesis in mammalian cells induced by triple helix formation and transcription-coupled repair. Science 271:802–805
11. Wang G, Levy DD, Seidman MM, Glazer PM (1995) Targeted mutagenesis in mammalian cells mediated by intracellular triple helix formation. Mol Cell Biol 15:1759–1768
12. Summers WC, Sarkar SN, Glazer PM (1985) Direct and inducible mutagenesis in mammalian cells. Cancer Surv 4:517–528

13. Watson JD, Crick FHC (1953) Molecular structure of nucleic acids - a structure for deoxyribose nucleic acid. Nature 171:737–738
14. Felsenfeld G, Davies DR, Rich A (1957) Formation of a 3-stranded polynucleotide molecule. J Am Chem Soc 79:2023–2024
15. Moser HE, Dervan PB (1987) Sequence-specific cleavage of double helical DNA by triple helix formation. Science 238:645–650
16. Le Doan T, Perrouault L, Praseuth D, Habhoub N, Decout JL, Thuong NT et al (1987) Sequence-specific recognition, photocrosslinking and cleavage of the DNA double helix by an oligo-[alpha]-thymidylate covalently linked to an azidoproflavine derivative. Nucleic Acids Res 15:7749–7760
17. Faruqi AF, Krawczyk SH, Matteucci MD, Glazer PM (1997) Potassium-resistant triple helix formation and improved intracellular gene targeting by oligodeoxyribonucleotides containing 7-deazaxanthine. Nucleic Acids Res 25:633–640
18. Kalish JM, Seidman MM, Weeks DL, Glazer PM (2005) Triplex-induced recombination and repair in the pyrimidine motif. Nucleic Acids Res 33:3492–3502
19. Camerini-Otero RD, Hsieh P (1993) Parallel DNA triplexes, homologous recombination, and other homology-dependent DNA interactions. Cell 73:217–223
20. Seidman MM (2004) Oligonucleotide mediated gene targeting in mammalian cells. Curr Pharm Biotechnol 5:421–430
21. Demidov VV, Potaman VN, Frankkamenetskii MD, Egholm M, Buchard O, Sonnichsen SH et al (1994) Stability of peptide nucleic-acids in human serum and cellular-extracts. Biochem Pharmacol 48:1310–1313
22. Ray A, Norden B (2000) Peptide nucleic acid (PNA): its medical and biotechnical applications and promise for the future. FASEB J 14: 1041–1060
23. Rooney SM, Moore PD (1995) Antiparallel, intramolecular triplex DNA stimulates homologous recombination in human cells. Proc Natl Acad Sci U S A 92:2141–2144
24. Walter A, Schutz H, Simon H, Birch-Hirschfeld E (2001) Evidence for a DNA triplex in a recombination-like motif: I. Recognition of Watson-Crick base pairs by natural bases in a high-stability triplex. J Mol Recog 14: 122–139
25. Faruqi AF, Egholm M, Glazer PM (1998) Peptide nucleic acid-targeted mutagenesis of a chromosomal gene in mouse cells. Proc Natl Acad Sci U S A 95:1398–1403
26. Chin JY, Kuan JY, Lonkar PS, Krause DS, Seidman MM, Peterson KR et al (2008) Correction of a splice-site mutation in the beta-globin gene stimulated by triplex-forming peptide nucleic acids. Proc Natl Acad Sci U S A 105:13514–13519
27. Schleifman EB, Bindra R, Leif J, Del Campo J, Rogers FA, Uchil P et al (2011) Targeted disruption of the CCR5 gene in human hematopoietic stem cells stimulated by peptide nucleic acids. Chem Biol 18:1189–1198
28. Guieysse AL, Praseuth D, Francois JC, Helene C (1995) Inhibition of replication initiation by triple helix-forming oligonucleotides. Biochem Biophys Res Commun 217:186–194
29. Jain A, Magistri M, Napoli S, Carbone GM, Catapano CV (2010) Mechanisms of triplex DNA-mediated inhibition of transcription initiation in cells. Biochimie 92:317–320
30. Ferdous A, Akaike T, Maruyama A (2000) Inhibition of sequence-specific protein-DNA interaction and restriction endonuclease cleavage via triplex stabilization by poly(L-lysine)-graft-dextran copolymer. Biomacromolecules 1:186–193
31. Hanvey JC, Peffer NJ, Bisi JE, Thomson SA, Cadilla R, Josey JA et al (1992) Antisense and antigene properties of peptide nucleic-acids. Science 258:1481–1485
32. Koppelhus U, Zachar V, Nielsen PE, Liu XD, EugenOlsen J, Ebbesen P (1997) Efficient in vitro inhibition of HIV-1 gag reverse transcription by peptide nucleic acid (PNA) at minimal ratios of PNA/RNA. Nucleic Acids Res 25:2167–2173
33. Praseuth D, Grigoriev M, Guieysse AL, Pritchard LL, HarelBellan A, Nielsen PE et al (1996) Peptide nucleic acids directed to the promoter of the alpha-chain of the interleukin-2 receptor. Biochim Biophys Acta 1309:226–238
34. Nielsen PE, Egholm M, Berg RH, Buchardt O (1993) Sequence specific-inhibition of DNA restriction enzyme cleavage by PNA. Nucleic Acids Res 21:197–200
35. Mollegaard NE, Buchardt O, Egholm M, Nielsen PE (1994) Peptide nucleic-acid.DNA strand displacement loops as artificial transcription promoters. Proc Natl Acad Sci U S A 91:3892–3895
36. Bahal R, McNeer NA, Ly DH, Saltzman WM, Glazer PM (2013) Nanoparticle for delivery of antisense gammaPNA oligomers targeting CCR5. Artif DNA PNA XNA 4:49–57
37. Chin JY, Reza F, Glazer PM (2013) Triplex-forming peptide nucleic acids induce heritable elevations in gamma-globin expression in

hematopoietic progenitor cells. Mol Ther 21(3):580–587

38. Faria M, Wood CD, Perrouault L, Nelson JS, Winter A, White MRH et al (2000) Targeted inhibition of transcription elongation in cells mediated by triplex-forming oligonucleotides. Proc Natl Acad Sci U S A 97:3862–3867
39. Birg F, Praseuth D, Zerial A, Thuong NT, Asseline U, Ledoan T et al (1990) Inhibition of simian virus-40 DNA-replication in CV-1 cells by an oligodeoxynucleotide covalently linked to an intercalating agent. Nucleic Acids Res 18:2901–2908
40. Volkmann S, Jendis J, Frauendorf A, Moelling K (1995) Inhibition of HIV-1 reverse transcription by triple-helix forming oligonucleotides with viral RNA. Nucleic Acids Res 23: 1204–1212
41. Diviacco S, Rapozzi V, Xodo L, Helene C, Quadrifoglio F, Giovannangeli C (2001) Site-directed inhibition of DNA replication by triple helix formation. FASEB J 15:2660–2668
42. Maher LJ, Wold B, Dervan PB (1989) Inhibition of DNA-binding proteins by oligonucleotide-directed triple helix formation. Science 245:725–730
43. Francois JC, Saisonbehmoaras T, Thuong NT, Helene C (1989) Inhibition of restriction endonuclease cleavage via triple helix formation by homopyrimidine oligonucleotides. Biochemistry 28:9617–9619
44. Hanvey JC, Shimizu M, Wells RD (1990) Site-specific inhibition of EcoRI restriction modification enzymes by a DNA triple helix. Nucleic Acids Res 18:157–161
45. Mayfield C, Ebbinghaus S, Gee J, Jones D, Rodu B, Squibb M et al (1994) Triplex formation by the human Ha-Ras promoter inhibits Sp1 binding and in-vitro transcription. J Biol Chem 269:18232–18238
46. Havre PA, Gunther EJ, Gasparro FP, Glazer PM (1993) Targeted mutagenesis of DNA using triple helix-forming oligonucleotides linked to psoralen. Proc Natl Acad Sci U S A 90:7879–7883
47. Takasugi M, Guendouz A, Chassignol M, Decout JL, Lhomme J, Thuong NT et al (1991) Sequence-specific photoinduced cross-linking of the 2 strands of double-helical DNA by a psoralen covalently linked to a triple helix-forming oligonucleotide. Proc Natl Acad Sci U S A 88:5602–5606
48. Vasquez KM, Wensel TG, Hogan ME, Wilson JH (1996) High-efficiency triple-helix-mediated photo-cross-linking at a targeted site within a selectable mammalian gene. Biochemistry 35: 10712–10719
49. Vasquez KM, Narayanan L, Glazer PM (2000) Specific mutations induced by triplex-forming oligonucleotides in mice. Science 290:530–533
50. Wang X, Tolstonog G, Shoeman RL, Traub P (1996) Selective binding of specific mouse genomic DNA fragments by mouse vimentin filaments in vitro. DNA Cell Biol 15:209–225
51. Chan PP, Lin M, Faruqi AF, Powell J, Seidman MM, Glazer PM (1999) Targeted correction of an episomal gene in mammalian cells by a short DNA fragment tethered to a triplex-forming oligonucleotide. J Biol Chem 274:11541–11548
52. Luo ZJ, Macris MA, Faruqi AF, Glazer PM (2000) High-frequency intrachromosomal gene conversion induced by triplex-forming oligonucleotides microinjected into mouse cells. Proc Natl Acad Sci U S A 97:9003–9008
53. Rogers FA, Manoharan M, Rabinovitch P, Ward DC, Glazer PM (2004) Peptide conjugates for chromosomal gene targeting by triplex-forming oligonucleotides. Nucleic Acids Res 32:6595–6604
54. McNeer NA, Chin JY, Schleifman EB, Fields RJ, Glazer PM, Saltzman WM (2011) Nanoparticles deliver triplex-forming PNAs for site-specific genomic recombination in CD34+ human hematopoietic progenitors. Mol Ther 19: 172–180
55. Branden LJ, Mohamed AJ, Smith CIE (1999) A peptide nucleic acid-nuclear localization signal fusion that mediates nuclear transport of DNA. Nat Biotechnol 17:784–787
56. Vasquez KM, Dagle JM, Weeks DL, Glazer PM (2001) Chromosome targeting at short polypurine sites by cationic triplex-forming oligonucleotides. J Biol Chem 276:38536–38541
57. Lacroix L, Lacoste J, Reddoch JF, Mergny JL, Levy DD, Seidman MM et al (1999) Triplex formation by oligonucleotides containing 5-(1-propynyl)-2′-deoxyuridine: decreased magnesium dependence and improved intracellular gene targeting. Biochemistry 38:1893–1901
58. Puri N, Majumdar A, Cuenoud B, Natt F, Martin P, Boyd A et al (2002) Minimum number of 2′-O-(2-aminoethyl) residues required for gene knockout activity by triple helix forming oligonucleotides. Biochemistry 41:7716–7724
59. Chin JY, Schleifman EB, Glazer PM (2007) Repair and recombination induced by triple helix DNA. Front Biosci 12:4288–4297
60. Raha M, Lacroix L, Glazer PM (1998) Mutagenesis mediated by triple helix-forming oligonucleotides conjugated to psoralen: effects of linker arm length and sequence context. Photochem Photobiol 67:289–294
61. Raha M, Wang G, Seidman MM, Glazer PM (1996) Mutagenesis by third-strand-directed

psoralen adducts in repair-deficient human cells: high frequency and altered spectrum in a xeroderma pigmentosum variant. Proc Natl Acad Sci U S A 93:2941–2946

62. Wang G, Glazer PM (1995) Altered repair of targeted psoralen photoadducts in the context of an oligonucleotide-mediated triple-helix. J Biol Chem 270:22595–22601
63. Majumdar A, Puri N, Cuenoud B, Natt F, Martin P, Khorlin A et al (2003) Cell cycle modulation of gene targeting by a triple helix-forming oligonucleotide. J Biol Chem 278: 11072–11077
64. Macris MA, Glazer PM (2003) Transcription dependence of chromosomal gene targeting by triplex-forming oligonucleotides. J Biol Chem 278:3357–3362
65. Vasquez KM, Wang G, Havre PA, Glazer PM (1999) Chromosomal mutations induced by triplex-forming oligonucleotides in mammalian cells. Nucleic Acids Res 27:1176–1181
66. Sargent RG, Rolig RL, Kilburn AE, Adair GM, Wilson JH, Nairn RS (1997) Recombination-dependent deletion formation in mammalian cells deficient in the nucleotide excision repair gene ERCC1. Proc Natl Acad Sci U S A 94:13122–13127
67. Faruqi AF, Seidman MM, Segal DJ, Carroll D, Glazer PM (1996) Recombination induced by triple-helix-targeted DNA damage in mammalian cells. Mol Cell Biol 16:6820–6828
68. Sandor Z, Bredberg A (1995) Triple-helix directed psoralen adducts induce a low-frequency of recombination in an Sv40 shuttle vector. Biochim Biophys Acta 1263:235–240
69. Knauert MP, Kalish JM, Hegan DC, Glazer PM (2006) Triplex-stimulated intermolecular recombination at a single-copy genomic target. Mol Ther 14:392–400
70. Schleifman EB, Chin JY, Glazer PM (2008) Triplex-mediated gene modification. Methods Mol Biol 435:175–190
71. Knauert MP, Lloyd JA, Rogers FA, Datta HJ, Bennett ML, Weeks DL et al (2005) Distance and affinity dependence of triplex-induced recombination. Biochemistry 44:3856–3864
72. Shahid KA, Majumdar A, Alam R, Liu ST, Kuan JY, Sui XF et al (2006) Targeted cross-linking of the human beta-globin gene in living cells mediated by a triple helix forming oligonucleotide. Biochemistry 45:1970–1978
73. Friedman S, Schwartz E (1976) Hereditary persistence of foetal haemoglobin with β-chain synthesis in cis position (Gγ-β+-HPFH) in a negro family. Nature 259:138–140
74. Xu XS, Glazer PM, Wang G (2000) Activation of human gamma-globin gene expression via triplex-forming oligonucleotide (TFO)-directed mutations in the gamma-globin gene 5′ flanking region. Gene 242:219–228
75. Samson M, Libert F, Doranz BJ, Rucker J, Liesnard C, Farber CM et al (1996) Resistance to HIV-1 infection in caucasian individuals bearing mutant alleles of the CCR-5 chemokine receptor gene. Nature 382:722–725
76. Steinberger P, Andris-Widhopf J, Buhler B, Torbett BE, Barbas CF 3rd (2000) Functional deletion of the CCR5 receptor by intracellular immunization produces cells that are refractory to CCR5-dependent HIV-1 infection and cell fusion. Proc Natl Acad Sci U S A 97:805–810
77. Hutter G, Nowak D, Mossner M, Ganepola S, Mussig A, Allers K et al (2009) Long-term control of HIV by CCR5 Delta32/Delta32 stem-cell transplantation. N Engl J Med 360: 692–698
78. Anderson J, Akkina R (2007) Complete knockdown of CCR5 by lentiviral vector-expressed siRNAs and protection of transgenic macrophages against HIV-1 infection. Gene Ther 14:1287–1297
79. Gorman M, Glazer PM (2006) Directed gene modification via triple helix formation. Curr Mol Med 1:391–399
80. Schleifman EB, Glazer PM (2014) Peptide nucleic acid-mediated recombination for targeted genomic repair and modification. Methods Mol Biol 1050:207–222
81. Shen H, Cheng T, Preffer FI, Dombkowski D, Tomasson MH, Golan DE et al (1999) Intrinsic human immunodeficiency virus type 1 resistance of hematopoietic stem cells despite coreceptor expression. J Virol 73:728–737
82. McNeer NA, Schleifman EB, Cuthbert A, Brehm M, Jackson A, Cheng C et al (2013) Systemic delivery of triplex-forming PNA and donor DNA by nanoparticles mediates site-specific genome editing of human hematopoietic cells in vivo. Gene Ther 20:658–669
83. Reza F, Glazer PM (2014) Triplex-mediated genome targeting and editing. Methods Mol Biol 1114:115–142
84. Egholm M, Christensen L, Dueholm KL, Buchardt O, Coull J, Nielsen PE (1995) Efficient Ph-independent sequence-specific DNA-binding by pseudoisocytosine-containing Bis-Pna. Nucleic Acids Res 23:217–222
85. Sazani P, Kang SH, Maier MA, Wei CF, Dillman J, Summerton J et al (2001) Nuclear antisense effects of neutral, anionic and cationic oligonucleotide analogs. Nucleic Acids Res 29:3965–3974
86. Koppelhus U, Awasthi SK, Zachar V, Holst HU, Ebbesen P, Nielsen PE (2002) Cell-

dependent differential cellular uptake of PNA, peptides, and PNA-peptide conjugates. Antisense Nucleic Acid Drug Dev 12:51–63

87. Karagiannis ED, Urbanska AM, Sahay G, Pelet JM, Jhunjhunwala S, Langer R et al (2013) Rational design of a biomimetic cell penetrating peptide library. ACS Nano 7:8616–8626
88. Sagan S, Burlina F, Alves ID, Bechara C, Dupont E, Joliot A (2013) Homeoproteins and homeoprotein-derived peptides: going in and out. Curr Pharm Des 19:2851–2862
89. Schwarze SR, Hruska KA, Dowdy SF (2000) Protein transduction: unrestricted delivery into all cells? Trends Cell Biol 10:290–295
90. Hauber J, Malim MH, Cullen BR (1989) Mutational analysis of the conserved basic domain of human immunodeficiency virus tat protein. J Virol 63:1181–1187
91. Ruben S, Perkins A, Purcell R, Joung K, Sia R, Burghoff R et al (1989) Structural and functional characterization of human immunodeficiency virus tat protein. J Virol 63:1–8
92. Schleifman EB, McNeer NA, Jackson A, Yamtich J, Brehm MA, Shultz LD et al (2013) Site-specific genome editing in PBMCs with PLGA nanoparticle-delivered PNAs confers HIV-1 resistance in humanized mice. Mol Ther Nucleic Acids 2:e135
93. McNeer NA, Schleifman EB, Glazer PM, Saltzman WM (2011) Polymer delivery systems for site-specific genome editing. J Control Rel 155:312–316
94. Hansen GI, Bentin T, Larsen HJ, Nielsen PE (2001) Structural isomers of bis-PNA bound to a target in duplex DNA. J Mol Biol 307:67–74
95. Orou A, Fechner B, Utermann G, Menzel HJ (1995) Allele-specific competitive blocker Pcr - a one-step method with applicability to pool screening. Hum Mutat 6:163–169
96. Parsons BL, McKinzie PB, Heflich RH (2005) Allele-specific competitive blocker-PCR detection of rare base substitution. Methods Mol Biol 291:235–245
97. Williams LD, Egli M, Ughetto G, van der Marel GA, van Boom JH, Quigley GJ et al (1990) Structure of 11-deoxydaunomycin bound to DNA containing a phosphorothioate. J Mol Biol 215:313–320
98. Radhakrishnan I, Patel DJ (1994) Solution structure of a pyrimidine.purine.pyrimidine DNA triplex containing T.AT, C+.GC and G.TA triples. Structure 2:17–32
99. Yeh JI, Pohl E, Truan D, He W, Sheldrick GM, Du S et al (2010) The crystal structure of non-modified and bipyridine-modified PNA duplexes. Chemistry 16:11867–11875

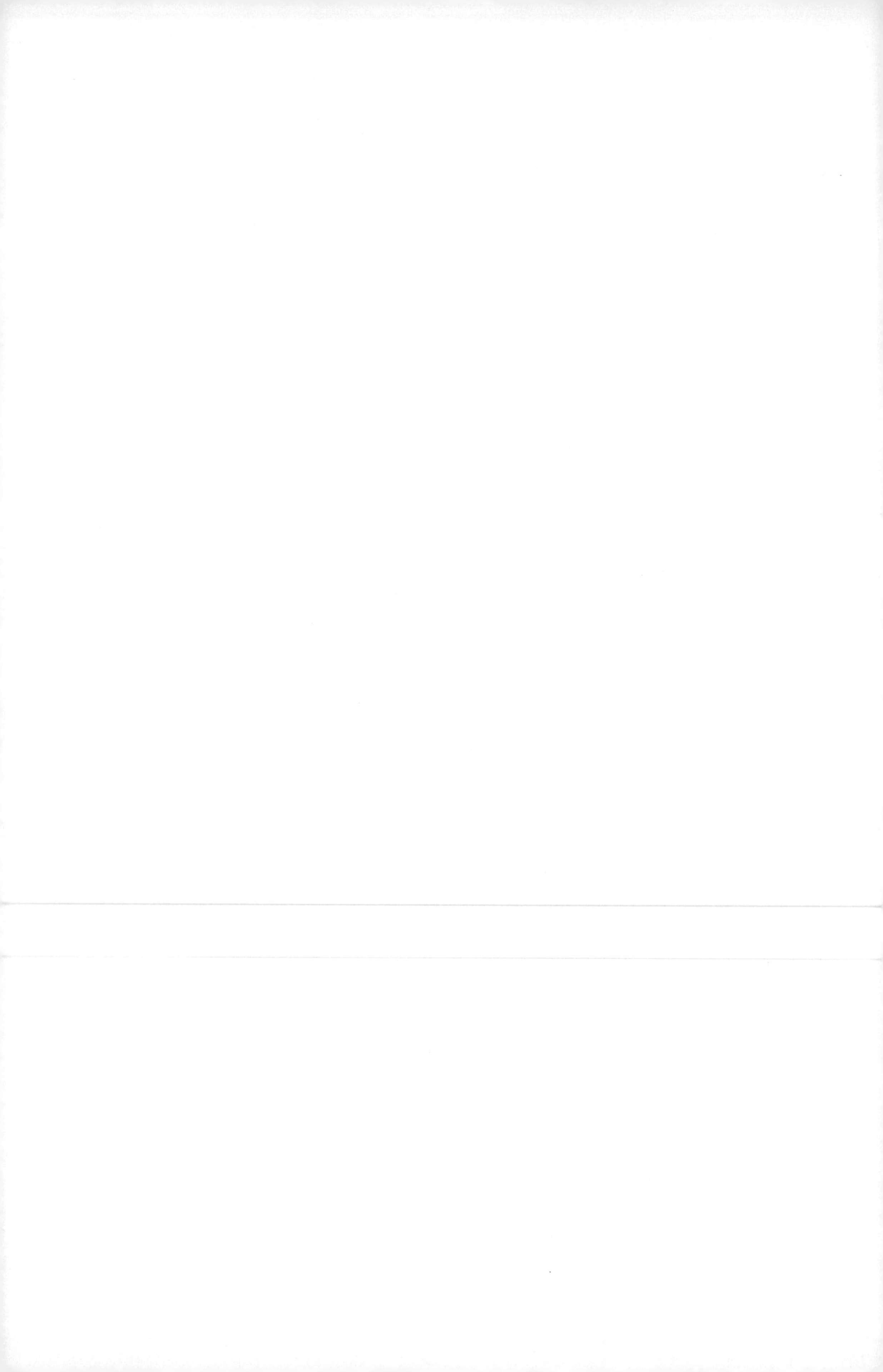

Chapter 5

Genome Engineering Using Adeno-Associated Virus (AAV)

Rob Howes and Christine Schofield

Abstract

The ability to edit the genome of cell lines has provided valuable insights into biological processes and the contribution of specific mutations to disease biology. These techniques fall into two categories based on the DNA repair mechanism that is used to incorporate the genetic change. Nuclease-based technologies, such as Zinc-Finger Nucleases, TALENS, and Crispr/Cas9, rely on non-homologous end-joining (NHEJ) and homology directed repair (HDR) to generate a range of genetic modifications. Adeno-Associated Virus (AAV) utilizes homologous recombination to generate precise and predictable genetic modifications directly at the target locus. AAV has been used to create over 500 human isogenic cell lines comprising a wide range of genetic alterations from gene knockouts, insertions of point mutations, indels, epitope tags, and reporter genes. Here we describe the generation and use of AAV gene targeting vectors and viruses to create targeted isogenic cell lines.

Key words Adeno-associated virus, AAV, Homologous recombination, Insertion, Disruption, Gene knockout, Human cell line, Precise, Flexible, Predictable, Gene targeting, Genome editing

1 Introduction

Adeno-associated virus (AAV) is a single stranded DNA parvovirus that has been demonstrated to have utility in generating human gene targeted cell lines [1]. AAV uses the homologous recombination DNA repair mechanism to introduce the desired genetic alterations (Fig. 1); however in human cells this mechanism is much less efficient than in bacteria or yeast [2]. In human cells, the alternative non-homologous end joining DNA repair pathway dominates which is utilized by other gene targeting technologies to create gene knockouts with high efficiency. This can complicate gene targeting using AAV as AAV genomes can integrate randomly throughout the genome at double-strand breaks [3].

The advantages of using AAV for chromosomal mutagenesis is that it is a predictable and flexible methodology [4]. As AAV creates a direct gene conversion event, one can design a particular modification in silico and this same modification will be present in the final targeted cell line without any confounding off-target modifications.

Shondra M. Pruett-Miller (ed.), *Chromosomal Mutagenesis*, Methods in Molecular Biology, vol. 1239,
DOI 10.1007/978-1-4939-1862-1_5,

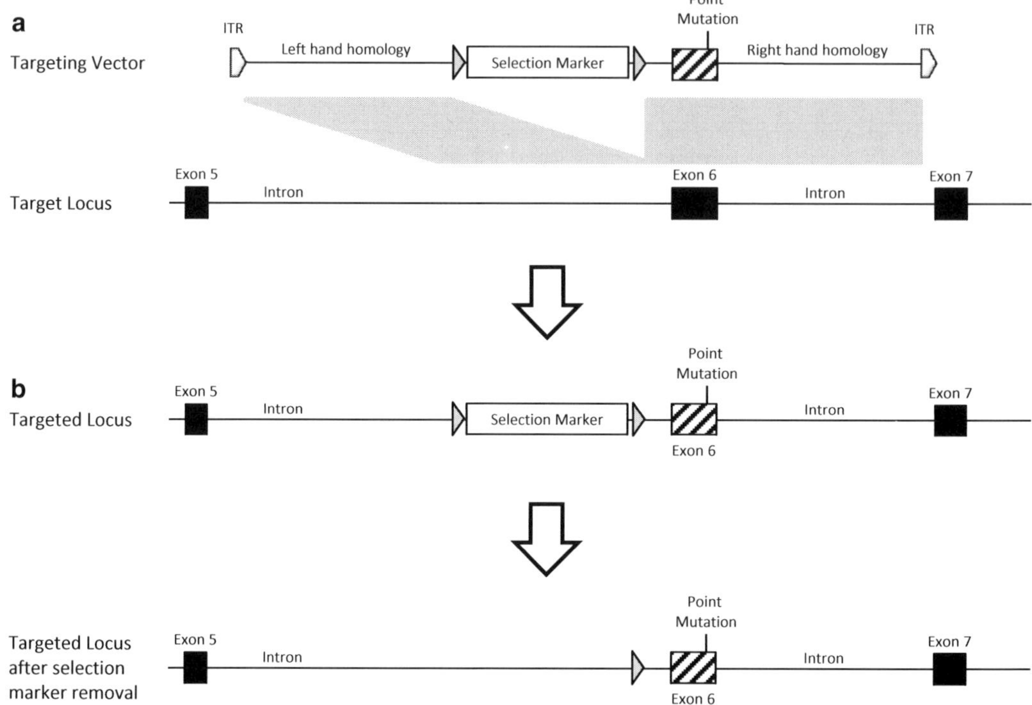

Fig. 1 Genome engineering using AAV-mediated homologous recombination. AAV Targeting Vectors (**a**) typically contain two regions of homology to the target locus (highlighted in *grey*) separated by a selection marker and flanked by the viral inverted terminal repeats (ITR's). The mutation to be introduced is located in one of the homology arms (in this case the right hand homology arm). The single-stranded targeting vector is delivered to the nucleus of the target cell where homologous recombination takes place to incorporate the selection marker and the mutation of interest (**b**). The selection marker is typically flanked by recombination sites (for example loxP or FRT sites) so by addition of the appropriate recombinase to the targeted cells, the selection marker can be removed resulting in a targeted cell line containing the mutation of interest and a recombination site in a nearby intron

The ability of AAV mutagenesis to introduce a wide range of genetic alterations has been previously demonstrated and allows the experimenter to choose the most appropriate modification for their particular purpose (Fig. 2) [5–8].

There are, of course, some drawbacks with using AAV as a gene targeting tool [9]. The limited packaging size of an AAV capsid of approximately 4.7 kb can limit its use in some applications such as the insertion of large cDNA's. AAV cannot efficiently target non-dividing cells and heterochromatic regions within the genome [10–13]. One of the complicating factors of using AAV is the highly efficient integration of the AAV genome at double-strand DNA breaks which can require screening of large numbers of colonies to identify the correctly targeted cells [2]. To overcome this, many AAV targeting approaches use a positive selection marker but this may be undesirable for some applications as the

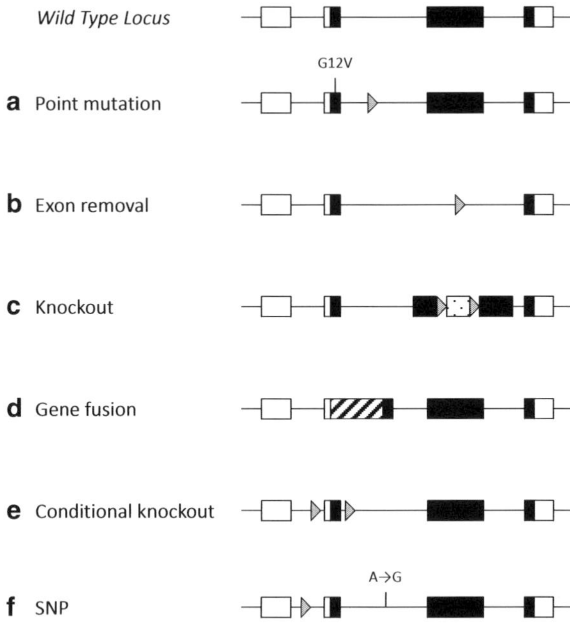

Fig. 2 Genomic modifications generated using AAV. AAV genome engineering has been used to generate a variety of modifications including single point mutations in coding exons (*a*); genetic knockouts such as removal of entire exons (*b*) or insertional mutagenesis (*c*); gene fusions such as reporter genes or gene tags (*d*); conditional gene knockouts (*e*); and noncoding changes such as SNP's (*f*)

selection marker or a recombination scar will be present in the final cell line [12]. AAV mutagenesis usually modifies a single allele in a targeting event so in order to target both alleles in a diploid cell line requires 2 rounds of gene targeting. However, this can be balanced by the flexibility of AAV so different modifications can be introduced at each locus.

The process of AAV chromosomal mutagenesis begins with characterization of the parental cell line. There are several characteristics that a cell line requires with the current AAV gene targeting system including the ability to proliferate indefinitely, tolerate single cell dilution cloning, transduction with the AAV virus, selection of the cell line with antibiotics and confirmation of the ploidy and mutation status of the target locus. Once this information is understood in a target cell line, this will help to increase the chances of a successful gene targeting experiment.

The next important stage in the gene targeting process is the targeting vector design and construction. The targeting vector design will be unique to the locus of interest and the mutation to be generated but there are several general principles that can be applied. These have been addressed in detail by others and so will not be repeated here [9, 12, 14]. The recombinant AAV particles are produced using a packaging cell line to encapsulate the recombinant gene targeting vector in the viral capsid. Purification of AAV can be achieved using many methods but the most common are based on ultracentrifugation and column purification. Once a target cell line has been infected with the recombinant AAV and placed under selection, cell colonies are screened for targeted integration events using genomic DNA PCR. Once positive recombinant cells have been identified, the targeted locus can be thoroughly characterized to confirm the integration and expression of the desired modification. Using this methodology, over 500 isogenic cell lines have been generated to date encompassing a wide range of genetic alterations [15].

2 Materials

2.1 Limiting Dilution of Parental Cell Line

1. 75 cm^2 tissue culture flasks.
2. 1.5 ml microcentrifuge tubes.
3. 15 and 50 ml conical bottom tubes.
4. 5, 10 and 25 ml serological pipettes.
5. 20 and 200 μl filter pipette tips.
6. 100 ml sterile reagent reservoir.

2.2 Infectability of Parental Cell Line with AAV

1. rAAV-LacZ (Vector Biolabs, Philadelphia, PA, USA); The most commonly used serotype of AAV that is used is serotype 2; however, other serotypes are available, including the chimeric pDJ serotype, which should be tested if there is a low level of transduction using serotype 2.
2. 38 % (w/v) formaldehyde.
3. 25 % (w/v) glutaraldehyde.
4. 50 mg/ml 5-Bromo-4-chloro-3-indolyl β-D-galactoside (X-Gal).
5. 500 mM ferricyanide.
6. 500 mM ferrocyanide.
7. 1 M $MgCl_2$.
8. 10 % (w/v) NP-40.
9. Phosphate Buffered Saline (PBS).
10. Fixing solution per ml: 52.6 μl 38 % (w/v) formaldehyde, 8 μl 25 % (w/v) glutaraldehyde and 939.4 μl PBS.
11. X-Gal staining solution per ml: 20 μl 5 mg/ml X-Gal,10 μl 500 mM ferricyanide, 10 μl 500 mM ferrocyanide, 2 μl 1 M $MgCl_2$, 2 μl 10 % NP-40, 956 μl PBS.

2.3 Antibiotic Selection Death Curve

1. 75 cm^2 tissue culture flasks.
2. 1.5 ml microcentrifuge tubes.
3. 5, 10 and 25 ml serological pipettes.
4. 15 and 50 ml conical bottom tubes.
5. 20, 200, and 1,000 µl filter tip pipette tips.
6. 1× TrypLE (Life Technologies, Carlsbad, CA, USA).
7. Phosphate Buffered Saline (PBS).
8. 0.4 % (w/v) Trypan Blue solution.
9. 50 mg/ml G418/geneticin.
10. 10 mg/ml puromycin.
11. 50 mg/ml hygromycin.
12. Cell culture media appropriate for the cell line.

2.4 Transfection of HEK293 Cells to Generate a Crude rAAV Virus Preparation

1. 75 cm^2, 175 cm^2 tissue culture flasks.
2. 1.5 ml microcentrifuge tubes.
3. 5, 10 and 25 ml serological pipettes.
4. 15 and 50 ml conical bottom tubes.
5. 200 and 1,000 µl filter tip pipette tips.
6. Dry ice/ethanol bath.
7. TrypLE (1×; Life Technologies).
8. Phosphate Buffered Saline (PBS).
9. 0.4 % (w/v) Trypan Blue solution.
10. pAAV targeting vector (*see* Subheading 2.1).
11. pDG helper plasmid (Plasmid Factory, Bielefeld, Germany).
12. Cell culture media: DMEM F12 with L-glutamine, 10 % FBS, 100 U/ml penicillin, 0.1 mg/ml streptomycin.

2.5 Adeno-Associated Virus (AAV) Purification

1. 50 ml conical bottom tubes.
2. Dry Ice/Ethanol bath.
3. Two T75 flasks or sterile containers with a capacity of greater than 100 ml.
4. One 20 ml luer-lock syringe per preparation.
5. Three 5 ml luer-lock syringes per preparation.
6. AAV purification kit (Virapur, San Diego, CA, USA).
7. 384 U/µl Benzonase nuclease (Sigma, St. Louis, MO, USA).
8. Phosphate Buffered Saline (PBS).
9. Lysis Buffer: 150 mM NaCl, 50 mM Tris–HCl pH 8.5.

2.6 AAV Quantification

1. Molecular biology grade water.
2. 10× DNaseI buffer (Sigma).
3. Amplification grade DNaseI (Sigma).
4. 10 % (w/v) SDS.
5. 1 M NaCl.
6. 20 mg/ml Proteinase K.
7. 2× Brilliant III Ultra-Fast QPCR Master Mix (Agilent Technologies, Santa Clara, CA, USA).
8. qRT-PCR optical grade plastic-ware.
9. NeomycinR TaqMan assay

(a)	Forward primer	Neo326F	ACCTTGCTCCTGCCGAG AAAGTAT
(b)	Reverse primer	Neo428R	CGATGTTTCGCTTGGTG GTCGAAT
(c)	FAM Probe	Neo366_FAM	[6-FAM]AATGCGGCGGCTGC ATACGCTTGAT[TAMRA].

10. PuromycinR SYBR assay

(a)	Forward primer	Puro96F21	CACGCGCCACACCGTCGACCC
(b)	Reverse primer	Puro283R22	GGGAACCGCTGAACTCGGCCA.

11. HygromycinR TaqMan assay

(a)	Forward primer	Hyg:14F18	AACTCACCGCGACGTCTG
(b)	Reverse primer	Hyg:111R20	GCCCTCCTACATCGAAGCTG
(c)	FAM Probe	Hyg:71L22	[6-FAM]CCCTCCGAGAGCTG CATCAGGT[TAMRA].

12. DNAse I mastermix: 5 μl 10× DNaseI buffer, 10 U amplification grade DNaseI, water to 50 μl.
13. 2× Proteinase K Buffer: 1 % SDS, 200 μM NaCl in water, must be kept DNAse-free.
14. Proteinase K mastermix: 75 μl 2× Proteinase K Buffer, 1 μl 20 mg/ml Proteinase K, 24 μl water.

2.7 Infection of Target Cell Line

1. 75 cm^2 tissue culture flasks.
2. 1.5 ml microcentrifuge tubes.
3. 5, 10 and 25 ml serological pipettes.

4. 15 and 50 ml conical bottom tubes.
5. 20, 200, and 1,000 μl filter tip pipette tips.
6. 1× TrypLE (Life Technologies).
7. Phosphate Buffered Saline (PBS).
8. 0.4 % (w/v) Trypan Blue solution.
9. Parafilm.
10. 50 mg/ml G418/geneticin.
11. 10 mg/ml puromycin.
12. 50 mg/ml hygromycin.
13. Cell culture media appropriate for the cell line of interest.

2.8 Plating Out of Infected Cell Line

1. 5, 10 and 25 ml serological pipettes.
2. 15 and 50 ml conical bottom tubes.
3. 200 μl/1,000 μl filter tip pipette tips.
4. Centrifuge (to hold 50 ml tubes).
5. 1× TrypLE (Life Technologies).
6. Phosphate Buffered Saline (PBS).
7. 0.4 % (w/v) Trypan Blue solution.
8. 96-well plates Tissue Culture sterilized.
9. Media with the appropriate level of antibiotic selection as determined in Subheading 3.4.

2.9 Screening for Homologous Recombinants by Genomic PCR

1. Inverted microscope.
2. 1× TrypLE (Life Technologies).
3. Phosphate Buffered Saline (PBS).
4. Appropriate standard media for cell line or 2× serum (Rescue Media), or conditioned media.
5. Direct PCR reagent (Viagen Biotech, Los Angeles, CA, USA) for gDNA Screen.
6. Proteinase K, >800 U/ml (Sigma).
7. 5× Trypsin Inhibitor solution (2.5 % Fetal Bovine Serum in PBS).
8. 1.5 ml microcentrifuge tubes.
9. 96-well V-bottomed polystyrene PCR plate.
10. V-bottomed 96-well plate (maximum volume no less than 300 μl).
11. 20, 200, and 1,000 μl filtered pipette tips.
12. Heat seals or strip caps.
13. Direct PCR Lysis Buffer for 110 reactions (2,100 μl Direct PCR Buffer and 21 μl Proteinase K).

2.10 PCR Screen of Targeted Cells

1. 1.5 ml microcentrifuge tubes.
2. 96-well V-bottom polystyrene PCR plate.
3. 15 and 50 ml conical bottom Falcon tubes.
4. 20, 200, and 1,000 μl filter tip pipette tips.
5. Heat seals or strip caps.
6. Thermal cycler/PCR machine.
7. Horizontal electrophoresis gel rig.
8. 100 μM Unique primer set as defined by the vector design.
9. Sterile distilled water PCR grade.
10. DMSO, PCR grade.
11. 25 mM each dNTP Mix.
12. 25 mM $MgCl_2$.
13. GoTaq 5× Buffer with Green loading dye (Promega, Madison, WI, USA).
14. GoTaq Hot Start enzyme (Promega).
15. Agarose.
16. 1× Tris Acetate EDTA (TAE) or Tris Borate EDTA (TBE) Buffer.
17. Loading Dye: 5 ml Glycerol, 25 mg bromophenol blue, 25 mg xylene cyanol, 5 ml sterile water.
18. 1 Kb DNA ladder marker.
19. PCR Mastermix (for 100 samples): 400 μl 5× PCR Buffer, 15 μl 100 μM Forward primer, 15 μl 100 μM Reverse primer, 20 μl 25 mM dNTP's, 120 μl 25 mM $MgCl_2$, 120 μl DMSO, 1,090 μl water (PCR grade), 20 μl 5 U/μl GoTaq Hot Start enzyme.

2.11 cDNA Genotyping of Targeted Cells

1. 1.5 ml RNase-free microcentrifuge tubes.
2. 96-well PCR plate or 0.2 ml PCR tubes with caps.
3. 20, 200, and 1,000 μl filter tip pipette tips.
4. Heat seals or strip caps (for 96-well plates).
5. Heat sealer.
6. Horizontal electrophoresis gel rig.
7. 100 ng/μl random hexamers.
8. dNTP Mix, 10 or 25 mM.
9. H_2O, PCR grade RNAse and DNAse free.
10. Superscript III 5× First Strand Synthesis Buffer.
11. 100 mM DTT.
12. RNase Inhibitor, 40 U/μl.
13. 200 U/μl Superscript III RT enzyme.

14. DMSO, PCR grade.
15. 25 mM $MgCl_2$.
16. GoTaq 5× Buffer with Green loading dye (Promega).
17. GoTaq Hot Start enzyme (Promega).
18. Agarose.
19. 1× Tris Acetate EDTA (TAE) or Tris Borate EDTA (TBE) Buffer.
20. Loading Dye: 5 ml Glycerol, 25 mg bromophenol blue, 25 mg xylene cyanol, 5 ml sterile water.
21. cDNA PCR Mastermix 1 (per sample): 1 μl 10 mM dNTP's, 1 μl 100 ng/μl Random Hexamers, 10 μl water.
22. cDNA PCR Mastermix 2 (per sample): 30 μl 5× Buffer, 0.3 μl 100 μM Forward Primer, 0.3 μl 100 μM Reverse Primer, 0.0.5 μl 25 mM dNTP's, 3 μl 25 mM $MgCl_2$, 3 μl DMSO, 30 μl water, PCR grade and 0.0.5 μl 5 U/μl DNA polymerase.

3 Methods

3.1 AAV Targeting Vector Design and Construction

Optimal vector design is essential for the successful outcome of a gene targeting experiment. There are many genetic alterations that can lead to the same phenotypic outcome and the choice depends on many factors such as the number and complexity of the transcripts of the target gene, multiple activities of the target protein and known effects of altering the target protein level such as lethality. For example, generation of a phenotypically null allele can be achieved in many ways depending on the outcome that you desire to be achieved. Optimal targeting vector design has been addressed by many other groups [9] but the key features include:

- Presence of any SNP's in the homology arms of the targeting vectors which can decrease the targeting efficiency
- Use of a selection cassette to increase the relative rates of gene targeting including the use of promoterless selection cassettes
- Positioning of the selection cassette relative to the targeted exon to allow correct mRNA splicing and expression

It is important to note that the specific targeting vector design will be unique to each experimental application but general principles will be applicable.

With the advent of cheap and reliable gene synthesis technologies, we routinely use commercial providers, such as Genscript and GeneWiz, to synthesize the gene targeting vectors which are subcloned into the AAV plasmid backbone containing the inverted terminal repeats (ITR's) necessary for packaging of the targeting vector into the viral capsid.

3.2 Limiting Dilution of Parental Cell Line

1. Day 1: Warm media for at least 10 min at 37 °C in water bath.
2. Remove existing media from cells and dispose to waste.
3. Wash cells with 10 ml PBS.
4. Aspirate PBS from flask. Add 2 ml of TrypLE to cells in T75 flask (change volume depending on flask size). Incubate the cells for 5 min at 37 °C 5 % CO_2 until the cells detach.
5. Triturate cells 3–4 times with 5 ml Stripette to break up any cell clumps. Add 3 ml media to the flask to neutralize the TrypLE.
6. Transfer the cells to a 15 ml Falcon tube.
7. Add 20 μl 1× Trypan Blue to a microcentrifuge tube.
8. Add 20 μl of TrypLE treated cells to the same microcentrifuge tube. Mix cells by gentle pipetting.
9. Transfer 10 μl cell suspension to a well of hemocytometer. Count the number of live cells in three fields of the hemocytometer. Calculate the number of cells/ml in the cell suspension.
10. Dilute cells to 1×10^4 cells/ml in 1 ml of fresh media.
11. Add the following volume of cells to 110 ml media in a large reservoir and mix well using a Stripette. For 1 cell per well add 55 μl (1×10^4 cells/ml) cell suspension to 110 ml media; for 0.5 cell per well add 33 μl (1×10^4 cells/ml) cell suspension to 110 ml media; for 5 cells per well add 275 μl (1×10^4 cells/ml) cell suspension to 110 ml media.
12. Dispense 200 μl/well in all 12 columns of a 96-well plate using multichannel pipette.
13. Dispense cells to total of five 96-well plates manually. Dispose of excess cell suspension to waste.
14. Transfer plates to incubator and incubate the cells at 37 °C 5 % CO_2.
15. Day 14: Visually screen plates for clear single colonies (starting with the lowest cell density), mark these by circling the well on the lid and record in a spreadsheet. Identify at least 10 clones and note the concentration of cells/ml that generates the highest number of single cell clones per 96-well plate.

3.3 Infectability of Parental Cell Line with rAAV

The aim of the infection test is to indicate an optimum cell density for infection based on the detection of cells stained for Beta-galactosidase activity. This information is used to inform the number of cells to be infected and plated out for screen to generated a genetically modified cell line.

1. Day 0: Seed cells in a 12 well plate at varying densities in a total of 1 ml volume as per the layout shown in Table 1 using **steps 1–9** of Subheading 3.2.
2. Incubate plates at 37 °C, 5 % CO_2.

Table 1
Plate layout for testing infectability of cells with rAAV

Clone no	Well 1 (cells)	Well 2 (cells)	Well 3 (cells)	Well 4 (cells)
1	2,500	5,000	10,000	20,000
2	2,500	5,000	10,000	20,000
3	2,500	5,000	10,000	20,000

3. Day 1: Replace the media to remove any dead cells. Infect the cells with 2 μl of rAAV-LacZ virus (at $1e^{13}$ GC/ml) and incubate the plates for 2–3 days at 37 °C, 5 % CO_2.
4. Day 4: Make up a fresh stock of the Fixing solution and X-Gal staining solution.
5. When the solutions are made wash the cells once with PBS. Add 1 ml of fixing solution to each well and incubate for 5 min at room temperature.
6. Remove the fixing solution and wash three times with PBS. Store the cells in PBS at 4 °C until needed or stain immediately (**step** 7 below).
7. Remove the PBS and add 1 ml of X-Gal staining solution. Incubate at room temperature for 24–48 h.
8. Remove the X-Gal staining solution from the plate. Wash cells with PBS and store in fresh PBS.
9. Visualize the cells using a light microscope. Infected cells will be positive for a blue stain.
10. Make a record of images. The cells and conditions with the most stain and therefore the most infected cells will be selected for the project.

3.4 Antibiotic Death Curve

Stable cell lines are often generated by selection of antibiotic resistance cells in population of cells transfected or infected with DNA containing an antibiotic resistance gene. In targeting cells with rAAV an antibiotic resistance gene is included in the viral genome, either under the control of an exogenous promoter or a promoter-less cassette that relies on endogenous gene expression. In all cases to select a targeted cell line, it is necessary to determine the concentration of the antibiotic which kills wild-type cells, termed a death curve. The death curve will determine the minimum concentration of the antibiotic used in culturing potentially targeted cells to kill any cells which do not express the resistance gene and leave targeted cells unaffected. Target cell lines may be tested with just one of the antibiotics, generally G418, or all three of the antibiotics (Geneticin or G418, hygromycin, puromycin).

1. Day 1: Remove media from flask of cells to be tested. Wash cells with PBS.
2. Incubate cells with appropriate volume of TrypLE at 37 °C/5 % CO_2 until cells have detached.
3. Triturate cells 3–4 times with 5 ml Stripette to break up any cell clumps. Add media to the flask equal to at least the volume of TrypLE used (preferably an excess) to neutralize the TrypLE.
4. Pipette 10 μl of cell suspension in a 0.5 ml microcentrifuge tube. Add 10 μl of Trypan Blue to the cell suspension.
5. Pipette 10 μl of this suspension to the chamber of a hemocytometer.
6. Count the number of cells in the three fields of the hemocytometer under the inverted microscope.
7. Calculate the number of cells/ml in the cell suspension.
8. Death curves require 2,000 cells/well (10,000 cells/ml). Each well will contain 190 μl of media and 10 μl of antibiotic solution. In order to seed the cells at the correct density follow this calculation:

   10,000/cells/ml = volume of cell suspension (μl) to add to 1 ml of media to give 10,000 cells.
9. Two 96-well plates are required for one antibiotic type. 40 ml of media will sufficiently fill all of the wells required. Multiply the previous answer by 40 to give the volume of cell suspension required to give 10,000 cells/ml in 40 ml of media.
10. Prepare the cell dilution by adding cells and fresh medium to a 50 ml Falcon tube, mixing, and then pouring into a reservoir. Using a multichannel pipette add 190 μl of this cell dilution to each well of the 96-well plates.
11. Using a multichannel pipette add 10 μl of antibiotic solution to the cells using a 1 in 3 dilution series from the top concentrations indicated in Table 2. Apply each concentration in triplicate.

Table 2
Suggested ranges of antibiotics used in characterization of parental cell lines

Antibiotic	Typical range	Top concentration in death curve analysis (mg/ml)
Geneticin (G418)	0.01 mg–1 mg/ml	10
Hygromycin	1 μg–1 mg/ml	10
Puromycin	0.1 μg–100 μg/ml	1

12. Return the plate to incubator at 37 °C/5 % CO_2. Incubate the plates for 14 days.
13. Day 15: Using a microscope check the viability of the cells by looking at the 96-well plates under the inverted microscope using a 40× objective. Choose the lowest antibiotic concentration that kills 100 % of the cells in all three wells of the triplicate. This is the concentration of antibiotic to use in the subsequent experiments.
14. If determination of cell viability is difficult add 50 μl of 1:2 dilution of Trypan Blue and PBS to each well. Leave for 5 min, remove Trypan Blue, wash cells with 50 μl PBS and view again under the microscope.
15. Take images of the cells to show the optimum antibiotic concentration required for the cell line and the progressive cell death at lower antibiotic concentrations.

3.5 Parental Cell Line Copy Number and Mutation Status

As AAV gene targeting usually modifies a single allele in a typical gene targeting event, it is essential to understand the ploidy of the target locus within your chosen cell line. Many methods exist to confirm the ploidy of your target locus but the most common ones are fluorescence in situ hybridization (FISH) and single-nucleotide polymorphism (SNP) analysis. Many common cell lines have been analyzed using high density SNP arrays, for example SNP6.0, so the data for your target cell line may be found on publically available databases such as COSMIC (http://cancer.sanger.ac.uk/cancergenome/projects/cosmic/) and the Cancer Cell Line Encyclopaedia (http://www.broadinstitute.org/ccle/home). We recommend using this information as a guide, as the ploidy of your particular clone of a cell line may differ, due to the genetic instability common in many cell lines. To confirm the ploidy of a cell line we use the Genome-Wide Human SNP Array 6.0 (Affymetrix) which is provided as a service by many commercial providers.

It is essential to confirm the mutation status of your target locus prior to gene targeting which we complete by genomic PCR (*see* Subheading 3.5) and sequencing using genomic DNA purified using standard methods such as Qiagen QIAprep kits using the manufacturer's recommendations.

3.6 AAV Virus Preparation

1. Day 1: Obtain flasks with low passage number HEK-293 (*see* **Note 1**).
2. Aspirate media from flasks. Wash cells with 10 ml of PBS (*see* **Note 2**).
3. Add 3 ml 1× TrypLE to cells. Incubate the cells for 5 min at 37 °C 5 % CO_2 until the cells detach.
4. Triturate cells 3–4 times with 5 ml Stripette to break up any cell clumps. Add 3 ml DMEM-F12 media to the flask to neutralize the TrypLE. Transfer the cells to a 50 ml Falcon tube.

5. Add 20 µl 1× Trypan Blue to a microcentrifuge tube. Add 20 µl TrypLE treated cells to the same microcentrifuge tube. Mix cells by gentle pipetting.
6. Transfer 20 µl cell suspension to a hemocytometer. Count the number of live cells in four fields of the hemocytometer.
7. Calculate the number of cells/ml in the cell suspension and calculate the volume of cells required for 4×10^6 cells per T75 flask.
8. Add this volume of cells to 5× T75 flasks. Add 10 ml DMEM-F12 media to each flask and gently rock the flask to evenly distribute the cells.
9. Return flasks to incubator and incubate the cells for 18–24 h at 37 °C 5 % CO_2.
10. Day 2: Warm DMEM-F12 with and without serum media in a water bath for at least 10 min at 37 °C.
11. If T75 flasks of HEK-293 cells are 70–80 % confluent. Remove existing media from T75 flask containing the HEK-293 cells.
12. Add fresh 9 ml DMEM-F12 with FBS media to the 75 cm^2 flask. Return the flask to the CO_2 incubator.
13. Incubate the cells for at least 30 min at 37 °C 5 % CO_2.
14. Calculate the volume of transfection mix required per T75 flask (18.75 µg pAAV helper (pDG or pRC and pHelper (Aglient Technologies, Santa Clara, CA)), 18.75 µg pAAV (with gene of interest), 37.5 µl Plus reagent, 56 µl Lipofectamine LTX, Media without serum to 1 ml).
15. Add all reagents in the following order:
    (a) Media without serum.
    (b) Plasmid DNA.
    (c) Plus reagent.
    (d) Lipofectamine LTX.
16. Dilute vector DNA and Plus reagent in serum free media in a 15 ml conical tube. Incubate at room temperature for 5 min.
17. Add Lipofectamine LTX to DNA/Plus reagent mix. Incubate at room temperature for 30 min.
18. Remove flasks from CO_2 incubator and add 1 ml of the lipofection mix to each flask.
19. Rock flask gently to distribute transfection mix.
20. Return flasks to CO_2 incubator and incubate the cells for 18–24 h at 37 °C 5 % CO_2.
21. Day 3: Inspect flasks for toxicity and cytopathic effect 18–24 h post transfection. If the cells look healthy then return to the incubator and incubate for a further 48 h at 37 °C 5 % CO_2.

22. Day 4 or 5: Pre-warm suitable media in a water bath for at least 10 min.
23. Harvest culture medium from transfected HEK-293 cells to 50 ml conical tubes.
24. Centrifuge cell suspension at 1,000 × *g* for 5 min.
25. Add 3 ml PBS to HEK-293 cells in flask to remove remaining media.
26. Remove PBS wash and transfer to another 50 ml conical tube.
27. Add 2 ml TrypLE to HEK-293 cells in each flask.
28. Incubate cells for 5 min at 37 °C.
29. Once the incubation is complete check cells have dissociated from the culture surface.
30. Resuspend dissociated cells in 2 ml of DMEM with FBS.
31. Transfer HEK-293 cells to a 50 ml conical Falcon tube.
32. Centrifuge the PBS wash (**step 26**) at 3,000 × *g* for 5 min.
33. Remove and discard PBS from PBS wash—you should be left with a very small cell pellet.
34. Centrifuge cell suspension at 1,000 × *g* for 5 min.
35. Decant clarified supernatant (from **step 34**) to a clean 50 ml Falcon tube.
36. Resuspend cell pellets from; PBS wash, supernatant clarification and TrypLE treatment steps (**steps 26**, **33** and **34**) in 10 ml of Media with FBS.
37. Centrifuge cell suspension at 1,000 × *g* for 5 min, remove supernatant.
38. Store clarified supernatant and cell pellet at −80 °C.

3.7 Adeno-Associated Virus (AAV) Purification (See Note 3)

1. Remove crude lysate (supernatant and cell pellet) from −80 °C freezer and transfer them to 37 °C water bath and allow them to thaw. The tube containing the supernatant may take up to 30 min to thaw. During this time proceed with the freeze–thaw of the cell pellet outlined in the steps below.
2. After 10 min of thawing, take out the tube containing the cell pellet and resuspend the pellet in 3 ml of lysis buffer (150 mM NaCl, 50 mM Tris–HCl pH 8.5) (*see* **Note 4**).
3. Return the tube to dry ice ethanol bath for 10 min to freeze the contents. Once the contents have frozen thaw the cell suspension in a 37 °C water bath for approximately 10 min. During thawing vortex the cell pellet to aid disruption of the cells.
4. Repeat **step 3** twice.
5. Combine both the cell suspension and the defrosted supernatant and aliquot between 50 ml Falcon/centrifuge tubes (the number of tubes depends up on the total volume of crude lysate).

6. Centrifuge the tubes containing the crude lysate at 3,000 × g for 30 min in a table top centrifuge.
7. After centrifugation, transfer the supernatant into a sterile T75 flask or any sterile bottle which can handle more the 100 ml.
8. To the clarified supernatant add 2.8 μl of Benzonase nuclease for every 10 ml of virus (i.e., 14 μl/50 ml) supernatant in the flask and incubate the flask at 37 °C for 30 min. The addition of Benzonase nuclease removes any contaminating DNA in the supernatant (*see* **Note 5**).
9. Unwrap the bottle top filter and carefully place the glass fiber prefilter disk on top of the membrane in the top funnel of the filter unit. Attach the bottle top filter to a vacuum pump.
10. Wet the glass fiber prefilter disk with 4 ml of sterile PBS (*see* **Note 6**).
11. Apply the vacuum to filter bottle.
12. Take the flask with the Benzonase treated virus from the incubator and gently pour into the top of the filter unit, being careful not dislodge the prefilter disk. The filtrate containing the virus will collect in the reservoir bottle below the filter disk.
13. Measure the volume collected using the scale given on side of the reservoir and add 1 part Dilution Buffer 1 to every 9 parts of the filtered supernatant (i.e., add 10 ml of Dilution Buffer 1–90 ml of the filtered supernatant).
14. Attach the purification filter assembly to a 5 ml syringe filled with sterile PBS. Wet the purification filter assembly with sterile PBS from large filter through to the small filter.
15. Remove the syringe and attach the tubing to the purification filter assembly connecting the long tube to the larger filter and small tube to the smaller filter. Attach the 20 ml syringe to the other end of the small tube. Place the long tube into the virus supernatant collected in the reservoir.
16. Pull the virus supernatant through the purification filter assembly into the syringe at a rate of 20 ml/min.
17. When the syringe is full of flow-through, pinch the short section of tubing on the syringe side of the filter to stop the flow of fluid.
18. Remove the syringe and dispose of the contents. Reattach the syringe.
19. Repeat **steps 4** and **5** until the entire volume of virus supernatant has passed through filter unit.
20. Disassemble the purification filter assembly and discard the large filter only.
21. Reattach the long tube to the small filter in place of the large filter.

22. Aliquot 30 ml of Wash Buffer 2 into a sterile 50 ml falcon tube and place the end of long tube into the Falcon tube containing Wash Buffer 2.
23. Pull the entire volume of Wash Buffer 2 through the small purification filter at a rate of 20 ml/min.
24. Empty the contents of syringe to waste and repeat the **step 10** until the entire volume of Wash Buffer 2 has pulled through.
25. Remove the tubing from both sides of the small filter.
26. Using the female luer adaptor attach a 5 ml syringe with plunger depressed to 1 ml to the outlet of the small filter. Attach a 5 ml syringe with plunger removed to the inlet of the small filter.
27. Add 1.5 ml of Elution Buffer 3 in to the syringe attached to the inlet of small filter and attach the plunger (*see* **Note** 7).
28. Hold the small filter with the two syringes attached in a horizontal position and pass the Elution Buffer 3 from the inlet syringe in to the outlet syringe through the small filter.
29. Slowly pass the Elution Buffer 3 back in to the inlet syringe and push a little air through the small filter to collect all elution into one syringe.
30. Discard the empty syringe and small filter unit.
31. Aliquot the eluted virus in to 0.25 ml aliquots into labelled sterile 1.5 ml tubes.
32. Store two vials at −80 °C and one at 4 °C for virus titer determination (*see* **Note 8**).

3.8 Virus Quantification (See Note 9)

1. Treat virus with DNaseI to eliminate any residual contaminating DNA, and Proteinase K in order to release the encapsidated genome. For each virus mix together 5 μl purified virus (*see* **Note 10**) and 45 μl of DNAseI Mastermix.
2. Incubate for 30 min at 37 °C, then 5 min at 95 °C to denature the DNAse.
3. To each sample add 100 μl of Proteinase K solution.
4. Incubate at 56 °C for 1 h, then 10 min at 95 °C.
5. Prepare 2 dilutions of this mixture (*see* **Note 11**): 1:15 in water (5 μl in 70 μl water) and 1:30 in water (5 μl in 145 μl water).
6. Prepare a 20× PCR Mastermix for all reactions, by adding the following reagents for each sample: Taq Man Assay (Neo and Hyg) 0.225 μl Forward primer (100 μM), 0.225 μl Reverse primer (100 μM), 0.05 μl Probe (100 μM), water 0.75 μl; SYBR Assay (Puro) 0.225 μl Forward primer (100 μM), 0.225 μl Reverse primer (100 μM), water 0.75 μl.
7. Prepare qRT-PCR Mastermix per well for all reactions as follows: 1.25 μl 20× PCR Mastermix, 12.5 μl Brilliant III Ultra-FAST

QPCR Master Mix (Agilent) or Brilliant III SYBR Green Master Mix (Agilent), 10.25 μl water.

8. Dispense 24 μl of final qRT-PCR mix into each relevant well of a 96-well plate (*see* **Note 12**).
9. Add 1 μl treated and diluted viral DNA OR standard DNA (plasmid containing Neo or Hyg cassette). (Taq Man assay); plasmid containing Puro cassette (SYBR assay) OR water to each relevant well. Include all samples in triplicate, each standard in triplicate, and three blanks where water is added instead of template. If possible, also include a control sample which consists of a treated and diluted rAAV2 purified virus of known titer (in triplicate)
10. Run qRT-PCR with the following cycling conditions: Taq Man Assay (Neo and Hyg) 1 cycle of 95 °C for 3 min, 40 cycles of 95 °C for 15 s followed by 60 °C for 60 s. SYBR Assay (Puro) 1 cycle of 95 °C for 3 min, 40 cycles of 95 °C for 15 s followed by 60 °C for 20 s, 1 cycle of 95 °C for 60 s, 60 °C for 30 s, and 95 °C for 30 s.
11. Analyze the data. The general process for analyzing the data is as follows:
    (a) View the amplification plots.
    (b) Use the default setting for the software's algorithms for analysis, the threshold bar should be on the linear part of the curve View Ct (threshold cycle) values, the replicates should be tight, within 0.5 Ct.
    (c) Generate the standard curve. This should be linear over the entire range where the unknowns/samples fall.
    (d) The specifications of the standard curve within this range should be: (a) correlation coefficient ≥ 0.99; (b) efficiency: 95–105 %; (c) slope value ~−3.3.
    (e) For SYBR assays, a dissociation curve is performed. This is used to assess the homogeneity of the PCR products, including the presence of primer dimers, thereby determining the specificity of the PCR reaction. Only one peak should be present in any sample.
    (f) If Samples have an equivalent or higher Ct value than the water control, then no titer can be determined and the assay should be repeated.
12. Obtain a report of the experiment with Ct values and initial copies single stranded gene/ml (Neomycin^R, or Hyg^R or Puro^R) for each sample (virus).
13. In order to calculate the virus titer in genome copies (GC)/ml, the value of initial copies of single stranded Neomycin^R or Hyg^R or Puro^R gene/ml should be multiplied by the dilution

factor used in the initial preparation of the virus for qRT-PCR (e.g., DNaseI and Proteinase K treatment). In this protocol, the dilution factor is 450 (after DNAseI, Proteinase K and 1:15 dilution) and 900 (after DNAseI, Proteinase K and 1:30 dilution).

3.9 Target Cell Line Infection with AAV and Selection (See Note 13)

1. Day 1: Warm appropriate media in a water bath for at least >10 min at 37 °C.
2. Aspirate media from flask and wash the cells with PBS. Add TrypLE to the cells.
3. Incubate the cells at 37 °C 5 % CO_2 until the cells detach. Triturate cells 3–4 times with 5 ml Stripette to break up any cell clumps.
4. Add media to the flask equal to at least the volume of TrypLE used (preferably an excess) to neutralize the TrypLE.
5. Place 20 μl of cell suspension in a 1.5 ml microcentrifuge tube. Add 20 μl 1× Trypan Blue to a microcentrifuge tube and mix by gentle pipetting. Transfer 20 μl of the cell suspension to a well of disposable hemocytometer.
6. Count the number of live cells in three fields of the hemocytometer.
7. Calculate the number of cells/ml in the cell suspension.
8. From this value calculate the volume of cells required for 5×10^5 cells in a T75 flask.
9. Volume for 5×10^5 cells $= 5 \times 10^5/(\text{cells/ml}) \times 1{,}000 =$ volume in μl.
10. Add this volume to a new T75 flask. Plate cells in the required number of flasks.
11. Add appropriate media to each flask to give a total volume of 15 ml. Gently rock the flasks to evenly distribute the cells.
12. Return flasks to incubator and incubate the cells for 24 h at 37 °C/5 % CO_2.
13. Day 2: Warm appropriate media in water bath for at least 10 min at 37 °C.
14. Obtain purified rAAV from −80 °C freezer. Thaw on ice. Titurate with a pipette to distribute the virus. Note concentration (genome copies (GC)/ml) and gene targeting event.
15. Aliquot 5 ml of media into a 15 ml conical tube.
16. Add appropriate volume of rAAV to the 5 ml aliquot of media (in 15 ml tube) using the calculation below: Volume of virus to add (μl) $= (\text{MOI} \times (5 \times 10^5 \text{ cells}) \times 1{,}000)/$Virus concentration (GC/ml). The typical MOI for infection is between 10^4 and 10^5 GC per ml.

17. Aspirate media from cells and add diluted virus to cells.
18. Return flasks to CO_2 incubator.
19. 4–6 h post infection, add 5 ml of media to each infected flask.
20. Incubate the cells for up to 72 h at 37 °C 5 % CO_2.
21. Infected cell plate outs (pools and or clones): Label the required number of 96-well plates with cell line, gene and mutation, date and plate number.
22. Warm media for at least 10 min at 37 °C in a water bath.
23. Remove existing media from cells and dispose to waste.
24. Wash cells with 10 ml PBS. Aspirate PBS from flask and dispose to waste.
25. Add 3 ml TrypLE to cells in T75 flask (change volume depending on flask size).
26. Incubate the cells for 5 min at 37 °C 5 % CO_2 until the cells detach.
27. Triturate cells 3–4 times with 5 ml Stripette to break up any cell clumps.
28. Add 3 ml media to the flask to neutralize the TrypLE.
29. Transfer the cells to a 15 ml Falcon tube.
30. Add 20 μl 1× Trypan Blue to a microcentrifuge tube.
31. Add 20 μl of TrypLE treated cells to the same microcentrifuge tube.
32. Mix cells by gentle pipetting.
33. Transfer 10 μl cell suspension to a well of disposable hemocytometer.
34. Transfer the hemocytometer to an inverted microscope.
35. Count the number of live cells in three fields of the hemocytometer.
36. Calculate the number of cells per ml in the cell suspension.
37. Dilute the cells to the required density in the appropriate media with selection for the cell line. To obtain clones cells should be plated at 0.5–2 cells/well, and for pools (wells containing 2–10 clones) a range of densities is advised from 200 to 2000 cells/well. Data obtained in parental characterisation section 3.2 and 3.3 can be used as a guide for the number of cells to be plated.
38. Transfer cells to large reservoir.
39. Dispense 200 μl per well to 11 columns of 96-well plate using multichannel pipette.
40. Dispense cells to total number of 96-well plates manually.
41. Transfer plates to incubator at 37 °C 5 % CO_2. Dispose of waste.

42. Day 10–14: Visually screen the plates to identify which wells contain colonies and if necessary which cell dilution has given 5 or less colonies/well. If there are some wells with only 1 colony present then mark these, as if they are positive in the PCR screen, they are likely to be a clonal population rather than a pool.

3.10 Screening for Homologous Recombinants by Genomic PCR

1. Harvest of gDNA for genomic screen: Prepare Direct PCR Lysis buffer.
2. Aliquot 20 μl of Lysis buffer in to each well of 96-well PCR plate. 1× 96-well plate is required for each 96-well plate of cells to be screened.
3. Remove plates to be harvested from incubator and check colonies are of appropriate size (50–70 % confluent).
4. Remove media. Wash wells with PBS and add 25 μl of TrypLE. Incubate at 37 °C for 5–10 min.
5. Remove plates from incubator and confirm dissociation of cells from plate surface.
6. Mix cell suspension by gentle pipetting.
7. Transfer 5 μl of cell suspension to PCR plate with lysis buffer.
8. Add 200 μl of media to the TrypLE treated cells.
9. Return cell plates to incubator.
10. Take cell lysis suspension from **step** 7. Seal the plate using caps or a heat seal.
11. Incubate the lysis plate under the following conditions: 15 min at 55 °C followed by 45 min at 85 °C.
12. Transfer lysed cells to 4 °C for short term storage or use directly for PCR Screen.
13. PCR Screen of targeted cells (*see* **Note 14**): Obtain all reagents except GoTaq enzyme. Thaw reagents at room temperature.
14. Prepare PCR Mastermix on ice. Mix the reaction immediately after the addition of the enzyme.
15. Aliquot 18 μl of the PCR mastermix to each well of a 96-well PCR plate.
16. Add 2 μl of template DNA to the appropriate wells. Seal the PCR plate with a heat sealer or strip caps.
17. Briefly centrifuge the plate to collect contents in the bottom of the wells.
18. Transfer plates to a PCR machine. Use the following Touchdown PCR protocol (*see* **Note 15**): 1 cycle of 94 °C for 3 min, 3 cycles of 94 °C for 15 s, 64 °C for 30 s, and 70 °C for 60 s per kb of product required, 3 cycles of 94 °C for 15 s, 61 °C for 30 s, and 70 °C for 60 s per kb of product required,

3 cycles of 94 °C for 15 s, 58 °C for 30 s, and 70 °C for 60 s per kb of product required, 35 cycles of 94 °C for 15 s, 57 °C for 30 s, and 70 °C for 60 s per kb of product required, 1 cycle of 70 °C for 5 min with a final 4 °C hold.

19. Remove plate at the end of the run and store at 4 °C.
20. Prepare a 1 % Agarose gel.
21. Transfer 5 μl of PCR product to each well of the agarose gel. Load appropriate DNA markers on the gel.
22. Run the agarose gel until the DNA markers are well defined in the region of interest.
23. Confirm presence of positive controls and lack of an amplicon in negative control. Record whether a positive amplicon is present for all samples.
24. Wells that provided a positive result should be repeated to confirm presence of an amplicon.
25. Inspect positively identified wells for the presence of viable cells.
26. Positive samples should be grown according to standard culture methods until enough cells are available for banking for long-term storage and further experiments.

3.11 Confirmation of Genotype of Targeted Cells at gDNA

1. Obtain the following samples: Purified and quantified gDNA samples, Forward and reverse primers for the specific PCR amplification required, Control gDNA: positive control for the reaction to ensure PCR is working.
2. Make stocks of the gDNA for PCR by diluting a sample to between 10 and 100 ng/μl.
3. Run PCR on the samples to be tested according to the method in Subheading 3.10.
4. Analyze samples and confirm presence of positives according to the protocol in Subheading 3.10.
5. Products amplified by PCR can be purified from the reaction mix using standard procedures, e.g., Qiagen QIAquick PCR purification kit or similar.
6. Purified samples can be sent for sequencing with appropriate sequencing primers to analyze the targeted status of the cell line clone at the DNA or RNA level. If the introduced mutation is a heterozygous point mutation, one would expect to see a double peak at the position of the introduced mutation.

3.12 Confirmation of cDNA Genotype

1. Isolation of total RNA using the Qiagen RNeasy purification kit (*see* **Note 16**): Obtain cell pellets of approximately 1e^{6} cells. Resuspend the pellet in a small volume of PBS (e.g., 20 μl).

2. Add 350 µl of buffer RLT to the pellet and pipette or vortex to mix.
3. Pipette the lysate directly into a QIAshredder spin column placed in a 2 ml collection tube, and centrifuge for 2 min at full speed.
4. Add 1 volume of 70 % ethanol to the homogenized lysate, and mix well by pipetting. Do not centrifuge.
5. Transfer up to 700 µl of the sample, including any precipitate that may have formed, to an RNeasy spin column placed in a 2 ml collection tube (supplied in the RNeasy purification kit).
6. Close the lid gently, and centrifuge for 15 s at 8,000 × *g*. Discard the flow-through.
7. If the sample volume exceeds 700 µl, centrifuge successive aliquots in the same RNeasy spin column. Discard the flow-through after each centrifugation.
8. Add 700 µl Buffer RW1 to the RNeasy spin column. Close the lid gently, and centrifuge for 15 s at 8,000 × *g* to wash the spin column.
9. Add 500 µl Buffer RPE to the RNeasy spin column. Close the lid gently, and centrifuge for 15 s at 8,000 × *g*.
10. Add 500 µl Buffer RPE to the RNeasy spin column. Close the lid gently, and centrifuge for 2 min at 8,000 × *g*.
11. Place the RNeasy spin column in a new 1.5 ml collection tube (supplied in the RNeasy purification kit).
12. Add 30–50 µl RNase-free water directly to the spin column membrane. Close the lid gently, and centrifuge for 1 min at 8,000 × *g*.
13. The RNA can be quantified using a spectrophotometer and then stored at −80 °C.
14. Genotyping cDNA samples by RT-PCR: Obtain all reagents except reverse transcriptase enzyme. Thaw reagents at room temperature.
15. Prepare cDNA Mastermix 1 on ice. Per reaction 1µl of random hexamers, 1µl 10mM dNTP, 10 µl water (molecular biology grade).
16. Aliquot 12 µl of master mix to each well of 96-well PCR plate or PCR tubes. Remember to include a no reverse transcriptase control sample.
17. Add 1 µl template RNA at a concentration of 1 µg to the appropriate well/tube.
18. Seal the PCR plate with heat sealer or strip caps or close the tubes. Briefly centrifuge the plate to collect the contents in the bottom of the wells.

19. Run the following protocol; 1 cycle at 65 °C for 5 min followed by 1 cycle at 4 °C for 1 min.
20. Remove plate/tubes at the end of the run and collect the contents in the tube by brief centrifugation.
21. Prepare the cDNA Mastermix 2 on ice (sufficient for 10 reactions) with and without reverse transcriptase: 40 μl of 5× First-Strand Buffer, 10 μl DTT (0.05M), 10 μl RNAse Inhibitor, 10 μl Supersript III enzyme.
22. Transfer 6 μl to each reaction tube. Collect the contents in the tube by brief centrifugation.
23. Run the following protocol: 1 cycle at 25 °C for 5 min, 1 cycle at 50 °C for 60 min, 1 cycle at 70 °C for 15 min followed by a hold at 4 °C. Remove plate at the end of the run and collect the contents in the tube by brief centrifugation.
24. Store First-Strand cDNA at 4 °C for short term or −20 °C for long term storage.
25. Obtain all reagents except GoTaq polymerase enzyme and thaw at room temperature.
26. Obtain the following samples: cDNA produced from the first step and Forward and Reverse primers for the specific PCR amplification required.
27. Dilute the cDNA 1:5 in distilled water.
28. Prepare PCR mastermix on ice. Mix the reaction immediately after the addition of the enzyme.
29. Aliquot 48 μl of master mix to each well of 96-well PCR plate or PCR tubes.
30. Add 2 μl template cDNA to the appropriate well. Seal the PCR plate with heat sealer or strip caps. Briefly centrifuge the plate to collect contents in the bottom of the wells.
31. Run the following Touchdown PCR protocol (*see* **Note 15**): 1 cycle of 94 °C for 3 min, 3 cycles of 94 °C for 15 s, 64 °C for 30 s, and 70 °C for 60 s per kb of product required, 3 cycles of 94 °C for 15 s, 61 °C for 30 s, and 70 °C for 60 s per kb of product required, 3 cycles of 94 °C for 15 s, 58 °C for 30 s, and 70 °C for 60 s per kb of product required, 35 cycles of 94 °C for 15 s, 57 °C for 30 s, and 70 °C for 60 s per kb of product required, 1 cycle of 70 °C for 5 min with a final 4 °C hold.
32. Remove plate at the end of the run and store at 4 °C.
33. Analysis of PCR amplification products by agarose gel electrophoresis and sequencing: This can be completed using the procotols for gDNA analysis and sequencing.

4 Notes

1. Do not let the cells become more than 80 % confluent. They should be passaged regularly (twice weekly) and discarded after 15–20 passages.
2. 3× T175 at 70 % confluence will be required to plate 5× T75 for transfection.
3. Harvest and purification should be carried out in a class II containment laboratory. Purification of a rAAV preparation generated by transfection of HEK-293 cells results in a viral preparation that is free from cellular contaminants, any exogenous plasmid DNA and is more concentrated compared to a crude lysate. We use a membrane based filtration method for laboratory scale purification of up to 10^{12} virus particles (AAV) in a stable storage buffer. The virus preparation can be stored short term at 4 °C (1–2 weeks) and long term for increased stability at −80 °C. AAV can also be purified by other methods such as caesium or iodixanol gradient ultracentrifugation. In order to purify the virus from the lysate first the virus has to be released from the HEK-293 cells. This is achieved by repeat freeze–thaw cycles of the lysate.
4. Although this step is not essential we have found that addition of lysis buffer increases the yield of purified virus.
5. The virus purification is performed using the AAV VIRAKIT (www.virapur.com).
6. This pre wetting enables the glass fiber prefilter disk to adhere to the 0.45 μm bottle top filter unit.
7. 1.5 ml is sufficient to elute the entire bound virus on the small filter. Larger volumes can be used, but the virus will be less concentrated.
8. To avoid repetitive freeze thaw cycles it is advised to store the eluted virus at lower volume aliquots.
9. Purified rAAV2 is quantified via quantitative real-time PCR (qRT-PCR) using a TaqMan assay (for Neo and Hyg) or a SYBR assay (for Puro). After purification, at least 10 μl purified virus should be stored at 4 °C for this purpose.
10. Great care must be taken in order to keep the vial sterile, to have an homogeneous/representative aliquot by pipetting up/down with a 200 μl pipette first but avoiding foam and degradation of the viruses; then take carefully 5 μl off for the titration.
11. Dilution is important since undiluted samples fail to be amplified in qRT-PCR reaction (possible residual inhibitory substances). Try to keep the same volume pipetted between the

two dilutions, using same pipettes, to avoid variations due to pipettes (P20 for the virus prep, P200 for the H_2O). The standard curve is based on DR-HPRT Neo rAAV, whose titer has been given by David Russell as 2.09E+13 GC/ml. This titer was determined by Southern Blot quantification. This titer has been validated using our Neo-plasmid-Standards-based qPCR. This gave a titer of =8.2E+12 GC/ml for the David Russell virus preparation.

12. Ensure plastic-ware used is optically suitable for use with the qRT-PCR machine.
13. In order to obtain efficient infection, it is important to use cells of a low passage number which are growing exponentially, i.e., 30–70 % confluent. It is also useful to consider whether the parental cell line has a genetic background which might interact with the chosen mutation to produce any unwanted effect (e.g., severely compromised growth or lethality). For a new target line it is advisable to infect with a range of MOIs. To determine targeting efficiency an uninfected control or an infected and unselected population is required. A T75 plated in the same manner but not infected and then processed for plate out and selection as the infected cells is a suitable control. Target cells should be less than 40 % confluent at infection.
14. The following protocol provides an outline of the steps required to screen a 96-well plate of previously lysed cells. Promega's GoTaq Hot Start polymerase is recommended for the purposes of this procedure. The enzyme is very robust for genomic targets while being very stable during the preparation and handling of the reaction mix. Details of relevant controls are given within. Primer design. PCR primers are designed to generate amplicons of 1–2 kb. For PCR screening experiments, primers should be designed to amplify from the resistant cassette into the genomic DNA outside of the homology arms. One approach is to incorporate a synthetic sequence into the resistant cassette that is equivalent to one of the wild type primers. This allows a single primer set to amplify both the targeted allele and the wild-type allele in a multiplex reaction (Fig. 3). In design the two amplicons should be greater than 500 bp difference to allow good separation on agarose gel. A second approach is to use two independent primer sets: set 1—will be specific for the targeted allele and have one primer in the resistance cassette while the other primer is located genomic region outside of the homology arms; set 2—will be wild-type specific. When possible, design primers to avoid regions of secondary structure, and/or repetitive elements in the DNA. Controls—standard vector (no additional control primers in the cassette). A selection of lysate should be amplified using a set of wild-type primers to control for the quality

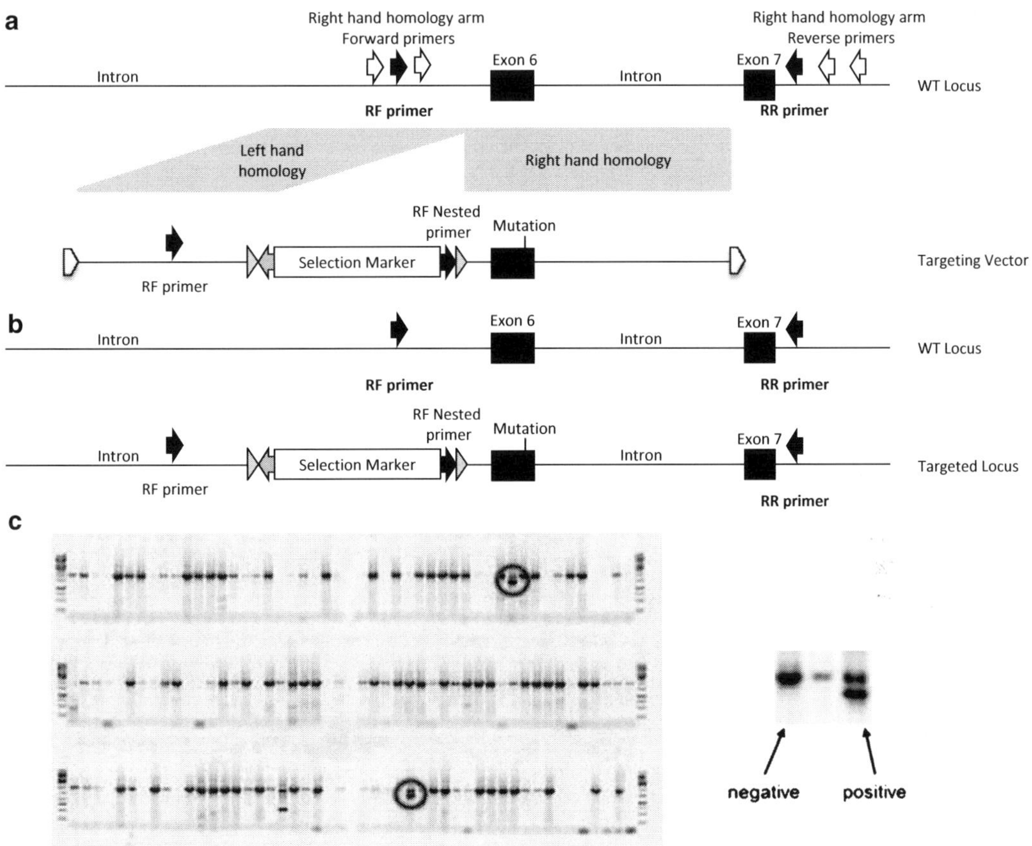

Fig. 3 Nested primer strategy to identify targeted loci. For each homology arm several sets of PCR primers are tested that amplify across the whole homology arm region (**a**) In this example, three forward primers are located in the center region of the left-hand homology arm are tested pairwise with three primers located in the target locus outside of the right-hand homology arm. Once a suitable pair of primers have been identified (*black arrowheads*), the forward primer is incorporated into the design of the target vector at the 3′ end of the selection marker cassette (RF Nested primer). Upon infection of the target cell line, targeted cells can be identified by the presence of the nested primer within the selection marker. Non-targeted loci (wild type) will produce a single PCR product (**b**), targeted loci, however, will produce a smaller product (from the nested forward primer) as well as a larger product—this larger product is usually not seen as it is not preferentially amplified during the PCR reaction. A typical agarose gel (**c**) of PCR products can easily identify the non-targeted cells from the targeted cells (*circled*). As AAV modifies a single allele in each targeting event, positive clones contain a double PCR product of the targeted (smaller) band and non-targeted (larger) band

of the DNA. DNA should be obtained from parental line. Amplification of parental cell line gDNA with wild-type primers will determine if the PCR mastermix is functional. Amplification of this control (parental gDNA) with neomycin specific primers will act as a negative control. If possible, a positive control should be included. This could be DNA taken from another cell line with the same targeting event or DNA from a targeted pool. A no template control (NTC) should be included to check for DNA contamination. Controls–control primers incorporated in to vector design. This PCR strategy contains an internal control for each sample enabling a wild-type and a targeted PCR product to be amplified in the same reaction. Lack of the wild-type band suggests the DNA is not suitable for efficient PCR amplification. In this instance assess the PCR using gDNA obtained from the parental line and repeat harvest of the targeted cell lines (Subheading 3.10). Amplification of this parental control gDNA with wild-type primers will confirm a functional PCR mastermix. Absence of targeted band will act as a negative control. A no template control (NTC) should be included to check for DNA contamination.

15. The annealing time should be altered dependent on the length of the product expected (60 s/kilobase). Review Primer test conditions for appropriate extension time for the primers being used.
16. If possible carryout all RNA work in a dedicated RNA area. Wipe down surfaces and pipettes with RNase away/RNA Zap before starting work. Using only filter tips and the dedicated reagents designated for RNA work.

Acknowledgements

The authors would like to thank all of the Cell Line Engineering team at Horizon Discovery Ltd for their work in developing and optimizing these protocols. We are also grateful to Prof. David Russell (University of Washington) and Prof. Eric Hendrickson (University of Minnesota) for their advice and constructive discussions during the optimization process. We would also like to thank Eric Rhodes for his helpful comments during writing of the manuscript.

References

1. Russell DW, Hirata RK (1998) Human gene targeting by viral vectors. Nat Genet 18:325
2. Vasileva A, Linden RM, Jessberger R (2006) Homologous recombination is required for AAV-mediated gene targeting. Nucleic Acids Res 34:3345–3360
3. Miller DG, Petek LM, Russell DW (2003) Adeno-associated virus vectors integrate at chromosome breakage sites. Nat Genet 36:767–773
4. Hendrickson EA (2007) Gene targeting in human somatic cells. In: Source book of models for biomedical research, vol 53. Humana Press, Totowa, NJ, pp 509–525
5. Traverso G, Bettegowda C, Kraus J, Speicher MR, Kinzler KW, Vogelstein B, Lengauer C (2003) Hyper-recombination and genetic instability in BLM-deficient epithelial cells. Cancer Res 63:8578–8581
6. Yun J, Rago C, Cheong I, Pagliarini R, Angenendt P, Rajagopalan H, Schmidt K, Willson JK, Markowitz S, Zhou S, Diaz LA Jr, Velculescu VE, Lengauer C, Kinzler KW,

Vogelstein B, Papadopoulos N (2009) Glucose deprivation contributes to the development of KRAS pathway mutations in tumor cells. Science 325:1555–1559

7. Hucl T, Rago C, Gallmeier E, Brody JR, Gorospe M, Kern SE (2008) A syngeneic variance library for functional annotation of human variation: application to BRCA2. Cancer Res 68:5023–5030
8. Konishi H, Karakas B, Abukhdeir AM, Lauring J, Gustin JP, Garay JP, Konishi Y, Gallmeier E, Bachman KE, Park BH (2007) Knock-in of mutant K-Ras in nontumorigenic human epithelial cells as a new model for studying K-Ras mediated transformation. Cancer Res 67:8460–8467
9. Khan IF, Hirata RK, Russell DW (2011) AAV-mediated gene targeting methods for human cells. Nat Protoc 6:482–501
10. Russell DW, Hirata RK (2008) Human gene targeting favors insertions over deletions. Hum Gene Ther 19:907–914
11. Hendrie PC, Russell DW (2005) Gene targeting with viral vectors. Mol Ther 12:9–17
12. Topaloglu O, Hurley PJ, Yildirim O, Civin CI, Bunz F (2005) Improved methods for the generation of human gene knockout and knockin cell lines. Nucleic Acids Res 33:e158
13. Miller DG, Petek LM, Russell DW (2003) Human gene targeting by adeno-associated virus vectors is enhanced by DNA double-strand breaks. Mol Cell Biol 23:3550–3557
14. Hirata RK, Russell DW (2000) Design and packaging of adeno-associated virus gene targeting vectors. J Virol 74:4612–4620
15. An up to date list of isogenic cell lines can be found at: http://www.horizondiscovery.com/

Chapter 6

Engineering of Customized Meganucleases via In Vitro Compartmentalization and In Cellulo Optimization

Ryo Takeuchi, Michael Choi, and Barry L. Stoddard

Abstract

LAGLIDADG homing endonucleases (also referred to as "meganucleases") are compact DNA cleaving enzymes that specifically recognize long target sequences (approximately 20 base pairs), and thus serve as useful tools for therapeutic genome engineering. While stand-alone meganucleases are sufficiently active to introduce targeted genome modification, they can be fused to additional sequence-specific DNA binding domains in order to improve their performance in target cells. In this chapter, we describe an approach to retarget meganucleases to DNA targets of interest (such as sequences found in genes and *cis* regulatory regions), which is feasible in an academic laboratory environment. A combination of two selection systems, in vitro compartmentalization and two-plasmid cleavage assay in bacteria, allow for efficient engineering of meganucleases that specifically cleave a wide variety of DNA sequences.

Key words Meganuclease, Protein engineering, Selection, Genome engineering

1 Introduction

Over the past decade, several technologies have been developed to introduce site-specific mutagenesis in highly complex genomes of various eukaryotes. Both oligonucleotide-based methods (such as triplex-forming oligonucleotides coupled to DNA modifying enzyme domains) and protein-based approaches (using sequence-specific nucleases and recombinases) have been shown to promote targeted genome modifications such as gene disruption, integration and sequence modification [1]. Engineered nucleases that have been most widely employed include zinc finger nucleases (ZFNs) [2, 3], transcription activator-like effector nucleases (TALENs) [4–6], RNA-guided CRISPR/Cas9 endonucleases (CRISPR) [7, 8], and LAGLIDADG homing endonucleases (also referred to as meganucleases) [9, 10]. These enzymes generate double strand breaks (DSBs) at their genomic target sites, and stimulate intrinsic DSB repair pathways. Template-independent repair, such as nonhomologous end joining, often gives rise to

Shondra M. Pruett-Miller (ed.), *Chromosomal Mutagenesis*, Methods in Molecular Biology, vol. 1239, DOI 10.1007/978-1-4939-1862-1_6, © Springer Science+Business Media New York 2015

small deletions and insertions, while homology-driven repair can lead to duplication of a donor template sequence (resulting in a specific sequence alteration, insertion or deletion) [3, 11, 12]. In addition, recent studies have demonstrated that simultaneous generation of DSBs at multiple genomic loci promotes more dynamic chromosomal mutations including deletion, inversion, and translocation [13–15].

Homing endonucleases ("meganucleases") are naturally occurring enzymes that are found in mitochondrial and chloroplast genomes (as well as archaea), and are often encoded in concert with surrounding intron or intein sequences [16]. When expressed in host cells, these enzymes generate strand breaks at alleles that lack the intervening sequence, leading to homology-driven repair that results in duplication of sequences carrying the meganuclease genes. As a result, intron- or intein-encoded homing endonucleases function as exceptionally efficient mobile elements that are inherited in a dominant, non-Mendelian manner.

Meganucleases that possess a single conserved LAGLIDADG motif per protein chain form homodimeric proteins that cleave palindromic and nearly palindromic DNA target sequences, while those that contain two such motifs per protein chain form larger, pseudo-symmetric monomers that can target completely asymmetric DNA sequences. To prevent cleavage at off-target sites that could otherwise compromise the viability of host cells, meganucleases recognize long (18–24 base pairs) target sequences, and thereby exhibit extremely high cleavage specificity. Because of their small size, stable folds and exceptional DNA cleavage specificities, several wild-type meganucleases (particularly I-SceI, a monomeric endonuclease that targets a 18 base-pair sequence), have been used to generate site-specific DSBs on integrated reporters in a variety of eukaryotic cells, and shown to do so without significant cytotoxicity [10, 12, 17]. More recently, these endonucleases have been redesigned to modify genomic target sites in mammalian and plant cells [9, 18–21].

Two recent studies have shown that by combining meganucleases with additional sequence-specific TAL effector DNA binding domains, highly active and specific genome engineering reagents can be created [22, 23]. However, this strategy still requires engineering of meganucleases to display a desired new specificity, corresponding to the genomic target site of interest. The complete redesign of meganuclease specificity toward such genomic sites has generally been accomplished in biotech settings that make use of industrial scale engineering pipelines [9, 19]. Here, we describe a detailed protocol to effectively obtain customized meganucleases that employs routine benchtop molecular biology protocols that are accessible for academic laboratories [23]. All steps required to initially redirect the specificity of meganucleases are conducted using a method termed in vitro compartmentalization (IVC). This approach, which was originally developed as described in [24–26], allows for expression and selection of engineered proteins in individual compartments generated in an oil–surfactant mixture (approximately 10^{10}/mL). Subsequently, the most active population of redesigned meganucleases are filtered for optimal gene targeting activity using a two-plasmid cleavage assay in bacteria [27] to select variant endonucleases that are significantly active *in cellulo.*

The use of in vitro compartmentalization, as a tool for protein engineering, facilitates construction of highly complex protein libraries, compared to previously described methods, and thereby allows investigators to query a significantly higher number of protein sequence variants at each stage of specificity redesign. While the methodology described in this chapter is specific for the alteration of nuclease specificity, the general approach is appropriate for the modification of any extensive protein recognition surface, and relies only on the development of a phenotypic selection strategy that can be utilized in the context of compartmentalized aqueous droplets.

2 Materials

2.1 Library Construction

1. PCR thermal cycler.
2. pET-21d(+) plasmid (EMD Millipore, Bedford, MA, USA) containing a meganuclease gene with codons optimized for bacterial expression (*see* **Note 1**).
3. Phusion High-Fidelity DNA Polymerase (Thermo Scientific, Pittsburg, PA, USA/New England Biolabs, Ipswich, MA, USA).
4. 10 mM dNTP mix.
5. 25 nmol scale degenerate PCR primers desalting grade.
6. PCR primers commonly used to construct a library: "Up1" (5′-CGT CCG GCG TAG AGG ATC GAG ATC-3′), "Up2" (5′-AGA TCT CGA TCC CGC GAA ATT-3′), UpRev (5′-CCA TGG TAT ATC TCC TTC TTA AAG TTA AAC-3′), "MidFwd" (5′-CAC TGA GAT CCG GCT GC TAA C-(tandem 2 copies of a target site)-CCG CTG AGC AAT AAC TAG-3′), "MidRev" (5′-GTT AGC AGC CGG ATC TCA GTG-3′), "Down1" (5′-CAC TCG TGC ACC CAA CTG ATC TTC-3′), and "Down2" (5′-TCA GGG TTA TTG TCT CAT GAG CG-3′).
7. 5 μg/mL ethidium bromide.
8. UltraPure Agarose.
9. 50× TAE buffer (pH 8.2–8.4 at 25 °C): 2.0 M Tris acetate, 0.05 M EDTA.

10. 6× Loading dye: 30 % glycerol, 6 mM EDTA, and 0.1 % orange G; filtered through a 0.2 μm PVDF membrane.
11. ≥98 % Guanosine.
12. 1 kb plus DNA ladder.
13. QIAquick Gel Extraction Kit (Qiagen, Valencia, CA, USA).

2.2 In Vitro Selection Using IVC

1. 4× annealing buffer: 40 mM Tris–HCl (pH 7.5), 600 mM NaCl (autoclaved).
2. Mineral oil for molecular biology (light oil).
3. ABIL EM90 (Evonik, Essen, Germany).
4. 98 % Triton X-100 for molecular biology, DNase, RNase, Protease free.
5. 1 M potassium glutamate (pH 7.5): 5.6 g of KOH is dissolved in 50 mL of water, and titrated with glutamate powder to adjust pH to 7.5. Water is added to make a final volume of 100 mL, and the solution is filtered through a 0.2 μm PVDF membrane.
6. 1 M magnesium acetate (pH 7.5): 14.2 g of magnesium acetate is dissolved in water, titrated with glacial acetic acid to adjust pH to 7.5, and filtered through a 0.2 μm PVDF membrane.
7. 1 M dithiothreitol (DTT) filtered through a 0.2 μm PVDF membrane.
8. 10 mg/mL bovine serum albumin (BSA) Fraction V filtered through a 0.2 μm PVDF membrane.
9. 14-mL polypropylene round-bottom tube.
10. Octagonal magnetic stir bar (Length: 13 mm; autoclaved for 15 min).
11. Magnetic stirrer.
12. PURExpress (New England Biolabs).
13. 40 units/μL Murine RNase inhibitor (New England Biolabs).
14. Phenol–Chloroform–Isoamyl alcohol (25:24:1, water-saturated, pH 7.9).
15. 3 M sodium acetate (pH 5.2): sodium acetate is dissolved in water, titrated with glacial acetate to adjust pH to 5.2 and autoclaved for 15 min.
16. Isopropanol.
17. 70 % ethanol.
18. RNase Cocktail Enzyme Mix (Life Technologies, Carlsbad, CA, USA).
19. QIAquick PCR Purification Kit (Qiagen).
20. 400,000 units/mL T4 DNA ligase.
21. DNA adaptors (*see* Table 1).

Table 1
DNA adaptors used to recover active meganuclease genes

Adaptor	Sequence
A	5′-GTTTGCTCAGGCTCTCCCCGTGGAGGTAATAATTG**NNNN**-3′ 3′-CAAACGAGTCCGAGAGGGGCACCTCCATTATTAAC-5′
B	5′-GATCTTACCGCTGTTGAGATCCAGTTCGATGTAACC**NNNN**-3′ 3′-CTAGAATGGCGACAACTCTAGGTCAAGCTACATTGG-5′
C	5′-CAAACGTCTGAACATCAATGCGGCCAAATCTTCATTCC**NNNN**-3′ 3′-GTTTGCAGACTTGTAGTTACGCCGGTTTAGAAGTAAGG-5′
D	5′-AGCTAACGTTGGTCCAAACAGGATACCTGCGGTGA**NNNN**-3′ 3′-TCGATTGCAACCAGGTTTGTCCTATGGACGCCACT-5′
E	5′-CCTAGACGGATAACGCGTACTCTTTCCTCCGATTGG**NNNN**-3′ 3′-GGATCTGCCTATTGCGCATGAGAAAGGAGGCTAACC-5′
F	5′-GCTCGAGACTCTCGCGAAAAGTAAGAAGGCTACATC**NNNN**-3′ 3′-CGAGCTCTGAGAGCGCTTTTCATTCTTCCGATGTAG-5′
G	5′-ACTGGTAGTCTCCGGCCATTTGTTCCTCAGCAAAGT**NNNN**-3′ 3′-TGACCATCAGAGGCCGGTAAACAAGGAGTCGTTTCA-5′
H	5′-GACTATATTCTCCAATCTCGGAGCAAAGGGGCTCGC**NNNN**-3′ 3′-CTGATATAAGAGGTTAGAGCCTCGTTTCCCCGAGCG-5′

Four-base, 3′ overhangs (bold) are complementary to those of target sites generated by variant endonucleases. Underlined are sequences of primers used for adaptor-specific PCR. Avoid ligation with adaptors used in the last two rounds of selection

22. PCR primers: "Up3" (5′-CGA AAT TAA TAC GAC TCA CTA TAG G-3′), "Up4" (5′-CCC TCT AGA AAT AAT TTT GTT TAA CTT-3′), primers specific to DNA adaptors (*see* Table 1).

2.3 Two-Plasmid Cleavage Assay in Bacteria

1. pEndo and pCcdB: these plasmids were originally described in ref. [27] (*see* Fig. 1).
2. DNA Clean & Concentrator-5 (Zymo Research, Irvine, CA, USA).
3. Restriction enzymes (AflIII, BglII, NheI, SacII, NcoI, and NotI) (New England Biolabs).
4. Quick Ligation Kit (New England Biolabs).
5. NovaXGF' competent cells (EMD Millipore).
6. DNA purification Miniprep Kit.
7. Oligonucleotides (used for cloning, PCR, and sequencing): "CcdB_seq1" (5′-GTT ATC GGG GAA GAA GTG GC-3′), "CcdB_seq2" (5′-CGG GTG ATG CTG CCA ACT TA-3′), "Endo_colonyPCR_fwd" (5′-CAC GGC AGA AAA GTC CAC ATT G-3′), Endo_colonyPCR_rev (5′-TGA GGG AGC CAC GGT TGA TG-3′), "Endo_seq_fwd" (5′-CGG CGT CAC ACT TTG CTA TG-3′), and "Endo_seq_rev" (5′-GAG CCA CGG TTG ATG AGA GCT TTG-3′).

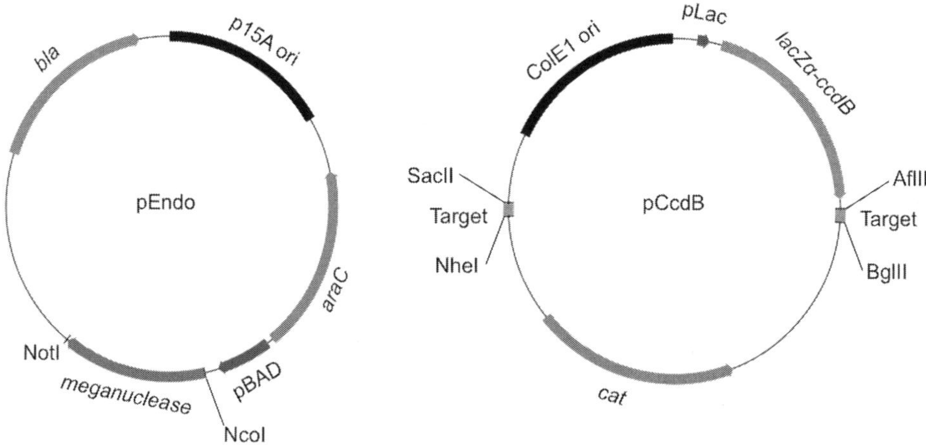

Fig. 1 The pEndo and pCcdB plasmids used for the two-plasmid cleavage assay in bacteria [27]. These plasmid maps were drawn using Flex Plasmid Draw (http://mavericktse.is-a-geek.com/PlasmidDrawv2/index.html)

8. SBG medium: 15 g tryptone, 10 g yeast extracts, 2.5 g NaCl, and 2.5 g glucose are dissolved in water and titrated with NaOH to adjust pH to 7.4. Water is added to make a final volume of 500 mL, and the medium is autoclaved for 15 min.
9. 1 mM Hepes (pH 7.0): Hepes is dissolved in water, titrated with NaOH to adjust pH to 7.0, and autoclaved for 15 min.
10. 10 % (v/v) glycerol (autoclaved for 15 min).
11. Gel DNA Recovery Kit (Zymo Research).
12. 1 kb DNA ladder.
13. 2× Gibson Assembly Master Mix (New England Biolabs).
14. DH5α and DH10B chemical competent cells.
15. Sterilized water.
16. Cell scraper.
17. Electroporation cuvette (0.2-cm path).
18. 2× YT medium.
19. 20 % L-arabinose filtered through a 0.2 μm PVDF membrane.
20. 10× M9 salts: 60 g/L Na_2HPO_4, 30 g/L KH_2PO_4, 5 g/L NaCl, and 10 g/L NH_4Cl.
21. 1 M Mg_2SO_4 autoclaved for 30 min.
22. 1 M $CaCl_2$ autoclaved for 30 min.
23. 1 % (w/v) thiamine filtered through a 0.2 μm PVDF membrane.
24. 1 M isopropyl β-D-1-thiogalactopyranoside (IPTG) filtered through a 0.2 μm PVDF membrane.
25. Control plate: 100 mL of 3 %(w/v) agar and 100 mL of 2× M9 salt supplemented with 2 %(v/v) glycerol and 1.6 %(w/v)

tryptone are separately autoclaved for 30 min, and combined. After the agar medium is cooled down to 50 °C, 200 μL of 1 M Mg_2SO_4, 200 μL of 1 M $CaCl_2$, 200 μL of 100 mg/mL carbenicillin, and 40 μL of thiamine are quickly added, and poured in 100-mm petri dishes.

26. Selective plate: the plates are made by further adding 200 μL of 20 % L-arabinose and 80 μL of IPTG to 200 mL of the control agar medium.
27. GoTaq DNA Polymerase (Promega, Madison, WI, USA).
28. GeneMorph II Random Mutagenesis Kit (Agilent Technologies, Santa Clara, CA, USA).
29. ElectroMax DH10B T1 phage-resistant competent cells (Life Technologies).
30. SOC medium.
31. 150-mm petri dish.
32. Glass beads autoclaved for 15 min.
33. Plasmid Maxi Kit.

3 Methods

The method described below is comprised of (1) a bioinformatic search for DNA sequences to which a meganuclease is to be retargeted (in this case, using the "LAHEDES" web server; http://www.homingendonuclease.net) [28], (2) engineering of a customized meganuclease using In vitro compartmentalization (IVC), and (3) filtering and optimization of final pool of active enzymes using two-plasmid cleavage assay in bacteria. To create variant endonucleases that display cleavage activity against a target site of interest (e.g., clinically relevant genome site), site-directed saturation mutagenesis and several rounds of selections using IVC are iterated. An overview of this step is depicted in Fig. 2. A pool of meganucleases redesigned through in vitro selections is further screened in bacteria so as to isolate substantially active enzymes *in cellulo*.

3.1 Determining Sequences to Be Targeted by Engineered Meganucleases

The previously developed "LAHEDES" web server is utilized to search for sequences that display high identities to the target sites for wild type meganucleases that have been crystallized in complexes with DNA [28]. In addition to the homodimeric I-CreI meganuclease that has previously been well described for engineering and gene targeting, crystal structures of eight monomeric meganucleases that recognize unique target sequences (I-SceI, I-OnuI, I-LtrI, I-LtrWI, I-PanMI, I-GzeMII, I-HjeMI, and I-SmaMI) are now available. A "Central Four Search", as described at that web server, is used to explore candidate sites that can be targeted by engineered meganucleases (Fig. 3). This search option

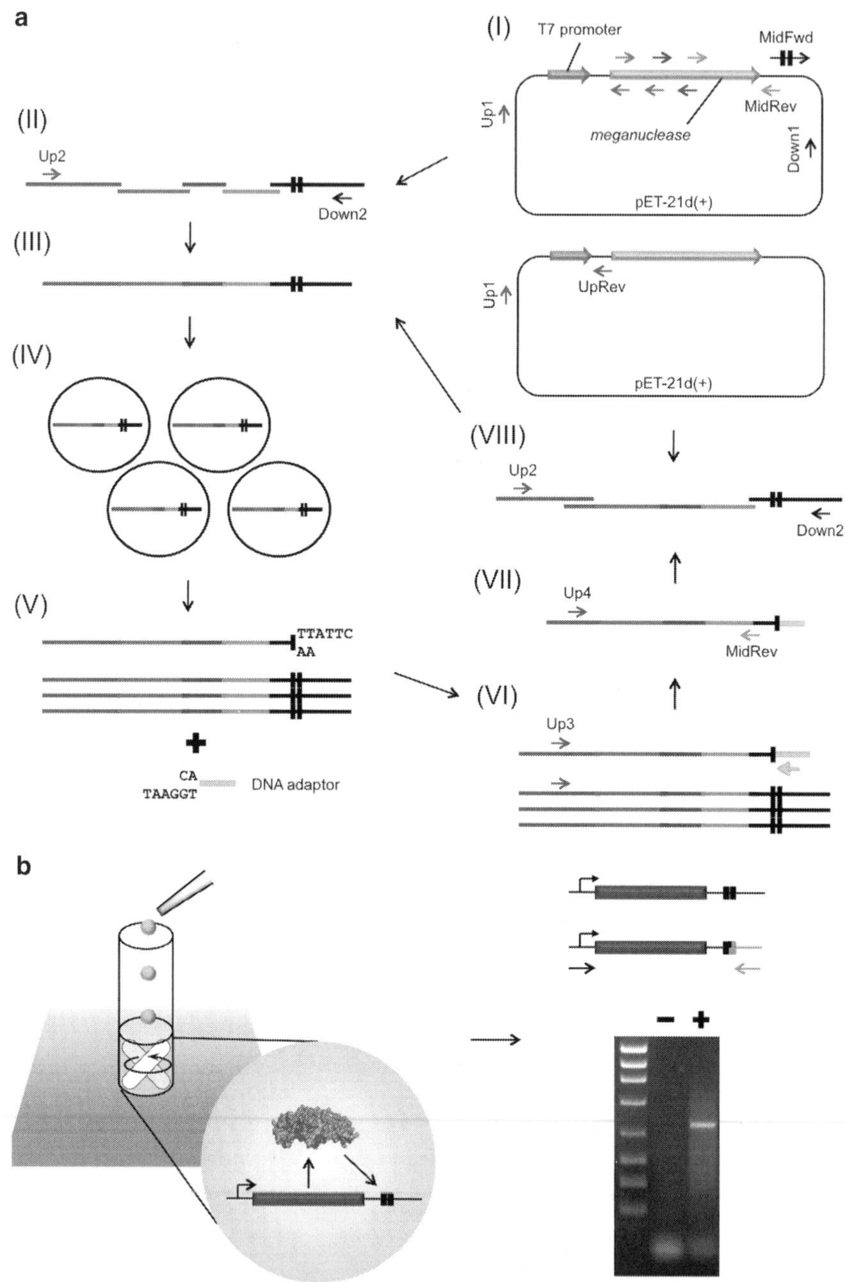

Fig. 2 Schematics of (**a**) PCR-based construction of a library and (**b**) in vitro selection using IVC. (**a**) A parental meganuclease gene cloned into pET-21d(+) is used as a PCR template to introduce site-directed mutagenesis. All PCR products are purified by gel extraction prior to subsequent rounds of PCR. In separate reaction mixtures, the T7 promoter and an ORF encoding an endonuclease are PCR-amplified with multiple sets of primers, a portion of which contain wobble bases to introduce saturation mutagenesis (*I*). In parallel, a fragment containing target sites are generated using the Down1 primer and the MidFwd primer (that contains tandem two copies of a target site (*black boxes*)). All sequences generated are assembled by overlap extension PCR using the Up2 and Down2 primers (*II*). The resulting library containing variant endonuclease genes and tandem two copies of a target site (*III*) are compartmentalized together with in vitro protein synthesis reaction mixture in an oil–surfactant mixture (*IV*).

identifies sites with few nucleotide mismatches at the central four base-pair positions, where sequence-dependent DNA bending appears to significantly influence the cleavage activity of meganucleases [29]. A target site for a redesigned meganuclease is chosen on the basis of its sequence identity to the original target sites for the wild type meganucleases and its location in a genomic target region. Using the approach described in this chapter, we have created variant endonucleases that target DNA sequences showing 50 % or higher identity to the original sites.

3.2 Retargeting Meganucleases to New Sequences Using IVC

3.2.1 Construction of a Library Encoding Variant Meganucleases Using Overlap Extension PCR

The DNA interface of meganucleases is typically composed of a pair of four antiparallel β-strands and their connecting loops, wherein clusters of 6–9 amino-acid residues dictate sequence specificity at 2–4 consecutive base-pair positions [10, 28]. Such a group of residues is termed "contact module", and can be defined for every three consecutive base-pair positions except the central four positions, where the DNA target is bent as a result of meganuclease binding, far fewer protein-DNA contacts are observed and we have found reprogramming of specificity to be more difficult.

A meganuclease is first retargeted towards two sites containing only one cluster of base pairs that differ from the original target site, in either of the two half sites (termed as "round 1" target sites) (Fig. 4). Up to nine residues to be randomized are identified within a "contact module" that covers these "round 1" base pair substitutions (*see* **Note 2**). To construct a library encoding variant meganucleases, DNA fragments containing a partial ORF of a meganuclease gene are PCR-amplified using degenerate primers, which shuffle amino-acid residues within the "contact module" against the altered DNA base pairs. A degenerate codon 5′-NNK-3′ is used to randomize amino-acid residues, but not included in approximately ten nucleotides from the 3′ termini of the primers (Fig. 5a). In addition, degenerate primers are designed to generate PCR products that share approximately 20 base-pair sequences with their adjacent fragments at both ends. In the second round of

Fig. 2 (continued) After extraction from emulsion, fragments with cohesive ends generated by meganucleases are ligated to DNA adaptors containing their complementary ends (*V*). Sequences coupled to adaptors are amplified using the Up3 primer and an adaptor-specific primer (*VI*). Meganuclease genes are amplified again using the Up4 primer and the MidRev primer to eliminate an adaptor sequence (*VII*). In parallel, a fragment containing the T7 promoter and the ribosome binding site is generated using the Up1 primer and the UpRev primer. Two copies of a target site are added by another round of PCR (*VIII*). (**b**) IVC allows endonucleases (expressed in compartmentalized droplets) to access only target sites coupled to its own gene (*black boxes*), thus maintaining genotype-phenotype linkage during selection. After a library encoding meganucleases is extracted from emulsion, genes associated with cohesive ends that are generated by endonucleolytic cleavage are ligated to a DNA adaptor. PCR using a pair of primers, one of which is specific to an adaptor sequence, give rise to a DNA band containing meganuclease genes from a library subjected to in vitro selection (lane "+"), but not from the corresponding naive one (lane "−")

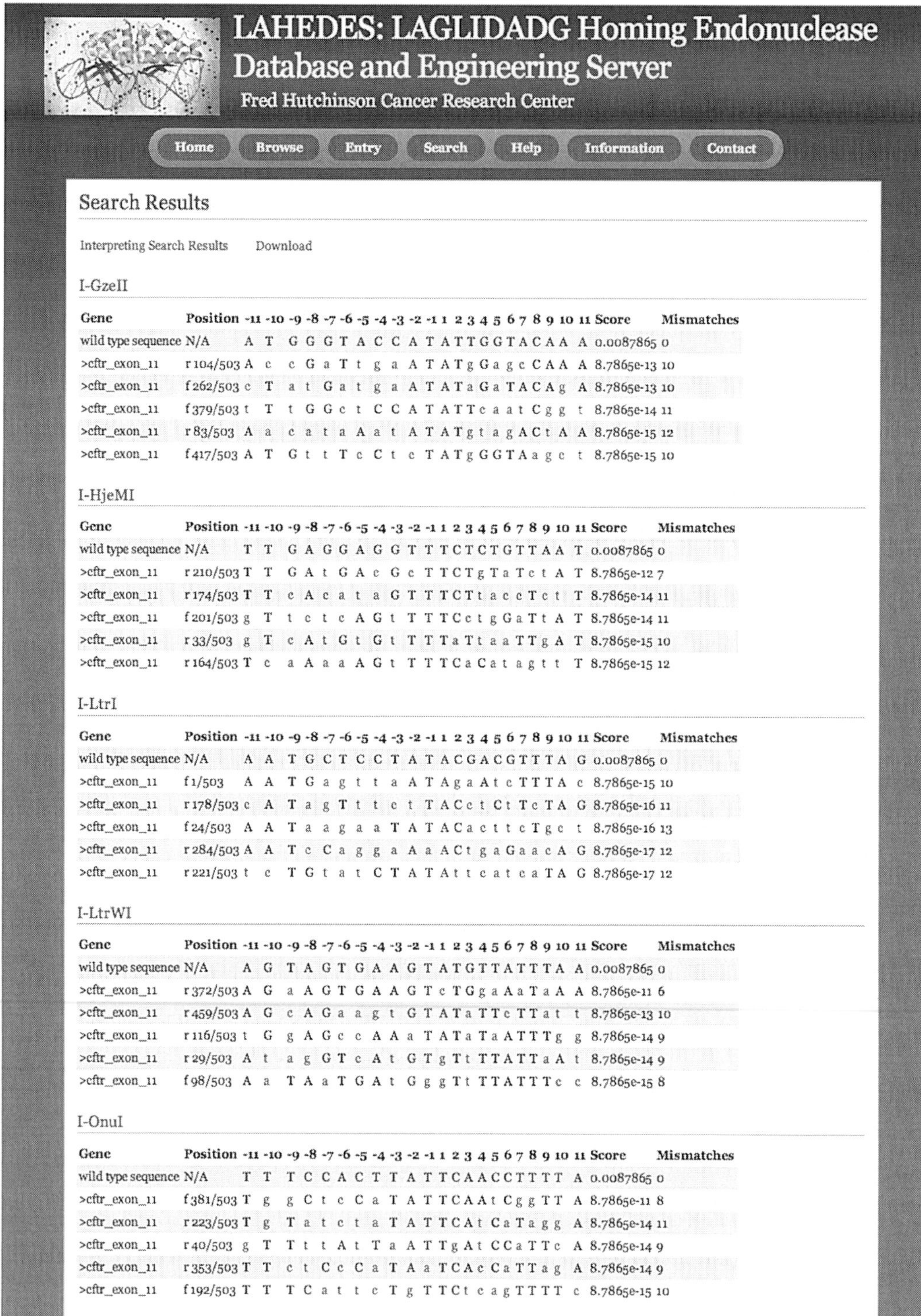

LAHEDES: LAGLIDADG Homing Endonuclease Database and Engineering Server

Fred Hutchinson Cancer Research Center

Home | Browse | Entry | Search | Help | Information | Contact

Search Results

Interpreting Search Results   Download

I-GzeII

Gene	Position	-11	-10	-9	-8	-7	-6	-5	-4	-3	-2	-1	1	2	3	4	5	6	7	8	9	10	11	Score	Mismatches
wild type sequence	N/A	A	T	G	G	G	T	A	C	C	A	T	A	T	T	G	G	T	A	C	A	A	A	0.0087865	0
>cftr_exon_11	r104/503	A	c	c	G	a	T	t	g	a	A	T	A	T	g	G	a	g	c	C	A	A	A	8.7865e-13	10
>cftr_exon_11	f262/503	c	T	a	t	G	a	t	g	a	A	T	A	T	a	G	a	T	A	C	A	g	A	8.7865e-13	10
>cftr_exon_11	f379/503	t	T	t	G	G	c	t	C	C	A	T	A	T	T	c	a	a	t	C	g	g	t	8.7865e-14	11
>cftr_exon_11	r83/503	A	a	c	a	t	a	A	a	t	A	T	A	T	g	t	a	g	A	C	t	A	A	8.7865e-15	12
>cftr_exon_11	f417/503	A	T	G	t	t	T	c	C	t	c	T	A	T	g	G	G	T	A	a	g	c	t	8.7865e-15	10

I-HjeMI

Gene	Position	-11	-10	-9	-8	-7	-6	-5	-4	-3	-2	-1	1	2	3	4	5	6	7	8	9	10	11	Score	Mismatches
wild type sequence	N/A	T	T	G	A	G	G	A	G	G	T	T	T	C	T	C	T	G	T	T	A	A	T	0.0087865	0
>cftr_exon_11	r210/503	T	T	G	A	t	G	A	c	G	c	T	T	C	T	g	T	a	T	c	t	A	T	8.7865e-12	7
>cftr_exon_11	r174/503	T	T	c	A	c	a	t	a	G	T	T	T	C	T	t	a	c	c	T	c	t	T	8.7865e-14	11
>cftr_exon_11	f201/503	g	T	t	c	t	c	A	G	t	T	T	T	C	c	t	g	G	a	T	t	A	T	8.7865e-14	11
>cftr_exon_11	r33/503	g	T	c	A	t	G	t	G	t	T	T	T	a	T	t	a	a	T	T	g	A	T	8.7865e-15	10
>cftr_exon_11	r164/503	T	c	a	A	a	a	A	G	t	T	T	T	C	a	C	a	t	a	g	t	t	T	8.7865e-15	12

I-LtrI

Gene	Position	-11	-10	-9	-8	-7	-6	-5	-4	-3	-2	-1	1	2	3	4	5	6	7	8	9	10	11	Score	Mismatches
wild type sequence	N/A	A	A	T	G	C	T	C	C	T	A	T	A	C	G	A	C	G	T	T	T	A	G	0.0087865	0
>cftr_exon_11	f1/503	A	A	T	G	a	g	t	t	a	A	T	A	g	a	A	t	c	T	T	T	A	c	8.7865e-15	10
>cftr_exon_11	r178/503	c	A	T	a	g	T	t	t	c	t	T	A	C	c	t	C	t	T	c	T	A	G	8.7865e-16	11
>cftr_exon_11	f24/503	A	A	T	a	a	g	a	a	T	A	T	A	C	a	c	t	t	c	T	g	c	t	8.7865e-16	13
>cftr_exon_11	r284/503	A	A	T	c	C	a	g	g	a	A	a	A	C	t	g	a	G	a	a	c	A	G	8.7865e-17	12
>cftr_exon_11	r221/503	t	c	T	G	t	a	t	C	T	A	T	A	t	t	c	a	t	c	a	T	A	G	8.7865e-17	12

I-LtrWI

Gene	Position	-11	-10	-9	-8	-7	-6	-5	-4	-3	-2	-1	1	2	3	4	5	6	7	8	9	10	11	Score	Mismatches
wild type sequence	N/A	A	G	T	A	G	T	G	A	A	G	T	A	T	G	T	T	A	T	T	T	A	A	0.0087865	0
>cftr_exon_11	r372/503	A	G	a	A	G	T	G	A	A	G	T	c	T	G	g	a	A	a	T	a	A	A	8.7865e-11	6
>cftr_exon_11	r459/503	A	G	c	A	G	a	a	g	t	G	T	A	T	a	T	T	c	T	T	a	t	t	8.7865e-13	10
>cftr_exon_11	r116/503	t	G	g	A	G	c	c	A	A	a	T	A	T	a	T	a	A	T	T	T	g	g	8.7865e-14	9
>cftr_exon_11	r29/503	A	t	a	g	G	T	c	A	t	G	T	g	T	t	T	T	A	T	T	a	A	t	8.7865e-14	9
>cftr_exon_11	f98/503	A	a	T	A	a	T	G	A	t	G	g	g	T	t	T	T	A	T	T	T	c	c	8.7865e-15	8

I-OnuI

Gene	Position	-11	-10	-9	-8	-7	-6	-5	-4	-3	-2	-1	1	2	3	4	5	6	7	8	9	10	11	Score	Mismatches
wild type sequence	N/A	T	T	T	C	C	A	C	T	T	A	T	T	C	A	A	C	C	T	T	T	T	A	0.0087865	0
>cftr_exon_11	f381/503	T	g	g	C	t	c	C	a	T	A	T	T	C	A	A	t	C	g	g	T	T	A	8.7865e-11	8
>cftr_exon_11	r223/503	T	g	T	a	t	c	t	a	T	A	T	T	C	A	t	C	a	T	a	g	g	A	8.7865e-14	11
>cftr_exon_11	r40/503	g	T	T	t	t	A	t	T	a	A	T	T	g	A	t	C	C	a	T	T	c	A	8.7865e-14	9
>cftr_exon_11	r353/503	T	T	c	t	C	c	C	a	T	A	a	T	C	A	c	C	a	T	T	a	g	A	8.7865e-14	9
>cftr_exon_11	f192/503	T	T	T	C	a	t	t	c	T	g	T	T	C	t	c	a	g	T	T	T	T	c	8.7865e-15	10

Fig. 3 Output of "Central Four Search" in the LAHEDES web server. Sites are selected, based on both the local sequence identity (at the base-pair positions −2 to +2) and the overall homology to the original target sites for chosen meganucleases (wild type sequences). Nucleotides that differ from the wild type sequences are shown in *lower cases*

Fig. 4 Schematic of sequential protein engineering to reprogram meganuclease specificity. Meganucleases are first retargeted to "round 1" targets, DNA sequences containing only one cluster of consecutive nucleotide substitutions from the original target (WT). Following rounds of in vitro selection, selected endonucleases are further mutated and screened to collect endonucleases that cleave next rounds of target sites, which contain additional base-pair mismatches. This iterative approach results in engineering of variant enzymes that target "half" altered sites. Half domains of such proteins are shuffled to create meganucleases that recognize "fully" altered targets

PCR, all sequences containing a partial ORF of a meganuclease gene are assembled with a fragment containing two copies of a "round 1" target site.

1. Design degenerate primers as described above (also *see* Fig. 5a).
2. To generate fragments containing a partial ORF of a meganuclease gene with targeted codons shuffled and a sequence containing two copies of a "round 1" target site, set up the following PCR reaction mixture:

5× Phusion HF Buffer	10 μL
10 mM dNTPs mix	1 μL
Phusion High-Fidelity DNA polymerase	0.5 μL
Water	32.5 μL
10 ng/μL pET-21d(+) plasmid (containing a meganuclease gene)	1 μL
A pair of 10 pmol/μL PCR primers	2.5 μL each

A pair of degenerate primers specific to a meganuclease gene, the "Up1" primer, the "MidRev" primer, the "MidFwd"

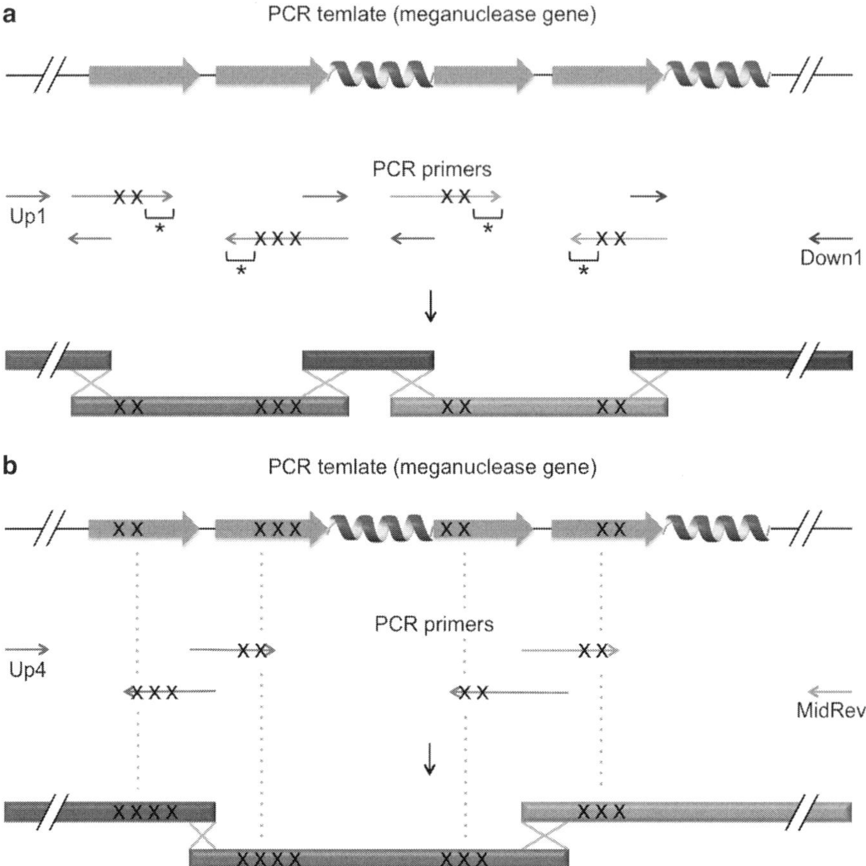

Fig. 5 General guidelines for design of degenerate primers (**a**) No degenerate codon is incorporated in approximately ten nucleotides from 3′ termini of primers (indicated by *asterisks*), and primers are designed such that resulting PCR fragments overlap at both ends with adjacent ones. Positions of degenerate codons are represented by "X"s. (**b**) In a case where degenerate primers including no wobble base in nine nucleotides from 3′ termini cannot be design, the 3′ termini are extended beyond the closest residues that have been shuffled (indicated by *dash lines*), and these residues are randomized again. Such residues are also counted in nine residues that are allowed to be randomized in one round of mutagenesis. Introduction of second-time saturation mutagenesis in 3′ terminal regions of degenerate primers ensures that at least a subset of degenerate primers stably form DNA duplexes with a PCR template, initiating PCR amplification

primer containing two copies of a "round 1" target site, and the "DownRev1" primer are added in each PCR tube (Fig. 2). The PCR conditions are: heat denaturation at 98 °C for 30 s, followed by 30 cycles of 98 °C for 30 s, 55 °C for 15 s, and 72 °C for 20 s, and the final extension at 72 °C for 1 min (*see* **Note 3**).

3. Add 10 μL of 6× the Loading dye in each PCR tube and separate 30 μL of each sample on an agarose-TAE gel containing 1 mM guanosine and 0.1 ng/mL ethidium bromide (*see* **Note 4**).

4. Cut out the PCR products, and combine 2–3 gel slices in one tube (except a sequence containing a "round 1" target site that can be reused to construct subsequent libraries) in order to save spin columns supplied with QIAquick Gel Extraction Kit.
5. Recover the DNA fragments using QIAquick Gel Extraction Kit, and elute them in 50 μL of supplied Buffer EB.
6. To generate a library encoding variant meganucleases, set up the following PCR reaction mixture:

5× Phusion HF Buffer	10 μL
10 mM dNTPs mix	1 μL
Phusion High-Fidelity DNA Polymerase	0.5 μL
Gel-extracted PCR fragments	1.5 μL each
10 pmol/μL the "Up2" and "Down2" primers	2.5 μL each
Water	Added to make a final volume of 50 μL

The PCR conditions are: heat denaturation at 98 °C for 30 s, followed by 30 cycles of 98 °C for 30 s, 55 °C for 15 s, and 72 °C for 30 s, and the final extension at 72 °C for 1 min (*see* **Note 3**).
7. Add 10 μL of 6× the Loading dye in a PCR product and separate 15 and 40 μL of the mixture in adjacent lanes on a 1 % agarose-TAE gel containing 0.1 ng/mL ethidium bromide.
8. To protect a PCR-generated library from UV-induced damage, divide two lanes where different volumes of the same PCR product are loaded, and mark the position where the library migrates only on a lane containing 15 μL of the sample under UV light. Match the two lanes again, and cut out DNA from the lane that is not exposed to UV (where 40 μL of a library is loaded).
9. Using QIAquick Gel Extraction Kit, recover and elute DNA in 30 μL of supplied Buffer EB (*see* **Note 5**). Load the collected elution on the same spin column and centrifuge it again.
10. Separate 1–2 μL of the purified PCR fragment on a 1 % agarose-TAE gel along with 2.5 and 5 μL of 1 kb plus DNA ladder, and quantify the library using a densitometry program such as ImageJ software (http://imagej.nih.gov/ij/).

3.2.2 First Round of In Vitro Selection Using IVC

Redesigned meganucleases are screened using IVC, where each of aqueous droplets that are compartmentalized in an oil phase contains an average of one DNA fragment encoding a variant endonuclease and an in vitro protein synthesis reaction mixture. A protein that is expressed in emulsion is retained in the same droplet throughout a time period for in vitro protein synthesis and selection, and thus has an opportunity to cleave only DNA target sites coupled to its own gene-coding sequence (Fig. 2b). After enzymatic reactions are terminated, extracted DNA fragments are subjected to a ligation reaction with a DNA adaptor containing cohesive ends that are compatible with a 3′, 4-base overhang generated by engineered meganucleases (Fig. 2). Genes linked to an adaptor are rescued by PCR using a pair of primers, one of which is specific to an adaptor (termed as adaptor-specific PCR). To eliminate the DNA adaptor from the adaptor-specific PCR product, another round of PCR is carried out and the resultant fragment is assembled with sequences containing the T7 promoter, the ribosome binding site, and two copies of "round 1" target sites (Fig. 2a).

1. To prepare a DNA adaptor, set up the following mixture:

4× Annealing buffer	4 μL
1 mM complementary oligonucleotides (*see* Table 1)	1.6 μL each
Water	8.8 μL

   Heat the sample at 95 °C for 5 min, and cool it down to 25 °C at a rate of 0.1 °C/s (using a thermal cycler). The double stranded DNA can be stored at −20 °C for a month.

2. Mix 40 μL of ABIL EM90 and 1 μL of Triton X-100 with 360 μL of light mineral oil in a 2-mL microcentrifuge tube (5× oil–surfactant mixture), and vortex it for 1 min (*see* **Note 6**).
3. Mix 560 μL of light mineral oil with 140 μL of the 5× oil–surfactant mixture in a 2-mL microcentrifuge tube, and vortex it for 1 min.
4. Add 140 μL of Saturation buffer (100 mM potassium glutamate (pH 7.5), 10 mM magnesium acetate, 1 mM DTT, and 5 mg/mL BSA) in the same tube, and vortex it for 1 min.
5. Incubate the tube at 37 °C for 20 min, and centrifuge it at 16,000 × *g* for 15 min at 4 °C.
6. Transfer 500 μL of the upper phase (white cloudy) into a 14-mL polypropylene round-bottom tube, and drop a stirrer bar in it.
7. Set up the following in vitro protein synthesis reaction mixture:

PURExpress solution A	10 μL
PURExpress solution B	7.5 μL
Murine RNase inhibitor (40 units/μL)	0.5 μL
Gel-extracted library (4 ng/μL)	2 μL
10 mg/mL BSA	3 μL
Water	7 μL

8. Stir the oil–surfactant mixture in ice water at 1,400 r.p.m, and slowly add 10 μL of the in vitro protein synthesis reaction mixture in the round-bottom tube (containing the oil–surfactant mixture) every 30 s. After all the aqueous mixture is added, continue stirring for additional 2.5 min.
9. Incubate the emulsified droplets at 30 °C for 4 h.
10. Heat the compartmentalized in vitro protein synthesis reaction mixture at 75 °C for 15 min, and cool it down on ice for 1 min.
11. Add 170 μL of supplied Buffer EB, vortex the emulsion for a few seconds and transfer it into a 1.5-mL microcentrifuge tube.
12. Centrifuge the tube at 16,000 × *g* for 20 min at 4 °C, and remove the upper (oil) phase.
13. Add 200 μL of Phenol–Chloroform–Isoamyl alcohol (pH 7.9) in the bottom (aqueous) phase, and mix them by vortexing.
14. Centrifuge the microcentrifuge tube again at 16,000 × *g* for 5 min at room temperature, and transfer the upper (aqueous) phase into a new 1.5-mL microcentrifuge tube.
15. Repeat **steps 13** and **14**.
16. Add 200 μL of isopropanol and 20 μL of 3 M sodium acetate (pH 5.2) in the aqueous phase, and centrifuge the mixture at 16,000 × *g* for 15 min at 4 °C.
17. Remove the supernatant, and add 500 μL of 70 % ethanol to the pellet.
18. After centrifugation at 16,000 × *g* for 5 min at 4 °C, remove the supernatant and dry the pellet at room temperature for 5–10 min.
19. Add 95 μL of supplied Buffer EB and 5 μL of RNase Cocktail Enzyme Mix in the tube, and incubate it at 37 °C for 30 min.
20. Clean up the library using QIAquick PCR Purification Kit, and elute it in 40 μL of supplied Buffer EB (*see* **Note 5**).
21. Set up the following ligation reaction mixture:

10× T4 DNA Ligase Reaction Buffer	1 μL
T4 DNA ligase (400,000 units/mL)	0.5 μL
0.2 pmol/μL DNA adaptor (diluted in water)	1 μL
Library extracted from emulsion *or* its corresponding naïve library (0.2 ng/μL)	8 μL

A naïve library is diluted with supplied Buffer EB, and used as a negative control. Its concentration is reduced in subsequent rounds of selection (second round, 0.025 ng/μL; third round, 0.0125 ng/μL).

22. Incubate the reaction mixture at 16 °C for 2 h to overnight.

23. Set up the following PCR premix for adaptor-specific PCR:

5× Phusion HF Buffer	40 μL
10 mM dNTPs mix	4 μL
Phusion High-Fidelity DNA Polymerase	2 μL
10 pmol/μL the Up3 primer and an adaptor-specific primer	2.5 μL each
Water	33.5 μL

Aliquot 48 μL of the PCR premix in four tubes, and add 2 μL of a ligation sample containing a naïve library into one tube and 2 μL of a ligation sample containing a library extracted from emulsion into each of the remaining tubes. The PCR conditions are: heat denaturation at 98 °C for 30 s, followed by 30 cycles of 98 °C for 30 s, 60 °C for 15 s, and 72 °C for 20 s, and the final extension at 72 °C for 1 min.

24. Analyze 5 μL of adaptor-specific PCR products on a 1 % agarose-TAE gel (*see* **Note 7**).
25. If a DNA fragment corresponding to meganuclease genes ligated to a DNA adaptor is specifically amplified from a library subjected to selection, combine 10 μL each of the PCR products that are amplified in the three tubes, and separate them on a 1 % agarose-TAE gel containing 1 mM guanosine and 0.1 ng/mL ethidium bromide.
26. Cut out a DNA band for variant endonuclease genes linked to a DNA adaptor, and purify the PCR product using QIAquick Gel Extraction Kit. Elute it in 50 μL of supplied Buffer EB.
27. To generate a sequence containing meganuclease genes but not an adaptor sequence, and to prepare a fragment containing the T7 promoter and ribosome binding site, set up the following two PCR reaction mixtures:

5× Phusion HF Buffer	10 μL
10 mM dNTPs mix	1 μL
Phusion High-Fidelity DNA Polymerase	0.5 μL
10 pmol/μL the Up4 primer and the MidRev primer	2.5 μL each
Gel-extracted DNA	3 μL
Water	30.5 μL

5× Phusion HF Buffer	10 μL
10 mM dNTPs mix	1 μL
Phusion High-Fidelity DNA Polymerase	0.5 μL
10 pmol/μL the Up1 primer and the UpRev primer	2.5 μL each
10 ng/μL pET-21d(+) plasmid	1 μL
Water	32.5 μL

The PCR conditions are: heat denaturation at 98 °C for 30 s, followed by 30 cycles of 98 °C for 30 s, 60 °C for 15 s, and 72 °C for 20 s, and the final extension at 72 °C for 1 min.

28. Separate the PCR products on a 1 % agarose-TAE gel containing 1 mM guanosine and 0.1 ng/mL ethidium bromide, and individually purify them as described in **step 26**.

29. To construct a new library for the next round of selection, set up the following PCR reaction mixture:

5× Phusion HF Buffer	10 μL
10 mM dNTPs mix	1 μL
Phusion High-Fidelity DNA Polymerase	0.5 μL
10 pmol/μL the "Up2" primer and the "Down2" primer	2.5 μL each
Sequence containing two copies of target sites (prepared in Subheading 3.2.1, **steps 2–5**)	1 μL
Two PCR fragments purified in **step 28**	1 μL each
Water	30.5 μL

The PCR conditions are: heat denaturation at 98 °C for 30 s, followed by 30 cycles of 98 °C for 30 s, 55 °C for 15 s, and 72 °C for 30 s, and the final extension at 72 °C for 1 min (*see* **Note 3**).

30. Purify and quantify a PCR-amplified fragment, as described in Subheading 3.2.1, **steps 7–10**.

3.2.3 Subsequent Rounds of Selection Using IVC

A reconstructed library is screened under more stringent conditions (by reducing a time period for in vitro protein synthesis and DNA target cleavage and/or by elevating temperatures). Increasing rounds of selection (up to six rounds) result in enrichment of stable, active enzymes in collected populations: total three rounds are generally sufficient.

1. Prepare the oil surfactant mixture as described in Subheading 3.2.2, **steps 2–6**.

2. Set up the following in vitro protein synthesis reaction mixture:

PURExpress solution A	10 μL
PURExpress solution B	7.5 μL
Murine RNase inhibitor (40 units/μL)	0.5 μL
Library prepared in Subheading 3.2.2, **step 30** (1 ng/μL)	2 μL
10 mg/mL BSA	3 μL
Water	7 μL

3. Compartmentalize the above mixture as described in Subheading 3.2.2, **step 8**, and incubate the emulsified droplets at 42 °C for 75 min (*see* **Note 7**).

4. Extract the library from emulsion, and ligate it to a different DNA adaptor from that used in the previous round of selection, as described in Subheading 3.2.2, **steps 10–22**.
5. Recover meganuclease genes by adaptor-specific PCR, and reconstruct a library, as shown in Subheading 3.2.2, **steps 23–30**. Set up an adaptor-specific PCR reaction mixture in two tubes for a sample extracted from emulsion (instead of three in the first round of selection).
6. In the third and later rounds of selection, add 0.5 ng of a library encoding variant meganucleases in an in vitro protein synthesis reaction mixture (shown in **step 2**), and carry out selection at 42 °C (or 37 °C; *see* **Note 8**) for 30 min as described above.
7. Extract a library from emulsion, and carry out ligation with a different DNA adaptor from that used in the last two rounds of selection, as shown in Subheading 3.2.2, **steps 10–22**.
8. Set up the adaptor-specific PCR reaction mixture in a single tube for each library (*see* Subheading 3.2.2, **step 23**), and purify an adaptor-specific PCR fragment (which is specifically amplified from a library subjected to selection), as described in Subheading 3.2.2, **steps 25** and **26**.

3.2.4 Selection of Meganucleases That Cleave a Target Site Containing More Than one Cluster of Base-Pair Substitutions

Additional base-pair substitutions are introduced to a "round 1" target site ("round 2" target site), and amino-acid residues of a "contact module" corresponding to newly altered base-pairs are shuffled in ORFs of variant endonucleases enriched in previous rounds of selection. A library is generated through two rounds of PCR amplification as described above. However, since neighboring residues have been shuffled in an earlier round of mutagenesis, degenerate primers that include no wobble base in nine nucleotides from 3′ termini often cannot be designed (Fig. 5b). If that is the case, 3′ termini of primers should be extended beyond one residue that have been randomized such that a subset of degenerate primers (containing a NNK codon near 3′ termini) form the correct Watson-Crick base pairs with a PCR template in 3′ terminal regions (which are required for DNA synthesis by a DNA polymerase). A residue to be shuffled again is also counted in up to nine positions to be randomized in a single round of mutagenesis. The resulting library is screened through a few rounds of selection against a "round 2" target site. This iterative approach is continued to obtain meganucleases that cleave a chimeric target site containing one half of the original target site for a parental enzyme and one half of a DNA sequence of interest ("half" altered target site) (Fig. 4). Half protein domains that are placed on half sites containing base-pair substitutions are shuffled to obtain variant meganucleases that target a "fully" altered site.

1. Design degenerate primers as described above (also refer to Fig. 5b).
2. To generate sequences containing partial ORFs of variant endonucleases with particular residues randomized, carry out

PCR as described in Subheading 3.2.1, **step 2** with the following modification: an adaptor-specific PCR product that is purified by gel extraction is used as a template (1–3 μL/reaction). A fragment containing two copies of a new target site and one containing the T7 promoter and the ribosome binding site are also PCR-amplified in parallel (refer to Subheading 3.2.1, **step 2** and Subheading 3.2.2, **step 27**).

3. Construct a library as described in Subheading 3.2.1, **steps 3–10**.
4. Carry out selection as described in Subheadings 3.2.2 and 3.2.3.
5. Repeat **steps 1–4** to create variant endonucleases that cleave each of two "half" altered target sites (*see* Fig. 4).
6. Using adaptor-specific PCR products that are recovered in the last round of selections against "half" altered target sites, PCR-amplify sequences encoding half domains of engineered enzymes (that are placed on half sites containing base-pair substitutions).

5× Phusion HF Buffer	10 μL
10 mM dNTPs mix	1 μL
Phusion High-Fidelity DNA Polymerase	0.5 μL
10 pmol/μL primers (*see* **Note 9**)	2.5 μL each
Gel-extracted DNA	3 μL
Water	30.5 μL

7. Construct a library as described in Subheading 3.2.2, **steps 28–30**.
8. Carry out selection as described in Subheadings 3.2.2 and 3.2.3.

3.3 Selection Using Two-Plasmid Cleavage Assay in Bacteria

A two-plasmid cleavage assay that was originally developed by Doyon et al. [27] is carried out to isolate variant endonucleases that display substantial cleavage activity against a "fully" altered target site in bacteria. In this assay, bacterial cells are transformed with two plasmids, pEndo and pCcdB: pEndo carries a redesigned meganuclease gene that is driven by the arabinose-inducible promoter, and pCcdB, which contains 2 copies of a "fully" target site in two different positions, encodes a bacterial toxin (CcdB) that is expressed in the presence of IPTG (Fig. 1). A DSB generated by an expressed endonuclease at a "fully" altered target site triggers degradation of the pCcdB plasmid by the endogenous RecBCD complex in transformed cells, resulting in colony formation on agar plates containing IPTG. Stringency can be modulated by varying both the temperature (30 or 37 °C) and the time period in which an engineered meganuclease is expressed prior to induction of the *ccdB* gene expression [21, 23].

3.3.1 Preparing Competent Cells Harboring a pCcdB Reporter Plasmid Containing "Fully" Altered Target Sites for Electroporation

1. Synthesize two complementary oligonucleotides and anneal to generate double-stranded fragments containing two copies of a "fully" altered target site with cohesive ends.

4× Annealing buffer	4 μL
1 mM complementary oligonucleotides (*see* **Note 10**)	1.6 μL each
Water	8.8 μL

2. Digest 1 μg of the pCcdB plasmid with AflIII and BglII, and purify it using DNA Clean & Concentrator-5. Elute it in 10 μL of supplied Elution Buffer.
3. 100-fold dilute the annealed DNA in water, and carry out a ligation in 10 μL of 1× Quick Ligation Buffer containing 1 μL of the diluted, double-stranded oligonucleotide, 1 μL of approximately 20–30 ng/μL the linearized pCcdB plasmid and 0.5 μL of Qiuck T4 DNA ligase at room temperature for 15 min.
4. Transform NovaXGF' competent cells with 1 μL of the reaction mixture, and spread cells on the LB plates containing 33 μg/mL chloramphenicol and 0.5 % glucose (*see* **Note 11**).
5. Purify pCcdB with two copies of a "fully" altered target site (*see* **Note 12**), and digest approximately 1 μg of the plasmid with NheI and SacII.
6. Clean up the linearized plasmid using DNA Clean & Concentrator-5, and carry out ligation and transformation as described in **step 3** in order to incorporate additional two copies of the same target site into the pCcdB plasmid.
7. Purify the pCcdB plasmid with four copies of an identical target site (*see* **Note 12**), and verify sequences of the target sites and their flanking regions using the CcdB_seq1 primer and the CcdB_seq2 primer.
8. Transformed NovaXGF' competent cells with 0.5 μL of a sequence-verified pCcdB plasmid, and spread cells on the LB plates containing 33 μg/mL chloramphenicol and 0.5 % glucose.
9. Inoculate a few colonies in 10 mL of the SBG medium supplemented with 33 μg/mL chloramphenicol, and grow cells at 37 °C overnight.
10. Transfer all cells grown overnight into 500 mL of the SBG medium supplemented with 33 μg/mL chloramphenicol, and continue to culture cells at 37 °C until $O.D._{600nm}$ reaches 0.6–1.0.
11. Incubate the medium on ice for 15 min, and harvest cells at 2,700 × *g* for 15 min and then resuspend in 40 mL of 1 mM Hepes (pH 7.0).

12. Harvest cells at 2,700 × *g* for 15 min at 4 °C, decant the supernatant, and gently suspend the pellet with 50 mL of 1 mM Hepes (pH 7.0).
13. Harvest cells at 2,700 × *g* for 15 min at 4 °C, decant the supernatant, and gently suspend the pellet with 20 mL of 10 % (v/v) glycerol.
14. Harvest cells at 2,700 × *g* for 15 min at 4 °C, decant the supernatant, and gently suspend the pellet with 1 mL of 10 % (v/v) glycerol.
15. Aliquot 50 μL of competent cells in each tube, which is frozen in liquid nitrogen, and stored at −80 °C until use.

3.3.2 Selection of Variant Endonucleases That Substantially Cleave a "Fully" Altered Target Site in Bacteria

1. PCR-amplify ORFs of meganuclease genes from a gel-extracted, adaptor-specific PCR product generated in the last round of selection against a "fully" altered target site.

5× Phusion HF Buffer	10 μL
10 mM dNTPs mix	1 μL
Phusion High-Fidelity DNA Polymerase	0.5 μL
10 pmol/μL PCR primers (*see* **Note 13**)	2.5 μL each
Adaptor-specific PCR product	3 μL
Water	30.5 μL

The PCR conditions are: heat denaturation at 98 °C for 30 s, followed by 30 cycles of 98 °C for 30 s, 60 °C for 15 s, and 72 °C for 20 s, and the final extension at 72 °C for 1 min.

2. Digest approximately 1 μg of the pEndo plasmid with NcoI and NotI.
3. Separate the PCR-amplified ORFs and the linearized plasmid on a 1 % agarose-TAE gel, and purify them using Gel DNA Recovery Kit. Elute them in 10 μL of Elution Buffer.
4. Separate 0.5 μL each of these two DNA fragments along with 5 μL of 1 kb DNA ladder on an agarose gel, and quantify the gel-extracted DNA samples using a densitometry software.
5. Set up the following reaction mixture to carry out Gibson Assembly:

2× Gibson Assembly Master Mix	5 μL
PCR fragment encoding variant endonucleases	30–40 ng
Linearized pEndo plasmid	60–75 ng
Water	Added to make a final volume of 10 μL

6. Incubate the reaction mixture at 50 °C for an hour, and transform 50 μL of chemical competent cells (such as DH5α and DH10B) with 4 μL of the Gibson Assembly reaction mixture. Spread all cells on a two LB plates containing 50 μg/mL ampicillin.
7. Add 1–2 mL of sterilized water on each plate (where several hundred colonies are expected to be observed), scrape off colonies using a cell scraper, and collect cells in two 2-mL microcentrifuge tube.
8. Purify the pEndo library using two spin columns supplied with QIAprep Spin Miniprep Kit, and combine DNA eluted from the two columns.
9. Thaw an aliquot (50 μL) of NovaXGF' competent cells that harbor pCcdB containing four copies of a "fully" altered target site on ice, and quickly transform the cells (50 μL) with 30–50 ng of the pEndo library by electroporation (*see* **Note 14**).
10. Immediately suspend the transformants with 1 mL of 2× YT medium, and transfer all the cell suspension into a 50-mL conical tube.
11. Incubate cells at 37 °C for 30 min, and add 10 mL of 2× YT medium containing 100 μg/mL carbenicillin and 0.02 %(w/v) L-arabinose to induce expression of meganucleases encoded on pEndo.
12. Culture cells at 30 °C for 4 h, and harvest them at 1,100 × *g* for 5 min.
13. Resuspend cells in 1 mL of sterilized water, and spread 50 μL of the cell suspension on the 2–3 selective plates (*see* **Note 15**). To calculate the transformation efficiency, dilute cells by 1,000-fold in sterilized water and spread 50 μL of the dilution on the control plate.
14. Incubate the plates at 30 °C for approximately 36 h.
15. Scrape off all colonies surviving on the selective plates, and purify plasmids using spin miniprep kit. If multiple spin columns are used, combine all eluted samples.
16. To generate a PCR fragment encoding variant meganucleases, set up the following PCR reaction mixture:

5× Phusion HF Buffer	10 μL
10 mM dNTPs mix	1 μL
Phusion High-Fidelity DNA Polymerase	0.5 μL
10 pmol/μL PCR primers used in **step 1**	2.5 μL each
Purified plasmid	1 μL
Water	32.5 μL

The PCR conditions are: heat denaturation at 98 °C for 30 s, followed by 30 cycles of 98 °C for 30 s, 60 °C for 15 s, and 72 °C for 20 s, and the final extension at 72 °C for 1 min.

17. To carry out selection under the same conditions, repeat **steps 3–14**.
18. To reconstruct the pEndo library, follow **steps 15** and **16**, then **steps 3–8**.
19. Carry out electroporation as described in **step 9**.
20. Suspend the transformants with 1 mL of 2× YT medium supplemented with 0.02 % (w/v) L-arabinose.
21. Transfer all the cell suspension into a 50-mL conical tube, and dilute it with 10 mL of the same medium.
22. Grow the cells at 37 °C for 1 h, and harvest cells from 1 mL of the culture at 1,100 × *g* for 5 min.
23. Resuspend cells in 0.5 mL of sterilized water, and spread 50 μL of the cell suspension on the selective plate (containing 0.4 mM IPTG and 0.02 % (w/v) L-arabinose). To calculate the transformation efficiency, dilute cells by 200-fold in sterilized water and spread 50 μL of the dilution on the control plate.
24. Incubate the plates at 37 °C for approximately 20 h.
25. To carry out the second round of selection under the same conditions, follow **steps 18–24**.
26. Randomly pick up colonies surviving on the selective plate, separately suspend them in 10 μL of water, and transfer 1 μL each of the cell suspension into a PCR tube.
27. Heat the PCR tubes at 95 °C on a thermal cycler for 5 min with the lids open.
28. Set up the following PCR premix (per six colonies):

5× Colorless GoTaq Reaction Buffer	20 μL
10 mM dNTPs mix	2 μL
GoTaq DNA polymerase (5 units/μL)	0.5 μL
10 pmol/μL the Endo_colonyPCR_fwd primer and the Endo_colony PCR_rev primer	2 μL each
Water	73.5 μL

Add 15 μL of the PCR reaction mixture in each PCR tube. The PCR conditions are: heat denaturation at 95 °C for 2 min, followed by 30 cycles of 95 °C for 30 s, 53 °C for 15 s, and 72 °C for 1 min 15 s, and the final extension at 72 °C for 2 min.

29. Sequence colony PCR products using the Endo_seq_fwd primer and the Endo_seq_rev primer.
30. Test individual pEndo plasmids under the same conditions used in the final round of selection.

3.3.3 Supplementary Protocol: Random Mutagenesis Using Error-Prone PCR

If few colonies survive on the selective plates in Subheading 3.3.2, **step 17** (*see* **Note 16**), a time period in which an engineered meganuclease is expressed prior to induction of the *ccdB* gene expression may be increased to 16 h in order to reduce a level of stringency. Then, random mutagenesis may be introduced across the entire ORFs of selected variant genes to increase protein stability and/or overall activity.

1. Transform NovaXGF' competent cells that harbor the pCcdB plasmid containing "fully" altered target sites (50 μL) with 30–50 ng of the pEndo library (prepared in Subheading 3.3.2, **step 8**) by electroporation.
2. Follow Subheading 3.3.2, **steps 10–11**, and incubate cells at 30 °C for 16 h.
3. Make a serial dilution of the culture with sterilized water ($1{:}10^2$–$1{:}10^5$), and spread 50 μL of each dilution on the selective plate. To calculate the transformation efficiency, spread 50 μL of the culture that is diluted by 10^6-fold on the control plate.
4. Incubate the plates at 30 °C for approximately 36 h.
5. Recover meganuclease genes as described in Subheading 3.3.2, **steps 15** and **16**, and reconstruct a pEndo library as described in and Subheading 3.3.2, **steps 3–8**.
6. Carry out selection under the same conditions as in the previous round (refer to **steps 1–4**).
7. Purify plasmids from cells surviving on the selective plates as shown in Subheading 3.3.2, **steps 15**.
8. To generate a PCR fragment containing meganuclease genes, set up the following PCR reaction mixture:

5× Phusion HF Buffer	10 μL
10 mM dNTPs mix	1 μL
Phusion High-Fidelity DNA polymerase	0.5 μL
10 pmol/μL PCR primers used in Subheading 3.3.2, **step 28**	2.5 μL each
Purified plasmids	1 μL
Water	32.5 μL

   The PCR conditions are: heat denaturation at 98 °C for 30 s, followed by 30 cycles of 98 °C for 30 s, 60 °C for 15 s, and 72 °C for 20 s, and the final extension at 72 °C for 1 min.

9. Clean up the PCR product using DNA Clean & Concentrator-5, and elute DNA in 20 μL of supplied Elution Buffer.

10. To carry out error-prone PCR using GeneMorph II Random Mutagenesis Kit, set up the following reaction mixtures with two different amounts of the PCR template purified above:

10× Mutazyme II reaction mixture	5 μL
40 mM dNTP mix (supplied with the kit)	1 μL
Mutazyme II DNA polymerase	1 μL
125 ng/μL PCR primers used in Subheading 3.3.2, **step 1**	1 μL each
Purified PCR fragment	20 or 100 ng
Water	Added to make a final volume of 50 μL

The PCR conditions are: heat denaturation at 95 °C for 2 min, followed by 30 cycles of 95 °C for 30 s, 55 °C for 15 s, and 72 °C for 1 min, and the final extension at 72 °C for 2 min.

11. Digest the pEndo plasmid and purify the above PCR samples and the linearized plasmid as described in Subheading 3.3.2, **steps 3–4**.

12. To carry out Gibson Assembly, set up the following reaction mixtures:

2× Gibson Assembly Master Mix	10 μL
PCR fragment amplified from 20 or 100 ng of a PCR template	60–80 ng
Linearized pEndo plasmid	120–150 ng
Water	Added to make a final volume of 20 μL

13. Incubate the reaction mixtures at 50 °C for an hour, and clean up the two reaction mixtures individually, using DNA Clean & Concentrator-5. Elute DNA in 10 μL of supplied Elution Buffer.

14. Transform 50 μL of ElectroMax DH10B T1 phage-resistant competent cells with 3 μL each of the purified Gibson Assembly samples by electroporation, and separately culture the cells in 2 mL of SOC medium at 37 °C for an hour.

15. To calculate the transformation efficiency, mix 2 μL of the culture with 100 μL of sterilized water and spread 50 μL of the dilution on a LB plate supplemented with 50 μg/mL ampicillin (100-mm dish). Then, spread the rest of the cells, using glass beads, on LB agar plates supplemented with 50 μg/mL ampicillin (160–200 μL/150-mm petri dish), and incubate all the plates at 37 °C overnight.

16. To calculate the total number of transformants, count colonies on a LB plate where 50 μL of 50-fold dilution of competent

cells are spread, and multiply the number by 2,000. Hundreds of thousands of transformants are generally obtained for each of two error-prone PCR samples. Scrape off all colonies that are formed on 150-mm plates, and purify a pEndo library plasmid using a plasmid maxiprep kit.

17. To screen meganuclease genes subjected to random mutagenesis, follow Subheading 3.3.2, **steps 9–30**.

4 Notes

1. We optimize the codon usage of a meganuclease gene for bacterial expression, using DNAworks (http://helixweb.nih.gov/dnaworks/), and synthesize the ORF (containing a start codon 5′-ATG-3′ and a stop codon 5′-TAA-3′) with extra sequences at both termini, which are required for insertion of the endonuclease gene into pET-21d(+) by Gibson Assembly. We purchase two gBlocks, each of which contains approximately a half of the ORF with approximately 20 base pairs of an overlap region (from Integrated DNA Technologies), and assemble these two fragments with the NcoI/NotI-digested pET-21d(+) plasmid using 2× Gibson Assembly Master Mix.
2. Amino-acid residues included in a "contact module" at every three consecutive base-pair positions are deposited in the LAHEDES web server (http://homingendonuclease.net), and can be found by clicking the name of a homing endonuclease in the "Homing Endonuclease Browser" under the "Browse" menu.
3. If PCR fails, increase a temperature during template-primer annealing up to 65 °C.
4. Use new TAE buffer when PCR fragments to be purified are separated on an agarose gel.
5. Washed a spin column twice (with 750 μL of Buffer PE and with 250 μL of the same buffer), and incubate it with supplied Buffer EB for 2 min at room temperature before centrifugation.
6. Use a positive-displacement pipettor (e.g., MICROMAN from Gilson) to add an accurate amount of surfactant.
7. If no DNA band is specifically amplified from a library that is subjected to in vitro selection, carry out ligation with different DNA adaptor(s). The efficiencies of PCR amplification of meganuclease genes linked to DNA adaptors are greatly dependent on sequences of adaptors and meganuclease genes.
8. If screening at 42 °C fails, reduce temperature to 37 °C.
9. To amplify 3′-terminal half of engineered meganuclease genes, use the MidRev primer and a forward primer that is annealed to a nucleotide sequence corresponding to a loop

connecting two half domains. To generate a fragment containing 5′-terminal half of variant endonuclease genes, use the Up4 primer and a primer that is reverse complement of the forward primer used to PCR-amplify 3′-terminal half genes. The resulting PCR products share an approximately 20 base-pair region that can be hybridized in the next round of PCR.

10. Sequences of oligonucleotides to be inserted into pCcdB are as follows: 5′-CGTGT-(2 copies of a target site)-A-3′ and 5′-GATCT-(two copies of a reverse complement target site)-A-3′ (to be inserted between AflII and BglII sites), and 5′-CTAGC-(two copies of a target site)-CCGC-3′ and 5′-GG-(two copies of a reverse complement target site)-G-3′ (to be inserted between NheI and SacII sites).
11. We use chemical competent cells of this *E. coli* strain that are prepared by a standard protocol.
12. We routinely purify pCcdB plasmids using spin miniprep kit from cells grown overnight in 3.5 mL of the LB medium containing 33 μg/mL chloramphenicol and 0.5 % glucose.
13. Primers used are as follows: 5′-CTT TAA GAA GGA GAT ATA CCC ATG-(5′ terminal sequence of a meganuclease gene sequence)-3′ and 5′-CCA ATT AAC CAA TTC TGA GCG GCC GCT TA-(reverse complement of a 3′ terminal meganuclease gene sequence)-3′. A start codon (5′-ATG-3′) and a stop codon (5′-TAA-3′) are underlined.
14. A cuvette is chilled on ice before electroporation.
15. Adjust the dilution rate and the volume of sterilized water to spread at least threefold excess of cells over the complexity of a library on the selective plates.
16. The survival rate of cells harboring the intact pCcdB plasmid on the selective plates is generally very low (<0.1 %). If a subset of variant endonucleases encoded by a library are capable of eliminating this reporter plasmid *in cellulo*, the survival rate (calculated by dividing the number of cells that form colonies on the selective plates by that on the control plates) is significantly high over the background level in the second round of selection.

References

1. Kolb AF, Coates CJ, Kaminski JM, Summers JB, Miller AD, Segal DJ (2005) Site-directed genome modification: nucleic acid and protein modules for targeted integration and gene correction. Trends Biotechnol 23:399–406
2. Carroll D (2011) Genome engineering with zinc-finger nucleases. Genetics 188:773–782
3. Urnov FD, Rebar EJ, Holmes MC, Zhang HS, Gregory PD (2010) Genome editing with engineered zinc finger nucleases. Nat Rev Genet 11:636–646
4. Bogdanove AJ, Voytas DF (2011) TAL effectors: customizable proteins for DNA targeting. Science 333:1843–1846
5. Christian M, Cermak T, Doyle EL, Schmidt C, Zhang F, Hummel A, Bogdanove AJ, Voytas DF (2010) Targeting DNA double-strand breaks with TAL effector nucleases. Genetics 186:757–761
6. Li T, Huang S, Jiang WZ, Wright D, Spalding MH, Weeks DP, Yang B (2011) TAL nucleases (TALNs): hybrid proteins composed of TAL

effectors and FokI DNA-cleavage domain. Nucleic Acids Res 39:359–372

7. Cong L, Ran FA, Cox D, Lin S, Barretto R, Habib N, Hsu PD, Wu X, Jiang W, Marraffini LA et al (2013) Multiplex genome engineering using CRISPR/Cas systems. Science 339:819–823
8. Mali P, Yang L, Esvelt KM, Aach J, Guell M, DiCarlo JE, Norville JE, Church GM (2013) RNA-guided human genome engineering via Cas9. Science 339:823–826
9. Arnould S, Perez C, Cabaniols JP, Smith J, Gouble A, Grizot S, Epinat JC, Duclert A, Duchateau P, Paques F (2007) Engineered I-CreI derivatives cleaving sequences from the human XPC gene can induce highly efficient gene correction in mammalian cells. J Mol Biol 371:49–65
10. Stoddard BL (2011) Homing endonucleases: from microbial genetic invaders to reagents for targeted DNA modification. Structure 19:7–15
11. Segal DJ, Meckler JF (2013) Genome engineering at the dawn of the golden age. Annu Rev Genomics Hum Genet 14:135–158
12. Silva G, Poirot L, Galetto R, Smith J, Montoya G, Duchateau P, Paques F (2011) Meganucleases and other tools for targeted genome engineering: perspectives and challenges for gene therapy. Curr Gene Ther 11:11–27
13. Brunet E, Simsek D, Tomishima M, DeKelver R, Choi VM, Gregory P, Urnov F, Weinstock DM, Jasin M (2009) Chromosomal translocations induced at specified loci in human stem cells. Proc Natl Acad Sci U S A 106:10620–10625
14. Lee HJ, Kweon J, Kim E, Kim S, Kim JS (2012) Targeted chromosomal duplications and inversions in the human genome using zinc finger nucleases. Genome Res 22:539–548
15. Sollu C, Pars K, Cornu TI, Thibodeau-Beganny S, Maeder ML, Joung JK, Heilbronn R, Cathomen T (2010) Autonomous zinc-finger nuclease pairs for targeted chromosomal deletion. Nucleic Acids Res 38:8269–8276
16. Stoddard BL (2005) Homing endonuclease structure and function. Q Rev Biophys 38: 49–95
17. Certo MT, Ryu BY, Annis JE, Garibov M, Jarjour J, Rawlings DJ, Scharenberg AM (2011) Tracking genome engineering outcome at individual DNA breakpoints. Nat Methods 8:671–676
18. Daboussi F, Zaslavskiy M, Poirot L, Loperfido M, Gouble A, Guyot V, Leduc S, Galetto R, Grizot S, Oficjalska D et al (2012) Chromosomal context and epigenetic mechanisms control the efficacy of genome editing by rare-cutting designer endonucleases. Nucleic Acids Res 40:6367–6379
19. Gao H, Smith J, Yang M, Jones S, Djukanovic V, Nicholson MG, West A, Bidney D, Falco SC, Jantz D et al (2010) Heritable targeted mutagenesis in maize using a designed endonuclease. Plant J 61:176–187
20. Grizot S, Smith J, Daboussi F, Prieto J, Redondo P, Merino N, Villate M, Thomas S, Lemaire L, Montoya G et al (2009) Efficient targeting of a SCID gene by an engineered single-chain homing endonuclease. Nucleic Acids Res 37:5405–5419
21. Takeuchi R, Lambert AR, Mak AN, Jacoby K, Dickson RJ, Gloor GB, Scharenberg AM, Edgell DR, Stoddard BL (2011) Tapping natural reservoirs of homing endonucleases for targeted gene modification. Proc Natl Acad Sci U S A 108:13077–13082
22. Boissel SJ, Astrakhan A, Jarjour J, Adey A, Shendure J, Stoddard BL, Certo MT, Baker D, Scharenberg AM (2014) MegaTALs: a rare-cleaving nuclease architecture for therapeutic genome engineering. Nucleic Acids Res 42:2591
23. Takeuchi R, Choi M, Stoddard BL (2014) Redesign of extensive protein-DNA interfaces of meganucleases using iterative cycles of in vitro compartmentalization. PNAS USA 111: 4061–4066
24. Miller OJ, Bernath K, Agresti JJ, Amitai G, Kelly BT, Mastrobattista E, Taly V, Magdassi S, Tawfik DS, Griffiths AD (2006) Directed evolution by in vitro compartmentalization. Nat Methods 3:561–570
25. Tawfik DS, Griffiths AD (1998) Man-made cell-like compartments for molecular evolution. Nat Biotechnol 16:652–656
26. Zheng Y, Roberts RJ (2007) Selection of restriction endonucleases using artificial cells. Nucleic Acids Res 35:e83
27. Doyon JB, Pattanayak V, Meyer CB, Liu DR (2006) Directed evolution and substrate specificity profile of homing endonuclease I-SceI. J Am Chem Soc 128:2477–2484
28. Taylor GK, Petrucci LH, Lambert AR, Baxter SK, Jarjour J, Stoddard BL (2012) LAHEDES: the LAGLIDADG homing endonuclease database and engineering server. Nucleic Acids Res 40:W110
29. Molina R, Redondo P, Stella S, Marenchino M, D'Abramo M, Gervasio FL, Epinat JC, Valton J, Grizot S, Duchateau P et al (2012) Non-specific protein-DNA interactions control I-CreI target binding and cleavage. Nucleic Acids Res 40:6936–6945

Chapter 7

Efficient Design and Assembly of Custom TALENs Using the Golden Gate Platform

Tomas Cermak, Colby G. Starker, and Daniel F. Voytas

Abstract

An important breakthrough in the field of genome engineering was the discovery of the modular *T*ranscription *A*ctivator-*L*ike *E*ffector (TALE) DNA binding domain and the development of TALE *n*ucleases (TALENs). TALENs enable researchers to make DNA double-strand breaks in target loci to create gene knockouts or introduce specific DNA sequence modifications. Precise genome engineering is increasingly being used to study gene function, develop disease models or create new traits in crop species. Underlying the boom in genome engineering is the striking simplicity and low cost of engineering new specificities of TALENs and other sequence-specific nucleases. In this chapter, we describe a rapid, inexpensive, and user-friendly protocol for custom TALEN construction based on one of the most popular TALEN assembly platforms, the Golden Gate cloning method. Using this protocol, ready-to-use TALENs with specificity for targets 13–32 bp long are constructed within 5 days.

Key words TALEN, Golden Gate assembly, Targeted mutagenesis, Genome engineering, Site-specific nuclease

1 Introduction

Genome engineering uses sequence-specific nucleases (SSNs) that introduce DNA double strand breaks (DSBs) in a gene to be modified [1]. Repair of the breaks proceeds by one of two widely conserved DNA repair pathways. Homologous recombination (HR)-based repair uses a DNA template with user-specified DNA sequence modifications that are copied into the chromosomal target locus. Induction of DSBs greatly increases the frequency of HR, as shown in both plants and animals. On the other hand, repair by the non-homologous end joining (NHEJ) pathway creates small deletions/insertions (indels) at the break site. Most of these indels cause frameshift mutations that may inactivate gene function. NHEJ is the preferred repair pathway in plant and animal cells, in contrast with yeast and bacteria, where HR predominates.

Shondra M. Pruett-Miller (ed.), *Chromosomal Mutagenesis*, Methods in Molecular Biology, vol. 1239,
DOI 10.1007/978-1-4939-1862-1_7, © Springer Science+Business Media New York 2015

The widespread use of genome engineering has been limited by the lack of SSNs that are reliable, easy to make, and inexpensive. This changed with the advent of *T*ranscription *A*ctivator-*L*ike *E*ffector *N*ucleases (TALENs) and, more recently, the discovery of CRISPR/Cas9 systems [2–5]. The ease and speed of making a custom TALEN or CRISPR/Cas9 reagents make it possible and affordable for most labs to introduce targeted genome alterations into essentially any site in a cell line or organism for which transformation methods are available. While TALENs and CRISPR/Cas9 SSNs have different characteristics that make them more or less suitable for specific applications, both are similarly efficient in inducing mutations, and engineering both to target a specific site in the genome is relatively easy. This chapter describes a platform for fast and efficient assembly of custom TALENs and discusses solutions to the problems one can encounter while using our method.

TAL effectors (TALEs) are virulence factors produced by plant pathogenic bacteria of the genus *Xanthomonas* [6]. The most significant finding in TALE research, with respect to genome engineering, was the discovery of the code that TALEs use to recognize their DNA targets in the promoters of plant genes [7, 8]. The central domain of a TALE is responsible for DNA binding (Fig. 1a).

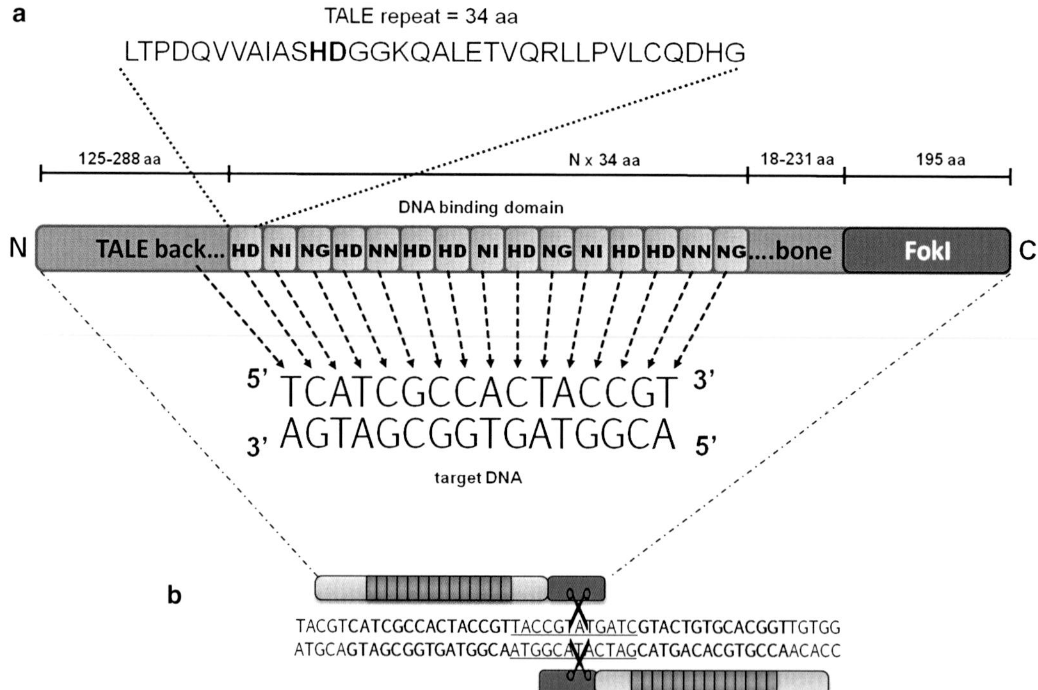

Fig. 1 Structure of a TALEN. The central repeat region, embedded in the TALE backbone, is fused to FokI nuclease to create a TALEN monomer (**a**). TALENs work as dimers: to make a DSB, two TALEN monomers are required to bind the DNA target in opposite orientations, separated by a DNA spacer (**b**)

It consists of nearly identical repeats of 34 amino acids. The last repeat in the array is identical to the rest of the repeats only in the first 20 amino acids. The 12th and 13th amino acid positions in the repeats are variable and are referred to as repeat variable diresidues (RVDs). There is a simple one-to-one correspondence between individual RVDs and the DNA bases they bind. The amino acid in position 13 makes specific interactions with a DNA base and the amino acid in position 12 stabilizes this interaction [9, 10]. Therefore, the type and order of RVDs in TALE repeat arrays determines TALE target specificity. New specificities are engineered by shuffling the TALE repeats with different RVDs. The specificity of three of the four most common RVDs (HD, NG, and NI) for their corresponding DNA bases (C, T, and A, respectively) is high. The fourth most common RVD, NN, has a similar affinity to both G and A [11]. Specificities and advantages of other, less frequent RVDs, including NH, NK, NS, and N* will be discussed below.

TALENs are fusions of the TAL effector DNA binding domain to the catalytic domain of the type II restriction endonuclease FokI [2, 3] (Fig. 1a). FokI, itself, has no sequence specificity and has to dimerize in order to create a DSB. A pair of TALEN monomers is therefore required to bind the DNA target to introduce the DSB (Fig. 1b). The binding sites of the two TALENs are separated by a spacer sequence that allows FokI to dimerize and cleave. In addition to the central repeats, a full length TALE DNA binding domain contains long N- and C-terminal extensions, parts of which are required for interaction with DNA. The N-terminal region adjacent to the repeats specifies interaction with a thymine at position −1 of the DNA target and is absolutely required for DNA binding. However, certain truncations of both N and C terminal regions have been shown to increase TALEN activity and influence the requirements for the length of DNA spacer [11–15]. Hereafter, different truncations of these regions will be referred to as "TALE backbones."

To engineer new specificity, the TALE repeats embedded in the TALE backbone are reorganized such that the RVD sequences correspond to the sequence of the DNA target site. To do this, our TALEN assembly platform uses a technique called Golden Gate cloning [16, 17] (Fig. 2a). This method offers the possibility of joining several DNA fragments in an ordered fashion in one simple reaction, based on the annealing of user-specified 4 bp overhangs created by Type IIs restriction enzymes (Fig. 2b). The protocol for TALEN assembly described below uses a library of 86 plasmids. This library includes ten single TALE repeat-containing module vectors and one last repeat-containing vector for each of the six different RVD types. Also included are 13 intermediate cloning vectors and a number of expression vectors. The expression vectors contain the TALE backbone fused to the FokI catalytic domain

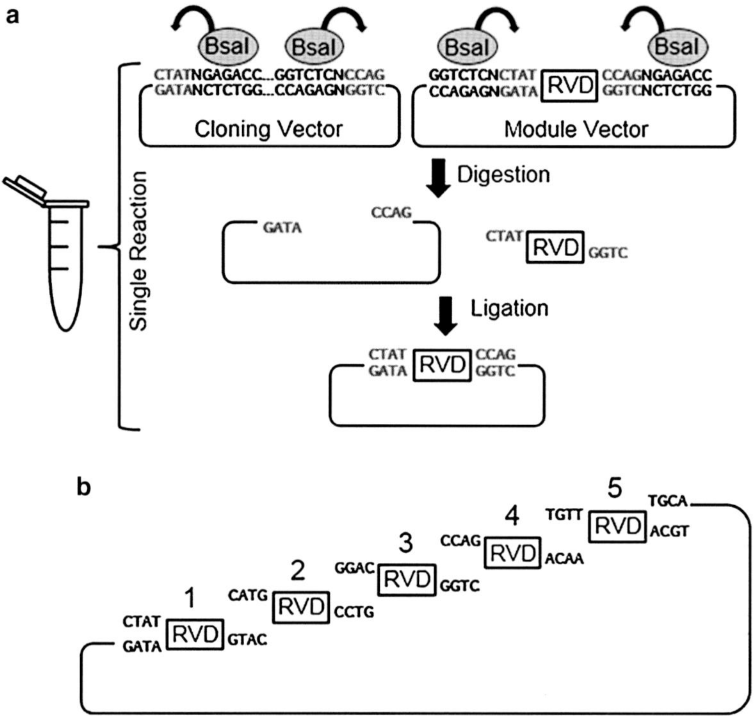

Fig. 2 Golden Gate cloning. A TALE repeat-containing fragment can be cloned from a module vector into an intermediate vector backbone in a single reaction. Cleavage by a Type IIs restriction enzyme (BsaI) releases the repeat fragment from a module vector and linearizes the intermediate vector backbone. Complementary 4 bp overhangs are left by BsaI cleavage on the fragment and the backbone, which are then ligated by DNA ligase. BsaI recognition sites are lost, and the final ligation product is stable (**a**). The reaction can be extended to multiple fragments. Fragments are cloned into a vector backbone in a desired order specified by the complementarity of their 4 bp overhangs (**b**)

and accommodate the full-length TALE repeat array, thereby creating a functional TALEN. Based on the principle of Golden Gate cloning, repeat-containing fragments released from the ten different types of module plasmid backbones upon cleavage with a Type IIs restriction enzyme will have ten different combinations of 4 bp overhangs specific for each position in the repeat array. Up to ten module plasmids, one specific for each position in the repeat array, are mixed with the appropriate intermediate vector in a single tube. After addition of the Type IIs restriction enzyme and DNA ligase, individual TALE repeats are released from the module backbones and fused into the vector in the specified order dictated by the combination of overhangs on each of the fragments. In the next step, arrays of up to ten repeats are fused together with the last repeat into the final expression vector, creating a functional

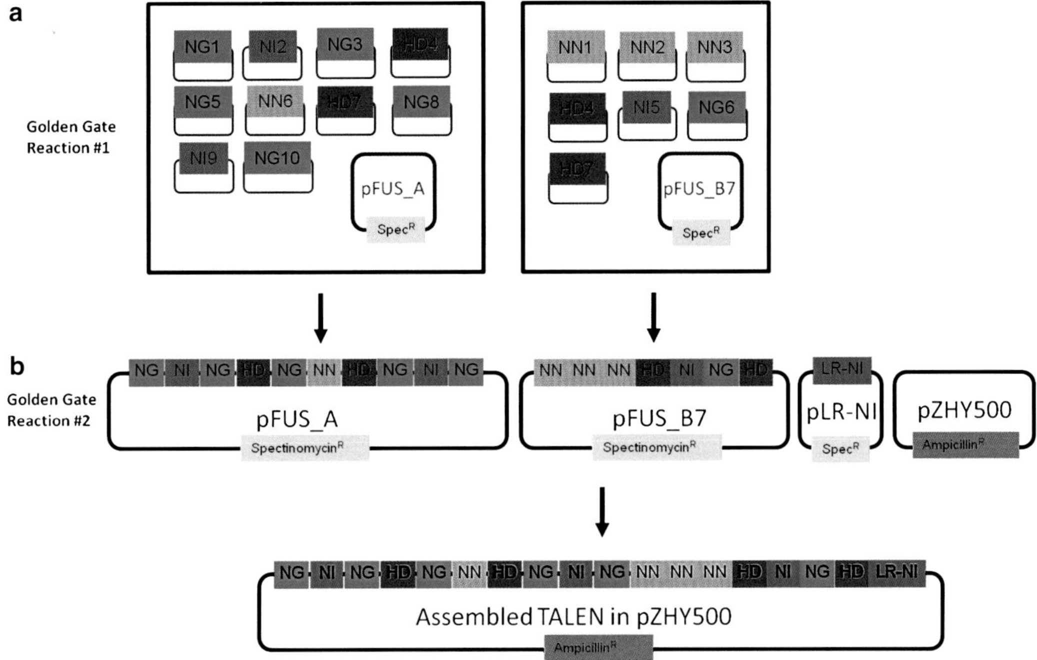

Fig. 3 Assembly of custom TALENs using the Golden Gate platform. Assembly of TALENs using the Golden Gate TALEN and TAL effector kit is divided into two steps. First, two arrays of ten or fewer repeats are cloned into intermediate cloning backbones, pFUS_A and pFUS_B (**a**). pFUS_A and pFUS_B arrays are then fused together with the last repeat-containing module into a final expression vector in the second Golden Gate reaction (**b**)

nuclease with custom specificity. Using our kit, custom TALE domains with 12–31 repeats can be constructed in two steps in 1 week (Fig. 3).

A number of different protocols for TALEN assembly have been developed since the construction of the first custom TALEN. Many of them use Golden Gate cloning or its variations [18–28], some involve PCR [25–29], some are based on ligation-independent cloning [30] and yet others are ligation-mediated approaches [29, 31, 32]. Several protocols enable automation of the process and construction of custom TALENs in high throughput [24, 30–32]. The platform described below gained popularity thanks to its simplicity and availability of the reagents from the plasmid repository Addgene (www.addgene.org). At the time of this submission, the kit has been distributed to more than 1,300 labs around the world, making it likely the most widespread TALEN assembly platform. Thanks to the high number of users, the number of add-on plasmids compatible with the original kit is growing, thereby increasing its applicability. These add-ons include various expression vectors with different TALE backbones that can be used for expression of TALENs or TALE fusions to other proteins in different organisms or for transcription in vitro, as well as module plasmids with new RVD types. Many of these are available from Addgene as well. The original kit described in Cermak et al. [18] has also been updated by its authors in light of the discovery of TALE backbones with higher activity and RVDs with better specificity, and is now available at Addgene as V2.0. Although not high-throughput, this tool will remain a good choice for both first-time and regular users of the TALEN technology thanks to its convenience.

The procedure for generating custom TALENs involves several steps: (1) preparing the Golden Gate TALEN and TAL Effector library of plasmids from glycerol stocks (this step is only done once and does not need to be repeated each time a TALEN is made); (2) selection of a target site in the target gene/locus; (3) assembly of the custom TALEN over five subsequent days (Subheadings 3.3, 3.4, 3.5, 3.6, and 3.7). This chapter provides instructions and troubleshooting for each step and can serve as a manual for TALEN assembly using the Golden Gate TALEN and TAL Effector Kit V2.0.

2 Materials

1. Golden Gate TALEN and TAL Effector Kit 2.0 (Addgene, Cambridge, MA, USA) (http://www.addgene.org/TALeffector/goldengateV2/).
2. Restriction enzyme BsaI. (BsaI-HF has lower activity with the experimental conditions used and is not recommended).
3. Restriction enzyme Esp3I.
4. T4 DNA ligase and 10× T4 DNA ligase buffer.
5. 100 mM DTT.
6. Plasmid-Safe nuclease (Epicentre Biotechnologies, Madison, WI, USA).
7. 25 mM ATP.
8. Taq DNA polymerase and buffer.
9. dNTPs.
10. Gel electrophoresis equipment.
11. DH5α chemically competent cells.
12. SOC medium: 5 g/l yeast extract, 20 g/l tryptone, 20 mM dextrose, 10 mM sodium chloride, 2.5 mM potassium chloride, 10 mM magnesium chloride.
13. 10 mg/ml Tetracycline stock. Dissolve 200 mg of tetracycline in 20 ml of 95 % ethanol. Aliquot into microcentrifuge tubes and store at −20 °C.
14. 50 mg/ml Spectinomycin stock. Dissolve 1 g of spectinomycin in 20 ml of ddH_2O. Filter-sterilize, aliquot into microcentrifuge tubes, and store at −20 °C.

15. 50 mg/ml Carbenicillin stock. Dissolve 1 g of carbenicillin in 20 ml of ddH_2O. Filter-sterilize, aliquot into microcentrifuge tubes, and store at −20 °C. Ampicillin may be used in the place of carbenicillin.
16. LB plates with 10 mg/l tetracycline, 50 mg/l spectinomycin, or 50 mg/l carbenicillin. LB medium is made by mixing 10 g tryptone, 5 g yeast extract, and 10 g sodium chloride in 1 l of ddH_2O. Then add 15 g granulated agar, autoclave, cool to 65 °C and add 1 ml of tetracycline, spectinomycin, or carbenicillin (ampicillin) stock before pouring the plates.
17. LB liquid medium with 10 mg/l tetracycline, 50 mg/l spectinomycin, or 50 mg/l carbenicillin. LB medium is made by mixing 10 g tryptone, 5 g yeast extract, and 10 g sodium chloride in 1 l of ddH_2O. Autoclave the medium, cool to at least 65 °C, and add 1 ml of tetracycline, spectinomycin, or carbenicillin (ampicillin) stock.
18. 40 mg/ml X-gal dissolved in either dimethylsulfoxide or *N*′,*N*-dimethylformamide.
19. 100 mM IPTG (isopropyl β-D-1-thiogalactopyranoside) dissolved in water and filter-sterilized.
20. QIAprep Spin Miniprep Kit (Qiagen).
21. Primers for correct clone identification and sequencing:

    pCR8_F1: 5′-TTGATGCCTGGCAGTTCCCT

    pCR8_R1: 5′-CGAACCGAACAGGCTTATGT

    TAL_F1: 5′-TTGGCGTCGGCAAACAGTGG

    TAL_R2: 5′-GGCGACGAGGTGGTCGTTGG

    TAL_Seq_5-1: 5′-CATCGCGCAATGCACTGAC

    TAL_R3: 5′-GGCTCAGCTGGGCCACAATG

    TC14_F1: 5′-CCTACTCAGGAGAGCGTTCA

    TC14_F2: 5′-CAAGCCTGATTGGGAGAAAA.
22. *Optional* restriction enzymes AatII, AflII, BspEI, BstAPI, or StuI and respective buffers for diagnostic digestion to confirm correctly assembled clones; restriction enzymes BglII or XhoI or XbaI and SacI or Eco53kI or AflII and respective buffers for sub-cloning TALENs to alternative expression vectors (*see* Subheading 3.7).

3 Methods

3.1 Preparing the Library of Plasmids from Glycerol Stocks

Prepare the Golden Gate TALEN and TAL Effector plasmid library by purifying plasmids from 5 ml LB cultures using the QIAprep Spin Miniprep Kit (*see* **Note 1**). Adjust concentration of each expression plasmid to 75 ng/μl and each module plasmid, including the last repeat plasmids, to 150 ng/μl (*see* **Note 2**).

Store the library of plasmid DNA at −20 °C. We highly recommend sequencing each plasmid prep of the Golden Gate module plasmids (*see* **Note 3**). Primer TC14_F2 can be used for sequencing all module plasmids except for pLR plasmids, for which TC14_F1 should be used instead. Once the library is ready, it can be used to make tens to hundreds of TALENs. Frequently used plasmids can be re-prepped individually. *See* **Notes 4–6** for known issues with several plasmids in the kit.

3.2 Using TAL Effector Nucleotide Targeter 2.0 to Select TALEN Target Sites

1. Download the sequence of your target gene/locus from Genbank (http://www.ncbi.nlm.nih.gov/genbank/) or other relevant database in FASTA format. Make sure you are downloading genomic DNA sequence, not mRNA sequence (*see* **Note 7**). Open the FASTA file in a text editor or DNA analysis software and highlight and copy the sequence.
2. Go to https://tale-nt.cac.cornell.edu/ [33] and select "TALEN targeter." Paste the sequence of the target gene in FASTA format into the "Sequence" window. Alternatively, the FASTA sequence file can be directly uploaded by using the "Choose file" button. Next, the user has two options:
   (a) Select "Use a preset architecture" tab and pick "Cermak et al., 2011 (spacer 15–24, RVDs = 15–20)" from the slide down menu. With this setting, the software will list all target sites that match the TALE repeat array length and DNA spacer length requirements compatible with all expression vectors included in the kit. Although there are vectors with two different TALE backbones, every target site identified with this setting will be compatible with either of them. The optimal range of spacer lengths, however, is narrower and different for each of the two TALE backbones. *See* **Note 8** for a description of the two TALE backbone architectures available in the kit.
   (b) To determine the optimal target sites for the most active TALE backbone, namely NΔ152/C63, we recommend selecting the "Provide Custom Spacer/RVD Lengths" tab and using following settings: "Minimum Spacer Length" = 13, "Maximum spacer length" = 16 (*see* **Note 9**), "Minimum Repeat Array Length" = 15, "Maximum Repeat Array Length" = 21 (*see* **Note 10**), "G substitute" = NN (*see* **Note 11**), "Filter Options" = "Show TALEN pairs (hide redundant TALENs)" (*see* **Note 12**), "Streubel et al. guidelines" = ON (*see* **Note 13**), "Upstream base" = "T only (Recommended)" (*see* **Note 14**). The other settings can be left as default.
3. Click on the "Submit" button. The program will provide a list of TALEN target sites matching the preset criteria. Unless you

are using a very short target sequence, this will be an exhaustive list of tens to hundreds of potential targets. Although it is possible to browse the targets directly on the website, we recommend creating an Excel file with the list of sites. In our experience, this makes the sorting and analysis easier. To create an Excel file, click on the file name under "Result File (Tab-Delimited)". The list of targets will appear in Tab-Delimited format. Use Ctrl + A (command + A for Mac) to select all text and then Ctrl + C (command + C for Mac) to copy the information. Open Excel. Right click on the cell in the top left corner (control plus click for Mac) and select "Paste special". Next, select "text" and hit OK.

4. The next step is to pick the one or two target sites best suited for your experiment (*see* **Note 15**). As all the target sites in the list already match the criteria for a TALEN pair with the NΔ152/C63 TALE backbone and optimal activity, the next criteria to consider are the following (from most important to least): (1) proximity of the DSB (which will be introduced in the spacer of the target site) to the desired site of mutation (*see* **Note 16**); (2) unique restriction sites in the spacer (*see* **Note 17**); (3) the percentage of HD RVDs and NN/NH RVDs; (4) size of TALENs (# of RVDs) (*see* **Note 18**).
5. Pick two target sites per gene/locus (*see* **Note 19**). If the genomic target is likely methylated, *see* **Note 20**. To include degenerate positions in the RVD array, *see* **Note 21**. Proceed to TALEN assembly.

3.3 Golden Gate TALEN Assembly: Day 1

1. Copy the TALEN RVD sequences and other specifics of the TALEN pairs into a separate word/text file. For TALENs up to 21 RVDs, divide the RVD array for each TALEN into the first ten RVDs, second ten RVDs (or fewer, depending on the length of the TALEN) and the last RVD, as shown in Fig. 4 (*see* **Note 22**). Assembling TALENs of up to 31 RVDs is possible; however, as there is no significant advantage in using TALENs with >21 RVDs (*see* **Note 10**), hereafter all steps in the protocol will describe the assembly procedure for TALENs up to 21 RVDs in length. For modifications to the protocol for longer TALENs, *see* **Notes 22**, **24**, **25**, **35** and **38**.
2. Thaw plasmid DNA for each of the module plasmids for positions 1 to $N-1$ (N is the total length of TALEN = number of RVDs) (*see* **Note 23**). Thaw the plasmid DNA for the intermediate vectors pFUS_A and pFUS_B($N-11$) (for TALENs >21 RVDs, *see* **Note 24**). To assemble the example TALEN in Fig. 4, the plasmids needed are pNG1, pNI2, pNG3, pHD4, pNG5, pNN6, pHD7, pNG8, pNI9, pNG10, and pFUS_A for the RVD array 1–10 and pNN1, pNN2, pNN3, pHD4, pNI5, pNG6, pHD7, and pFUS_B7 for the RVD array 11 to 17

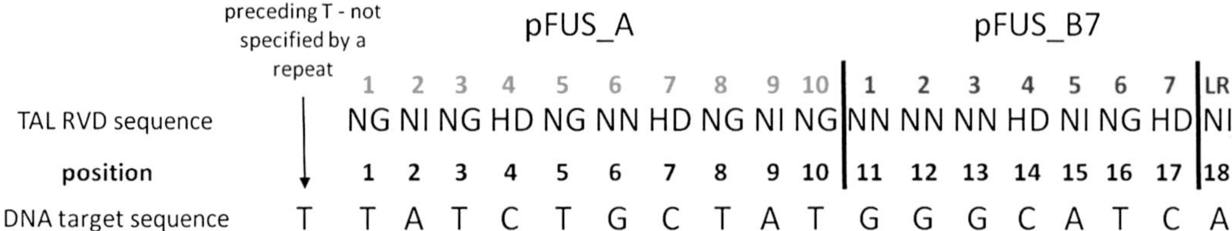

Fig. 4 Dividing repeats of a TALEN for assembly into pFUS_A and pFUS_B vectors. The RVDs of an exemplary TALEN are shown, divided into the first ten, the next seven and the last repeat. The first ten RVDs, highlighted in *green*, are assembled into pFUS_A; the next seven, highlighted in *blue*, are assembled into pFUS_B7; the last repeat (number *18*) is assembled in Golden Gate reaction #2 using the last repeat plasmid pLR-NI

(the last RVD will be added in the second Golden Gate reaction). Prepare Golden Gate reaction #1 for RVD array 1 to 10; another Golden Gate reaction will be prepared for RVDs 11 to ($N-1$) (for TALENs >21 RVDs, *see* **Note 25**). For RVD array 1 to 10, the reaction will include (*see* **Note 26**):

(a) 150 ng (1 μl) of each module vector (RVDs 1–10).

(b) 75 ng (1 μl) of pFUS_A vector.

(c) 1 μl BsaI (*see* **Note 27**).

(d) 1 μl T4 DNA ligase (*see* **Note 28**).

(e) 2 μl 10× T4 DNA ligase buffer (to final concentration of 1×) (*see* **Note 29**).

(f) H_2O to a total reaction volume of 20 μl (*see* **Note 30**).

3. For RVD array 11 to ($N-1$), the reaction will include:

(a) 150 ng (1 μl) of each module vector [RVDs 11 to ($N-1$)].

(b) 75 ng (1 μl) of pFUS_B($N-11$) vector.

(c) 1 μl BsaI (*see* **Note 27**).

(d) 1 μl T4 DNA ligase (*see* **Note 28**).

(e) 2 μl 10× T4 DNA ligase buffer (to final concentration of 1×) (*see* **Note 29**).

(f) H_2O to a total reaction volume of 20 μl (*see* **Note 30**).

4. Place the Golden Gate reactions in a PCR machine and run the following cycle: 10× (37 °C/5 min + 16 °C/10 min) + 50 °C/5 min + 80 °C/5 min.

5. Plasmid-Safe nuclease treatment:

(a) Cool the Golden Gate #1 reactions on ice and add:

- 1 μl 25 mM ATP.
- 1 μl Plasmid-Safe nuclease.
- Incubate at 37 °C/1 h (*see* **Note 31**).

This step destroys all unligated linear dsDNA fragments, including incomplete ligation products with less than the desired number of repeats, and linearized, unligated vector backbones. If not removed, the incomplete, shorter fragments will become inserted into the pFUS vector in vivo by recombination in the bacterial cell (sequences at the start of the first repeat and the end of the last repeat are present in the pFUS vector backbone).

6. Transform each Golden Gate reaction into *E. coli* (DH5α) by adding 5 μl of the reaction to 50 μl of chemically competent cells in a 1.5 ml microcentrifuge tube. Incubate on ice for at least 10 min and heat-shock for 70 s at 42 °C. Place on ice immediately and within 2 min add 200 μl of room temperature SOC medium. Allow the cells to recover by shaking at 37 °C for 1 h. Spread the entire volume (*see* **Note 32**) of the transformation mixture on LB agar plates with 50 mg/l spectinomycin, supplemented with X-gal and IPTG (*see* **Note 33**). Incubate overnight at 37 °C.

3.4 Golden Gate TALEN Assembly: Day 2

1. The following day, select three or more white colonies (*see* **Note 34**) and screen for correct array assembly by colony PCR. Use primers pCR8_F1 and pCR8_R1 for both pFUS_A and pFUS_B assemblies (for TALENs >21 RVDs, *see* **Note 35**). Save the screened colonies for subsequent inoculation (*see* **Note 36**). Per 25 μl PCR reaction, use: 2.5 μl 10× standard *Taq* reaction buffer; 0.5 μl 10 mM dNTPs; 0.5 μl each primer (10 μM); 0.1 μl Taq DNA polymerase (5 U/μl); 20.9 μl ddH_2O. PCR cycling conditions are as follows: 95 °C for 60 s; 30× (95 °C for 20 s, 55 °C for 30 s, 68 °C for 60 s); 68 °C for 5 min.
2. Upon completion, run the PCR reactions on an agarose gel. The products from correct clones will form a ladder of bands, where each band represents one repeat; this result is expected because of the repeated nature of the RVD array. The full length product of the expected size will be the strongest band. The size of this band should be 1,249 bp for 10 repeat arrays in pFUS_A vectors and approximately 1,271 – [(10 – N) × 102] bp (N is the number of the pFUS_B vector used, i.e., if pFUS_B7 is used, N = 7) for repeat arrays in pFUS_B vectors. An example of colony PCR results is shown in Fig. 5a. Inoculate clones with the desired inserts into 3 ml of LB supplemented with 50 mg/l spectinomycin. Incubate and shake overnight at 220 rpm, 37 °C.

3.5 Golden Gate TALEN Assembly: Day 3

1. On the following day, purify plasmids from the LB cultures using a QIAprep Spin Miniprep kit. Optionally, correct clones can be confirmed by diagnostic digestion of purified plasmids. Restriction enzymes AflII and XbaI will release the array of repeats from all pFUS backbones. Mix about ~600 ng of the

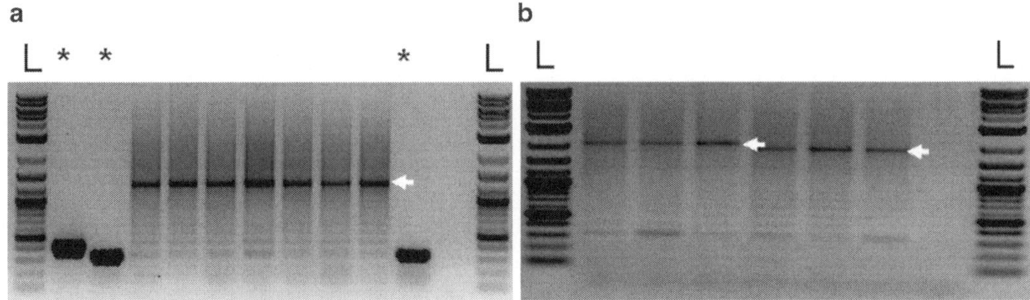

Fig. 5 Sample colony PCR results from the first and second Golden Gate reactions. Typical results of colony PCR using pFUS_A clones as templates (**a**). Typical results of colony PCR with fully assembled TAL DNA binding domains. The three on the left are 19-repeat arrays, and the three on the right are 16-repeat DNA arrays (**b**). *Arrowheads* indicate full-length PCR amplicons; *asterisks* indicate failed reactions; L = DNA ladder 2-Log (New England Biolabs)

plasmid DNA with 2 μl of CutSmart™ buffer and 0.5 μl of each enzyme and add ddH_2O to 20 μl. Incubate the reaction at 37 °C for 1 h and run on an agarose gel. The size of the released fragment will be 1,048 bp for pFUS_A vectors and approximately $1{,}070-[(10-N)\times 102]$ bp (N is the number of the pFUS_B vector used, i.e., if pFUS_B7 is used, $N=7$) for pFUS_B vectors. Clones that release fragments that are shorter than expected should not be used in further steps. Additionally, the RVD sequence can be at least partially verified by restriction digest with BspEI (*see* **Note 37**) or confirmed by DNA sequencing with primers pCR8_F1 and pCR8_R1.

2. Mix Golden Gate reaction #2 to fuse the two products of the first Golden Gate reactions and create the final full length TALEN. Prepare plasmids for each repeat array being fused—ten repeats in pFUS_A vector and ten or fewer repeats in a pFUS_B vector (for TALENs >21 RVDs, *see* **Note 38**). From the Golden Gate kit, thaw the plasmid DNA for the last repeat vector (this is the last repeat in the array that was not included in the Golden Gate reaction #1). The example TALEN in Fig. 4 would require the RVD NI, and so the vector pLR-NI would be used. Also thaw the final expression vector of your choice—this can be one of the vectors included in the kit—pTAL1, pTAL2, pTAL3, pTAL4, pZHY500, pZHY501, or one of the add-on vectors available from Addgene or from other research groups (for choice of the final expression vector *see* **Notes 8** and **39**). For each TALEN, prepare the following reaction (*see* **Note 26**):
   (a) 150 ng of each array vector (pFUS_A with ten repeats cloned and pFUS_B with ten or fewer).
   (b) 150 ng (1 μl) of the last repeat vector.

(c) 75 ng (1 μl) of the final expression vector.

(d) 1 μl Esp3I (*see* **Note 40**).

(e) 1 μl T4 DNA ligase (*see* **Note 28**).

(f) 2 μl 10× T4 DNA ligase buffer (*see* **Note 29**).

(g) H_2O up to 20 μl total reaction volume (*see* **Note 30**).

3. Place the Golden Gate reactions in a PCR machine and run the following cycles: 10× (37 °C/5 min + 16 °C/10 min) + 37 °C/15 min + 80 °C/5 min (for shorter cycles *see* **Note 41**). Plasmid-Safe nuclease treatment is not necessary in Golden Gate reaction #2, because the final expression vector has no homology with the inserted repeats.

4. Transform each Golden Gate reaction into *E. coli* (DH5α) by adding 5 μl of the reaction to 50 μl of chemically competent cells in a 1.5 ml microcentrifuge tube. Incubate on ice for at least 10 min and heat-shock for 70 s at 42 °C. Place on ice immediately and within 2 min add 200 μl of room temperature SOC medium. Recover the cells by shaking at 37 °C for 1 h. Plate the full volume (*see* **Note 32**) of the transformation mixture onto LB agar plates with 50 mg/l carbenicillin (or appropriate antibiotic for the expression vector, if vectors other than the ones included in the kit are used), supplemented with X-gal and IPTG (*see* **Note 33**). Incubate overnight at 37 °C.

3.6 Golden Gate TALEN Assembly: Day 4

1. On the following day, select three white colonies and screen for correct array assembly by colony PCR. Use primers TAL_F1 and TAL_R2. Save the screened colonies for inoculation (*see* **Note 36**). Per 25 μl PCR reaction, use 2.5 μl 10× standard *Taq* reaction buffer, 0.5 μl dNTPs, 0.5 μl each primer (10 μM), 0.1 μl Taq DNA polymerase (5 U/μl) and 20.9 μl ddH_2O. PCR cycling conditions are as follows: 95 °C for 60 s; 30× (95 °C for 20 s, 55 °C for 30 s, 68 °C for 2 min); 68 °C for 5 min.

2. Upon completion, run the PCR reactions on an agarose gel. The products from correct clones will form a ladder of bands, where each band represents one repeat. The full length product of expected size will be the most abundant band. The size of this band should be 2,357 bp for TALENs with 21 RVDs and $2{,}357-[(21-N)\times 102]$ bp or $2{,}357+[(N-21)\times 102]$ bp for shorter or longer TALENs, respectively (N is the number of RVDs of the full length TALEN). An example of colony PCR results for Golden Gate reaction #2 is in Fig. 5b. Inoculate clones with the desired inserts into 5 ml of LB supplemented with 50 mg/l carbenicillin (or appropriate antibiotic for the expression vector, if vectors other than the ones included in the kit are used). Incubate and shake overnight at 220 rpm, 37 °C.

3.7 Golden Gate TALEN Assembly: Day 5

1. On the following day, purify plasmids from the LB cultures using the QIAprep Spin Miniprep Kit. Optionally, correct clones can be confirmed by diagnostic digestion of purified plasmids. Restriction enzymes StuI or BstAPI and AatII will release the full length array of repeats from expression backbones pTAL1-4 and pZHY500-501. Mix ~600 ng of the plasmid DNA with 2 μl of CutSmart™ buffer and 0.5 μl of each enzyme and add ddH_2O to 20 μl. Incubate the reaction at 37 °C for 1 h and run on an agarose gel. If using StuI, the size of the released fragment will be 2,336 bp for TALENs with 21 RVDs and $2{,}336-[(21-N)\times 102]$ bp or $2{,}336+[(N-21)\times 102]$ bp for shorter or longer TALENs, respectively (N is the number of RVDs of the full length TALEN). Fragments will be 108 bp shorter if using BstAPI. Clones that release fragments of shorter than the expected size should not be further used. Additionally, the RVD sequence can be at least partially verified by restriction digest with BspEI (*see* **Note 37**). The sequence of the complete repeat array should be confirmed by DNA sequencing with primers TAL_F1 or TAL_Seq_5-1 and TAL_R2 or TAL_R3 (*see* **Note 42**). For hints about TALEN sequencing, *see* also **Note 43**. For tools to assemble TALEN sequences and build vector maps, *see* **Note 44**.
2. After sequence confirmation, TALENs are ready for use in the target organism or for sub-cloning into an expression vector of choice. To sub-clone TALENs from vectors pTAL3-4 and pZHY500-501, enzymes BglII, XhoI or XbaI (*see* **Note 45**) on the 5′ end of the TALEN are used in combination with SacI, Eco53kI (blunt end) or AflII on the 3′ end of the TALEN. To cut out only the TALE DNA binding domain (without FokI), BamHI can be used (cuts twice) or a combination of BglII or XhoI with EcoRV (blunt end) for pTAL3 and 4. For pZHY500 and 501, a combination of BglII, XhoI, or XbaI with BamHI can be used.

4 Notes

1. We have found that prepping the plasmids for Golden Gate reaction #1 with kits from other suppliers seems to reduce efficiency of the reaction. For the Golden Gate reaction #2, we routinely use miniprep kits from other suppliers with excellent success.
2. Provided are the concentrations in which the respective plasmids will be used in Golden Gate reactions. Adjusting the concentrations of all vectors in the library to a standard greatly simplifies the assembly procedure; the user simply needs to add 1 μl of each vector to each reaction.

3. We recommend DNA sequencing every time you re-prep a plasmid. After sequencing, carefully look at the sequencing trace from each of the repeat-containing plasmids, specifically the bases that encode the RVD. If you have a mix of two plasmids with the same number (e.g., pNN1 and pNG1), you can see double peaks at the bases that encode the RVD, even though base calling by a program may indicate only one type RVD.
4. *E. coli* containing pHD2 grows slowly. The basis for this phenotype is unknown.
5. Plasmids grown in DH5α yield significantly more plasmid DNA than those grown in DH10B. Plasmids obtained from Addgene are in one of these two *E. coli* strains. The basis for this plasmid yield difference is unknown.
6. Several plasmids are known to have rearrangements in the vector backbones. These rearrangements do not affect their use in the Golden Gate assembly protocol. The plasmids with known rearrangements are pNK1, pNK2, pNK3, pNK4, pNK5, pNK6, pNK7, pNK8, pNK9, pNK10, pNG9.
7. If mRNA or cDNA is used to design target sites, it may happen that a target site will be selected that overlaps two exons. As such a sequence doesn't exist in the genomic DNA due to the presence of an intron, such TALENs will not induce DSBs. To confirm the target sequence is correct, we highly recommend sequencing the target locus before designing TALEN target sites. There may be sequence differences between different cell lines/strains/varieties of the same species. Such differences, including single nucleotide polymorphisms, can prevent TALENs from binding.
8. The original TALEN backbone described in Christian et al. [2] and Cermak et al. [18] used a nearly full-length TALE protein fused to the FokI catalytic domain. This backbone had 288 aa and 231 aa on the N and C terminal ends, respectively, of the TALE repeat array. This TALE backbone, also called the "BamHI backbone" (it is the BamHI fragment of the TALE protein Tal1C) is used in vectors pTAL1-4. However, truncations of the N and C termini of the TALE backbone increase the activity of TALENs by several fold, and several truncated versions of the TALE backbones have been developed. Among these, the most frequently used is the Δ152/C63 backbone first described by Miller et al. [12]. The authors removed 152 amino acids from the N terminus and left 63 amino acids on the C terminus of the repeat array. The increased activity was confirmed in several organisms including plants [15]. Vectors pZHY500 and pZHY501 are derived from pTAL3 and pTAL4 and have the Δ152/C63 backbone. We highly recommend

using these vectors over pTAL1 and pTAL4, as the difference in activity is substantial, particularly for spacers with lengths under 20 bp (*see* **Note 9**). Most of the final expression plasmids available from Addgene have the Δ152/C63 TALE backbone. The Δ152/C18 backbone is used in vectors pZHY565 and 566 [15] (available as Addgene add-ons to the kit), which are derived from pTAL3 and pTAL4. Christian et al. [11] showed that the Δ152/C18 backbone has high activity over a much narrower range of spacer lengths (also *see* **Note 9**) than the BamHI or Δ152/C63 backbones. Therefore, it can be used (1) to increase specificity of TALENs by presumably reducing the number of off-target sites in which the two TALEN binding sites are properly spaced for cleavage and (2) to increase activity at sites with very specific spacer lengths.

9. The TALE backbones BamHI, Δ152/C63 and Δ152/C18 (for description of TALE backbones *see* **Note 8**) have different requirements for the DNA spacer length between the two TALEN binding sites. This is mainly due to the number of amino acids at the C terminus of the TALE array that separates the DNA binding domain from the FokI catalytic domain (hereafter referred to as the linker). The shorter the protein linker is, the closer the FokI domain is to the DNA binding domain of a TALEN. Therefore, a shorter spacer is needed to allow the FokI monomers to dimerize. The comparison of the spacer length optima for the three backbones was described in Christian et al. [11]. The BamHI backbone is active through a wide range of spacers, from 14 to 39 bp, with optimal activity on longer spacers of 21–33 bp. The Δ152/C63 backbone is also active through a wide range of spacers, from 13 to 39 bp, but highest activity is achieved with shorter spacers of 13–16 bp. The activity of the Δ152/C18 backbone is extremely low on spacers longer than 17 bp, and it is very high on spacers 13–15 bp in length. Based on this, for the widely used Δ152/C63 backbone, we recommend using spacer lengths from 13 to 16 bp. Although activity of the BamHI backbone is relatively high on spacers >21 bp, shorter spacers offer higher specificity (due to lower numbers of possible off-target sites) and easier screening for TALEN-induced mutations (*see* **Note 17**).

10. The minimum number of RVDs that can be used to build a functional TALEN using the method described here is 12. Although TALENs with 12 RVDs have been shown to be active with certain TALE architectures [31], in our experience, TALENs with 15 and more RVDs perform best at inducing DSBs in vivo. Adding more RVDs up to certain number may also increase activity and specificity. However, there might be a limit above which adding more RVDs may not increase activity as suggested by Meckler et al. [34].

They showed that there is a decrease in DNA binding affinity towards the C-terminal end of TALE repeat arrays; thus, the C-terminal TALE repeats do not contribute significantly to the overall binding affinity. Although the absolute number of repeats that contribute to specificity and/or activity has not been determined for TALENs, we hypothesize that adding RVDs above 20 will have little or no effect on TALEN efficacy. A TALEN with 21 RVDs is the longest TALEN that can be built using only two intermediate vectors (2 pFUS assemblies of 10 RVDs plus the last repeat). TALENs up to 31 RVDs can be assembled, but these require building an additional array of RVDs in Golden Gate reaction #1 (three pFUS assemblies). To identify the maximum number of optimal targets for TALENs that can be made with least effort, we recommend using a repeat array length of 15–21.

11. RVD NN binds to G and A with similar affinity. To increase the specificity of TALENs, RVDs that are strictly specific for G were sought. High specificity for G was reported for RVDs NK [12, 35] and NH [36]. Although highly specific for G, NK has low binding affinity, and higher numbers of NK repeats in an array causes TALEN activity to decrease [11, 37]. Although the RVD NH has high specificity for G, activities of custom TALE transcription activators with NH RVDs did not reach levels of their NN counterparts, and the binding affinity of the RVD NH was described as "intermediate" (as opposed to "strong" for NN) [36]. In our experience, we have not seen significant toxicity of NN-containing TALENs, suggestive of off-target cleavage, and we therefore recommend using NN as a default RVD for G to ensure TALENs of high activity. Use of NH should be considered in cases where specificity is extremely important (e.g., gene therapy). Alternatively, NH or NK can be used in positions where off-target cleavage is a concern. For example, if there is an off-target site in the genome that only differs from the primary target site by one or few G to A substitutions, NH or NK RVDs can be used in these positions to prevent off-target binding and cleavage. For the remaining Gs, NN can be used to maximize activity.

12. In "Filter options," the user can select from three settings. If you are targeting a specific base position in your sequence, you can select "Show TALEN pairs targeting a specific site" and set the "Cut Site Position." The output will contain only TALENs targeting that specific site (the base position you selected will be in the middle of the spacer). We recommend selecting one of the other settings that list TALEN pairs for all base positions in the provided sequence. With this type of setting, you will be able to see sites where your target base is not in the middle of the spacer, although such sites may be more

optimal based on the criteria described in **Notes 9–11**. For most applications, including gene knockouts and replacements, it is not necessary for the targeted position to be in the middle of the spacer. You can also choose to "show" or "hide redundant TALENs." For the latter, only one TALEN pair for each base position (centered in the spacer) in the sequence will be shown. The rest of the TALEN pairs, including the ones that have different repeat array or spacer lengths but target the same position, will be hidden.

13. Streubel et al. [36] described a set of guidelines for custom TALE and TALEN design based on binding affinities and specificities of individual RVDs. These data were derived from assays that measured transcriptional activity of TALEs in vivo. Based on their binding affinities, individual RVDs were divided into "weak," "intermediate," and "strong" groups. The number of strong RVDs (i.e., HD and NN) in the TALE repeat array was shown to positively correlate with TALE activity. On the other hand, TALEs with no strong RVDs had very low activity. A certain number of strong RVDs is therefore required for optimal activity. In addition, the position of strong RVDs in the array and avoidance of stretches of identical RVDs are also important. We recommend starting with the setting for the Streubel et al. design parameters in the "on" position, and if no satisfactory target sites are found, switch the setting to "off." Although these guidelines may help to identify more active TALENs in some cases, we find that TALENs that obey the guidelines work similarly well. This may be due to the effect of dimerization required for TALEN activity that can compensate for weak binding of TALEN monomers, unlike in TALEs where only one DNA binding monomer is used.

14. "Upstream base" is the base immediately preceding the first base of a TALEN target site contacted by the first RVD in the repeat array. We highly recommend using T only as a preferred "upstream" base. Most of the natural TAL effector target sites have a T in this position, and structural studies revealed interaction of the N-terminal region of the TALE protein specifically with this T [9]. The authors of this study speculate that a substitution for a cytosine (but not purine) could be accommodated by the structure of the protein, and some custom TAL effectors have been reported to be active on sites preceded by a C [12]; however, this is not an optimal configuration and should be avoided if possible.

15. For gene knockouts, the targeted region should be a conserved catalytic domain or critical amino acid residue. If no such sequences are known, an alternative is to target a region close to the 5′ end of the gene. Frameshift mutations near the 5′ end are more likely to cause null (knockout) phenotypes than

mutations in the 3′ end of the gene, where in some instances, functional protein can still be produced. One should also avoid targeting introns. For targeted gene modification, proximity of the DSB to the site of mutation should be considered (*see* also **Note 16**).

16. Generally, for gene replacement or targeted insertion by HR, DSBs within tens of base pairs on either side of the modification will yield comparable gene targeting efficiencies. Therefore, it is not absolutely necessary for the DSB to overlap exactly with the site of mutation. For NHEJ-mediated gene knockouts, most mutations near the 5′-end of the gene will likely have comparable effects. If aiming for a NHEJ-mediated deletion of a specific amino acid, DSBs within 5 bp of the targeted codon are likely to result in deletions that include that codon.
17. The most simple and reliable method to screen for TALEN induced mutations is "enrichment PCR." If a unique restriction enzyme binding site is present in the spacer of a TALEN target site, it is very likely to be disrupted upon TALEN cleavage and error-prone repair of the break by NHEJ. Genomic DNA isolated from a sample treated with TALENs can be predigested with this restriction enzyme and used as a template for PCR amplification of the TALEN target site. The PCR product is digested again, and the products with TALEN-induced mutations are detected as digestion-resistant bands on a gel. For semiquantitative assessment, genomic DNA predigestion may be skipped, and frequency of mutagenesis is determined by calculating the fraction of the digestion-resistant products divided by the total amount of PCR product. In order to use this convenient and sensitive screen for mutagenesis, we highly recommend selecting TALEN target sites that have a unique restriction enzyme site in the spacer. Restriction enzymes sites in the middle of the spacer region and those with longer recognition sites (e.g., 6–12 bp) are preferred, as they are more likely to detect mutations. One should also look at the reliability of the restriction endonuclease; in order to get clear results, unstable enzymes should be avoided when possible.
18. A higher percentage of RVDs HD and NN in the repeat array correlates with TALENs with higher activity (for reasoning, *see* **Note 13**). Longer TALENs may be more specific (*see* **Note 10**).
19. In our experience, >95 % of properly designed TALENs are active. However, there can be significant differences in the levels of activity between TALENs with different array lengths, spacer sizes, or RVD compositions [18]. In addition, even a TALEN that is highly active in cleaving episomal targets can be blocked from inducing mutations at the endogenous target

due to intrinsic characteristics of the target locus, such as DNA methylation or chromatin structure (for targeting methylated targets *see* **Note 20**). Based on this, we recommend making two TALEN pairs per target locus to ensure identifying one that will be highly active.

20. Methylated sites are not cleaved by TALENs designed using TAL effector targeter 2.0, because the RVD HD, used to recognize cytosine, does not interact with 5-methyl cytosine (5mC). However, it was shown that 5mC is efficiently recognized by the T binding RVDs NG [38] and N* [39]. While N* modules are not yet available in the Golden Gate kit, NG can be used for positions in the DNA target where 5mC is known to occur. TALENs modified in this way efficiently cleave methylated sites.
21. In some cases, targeting two or more different alleles of a gene at the same time may be desired. If the alleles differ by few SNPs, this can be achieved with the same pair of TALENs that have the degenerate RVDs NS or NP. The RVD NS was believed to be unspecific and interact with all four DNA bases; however, Streubel et al. [36] later showed that NS has a preference for G and A. The same authors reported specificity of NP for A, C, or T. NS modules are available as Addgene add-ons to the kit. Note that the NP RVD modules are not available at Addgene as of this submission.
22. For TALENs of 22–31 RVDs, separate the RVD array of each TALEN into the first ten RVDs, the second ten RVDs and the third ten RVDs (or fewer, depending on the length of TALEN), and the last RVD.
23. Module plasmids are labeled with numbers 1–10. These numbers correspond to the position of the RVD in the array of repeats. Using the method described here, only arrays of up to ten repeats (RVDs) can be assembled, and these are then fused into a longer array in a second Golden Gate reaction. An intermediate, separate array is built for RVDs 1–10, 11–20 and 21–30. The same plasmid modules are used to build all three arrays; that is, module plasmids 1–10 are also used when building arrays 11–20 and 21–30. For example, the RVD HD in position 14 in the example TALEN in Fig. 4 will be in position 4 of the intermediate array in pFUS_B7 vector, and therefore plasmid pHD4 will be used for this position.
24. For TALENs with >21 RVDs, intermediate vectors pFUS_A30A, pFUS_A30B, and pFUS_B($N-21$) must be used.
25. For TALENs with >21 RVDs, three Golden Gate #1 reactions are needed to assemble RVD arrays 1 to 10, 11 to 20, and 21 to ($N-1$).

26. It is very important to use the specified enzymes and buffers from suppliers as described in the Subheading 2. Although some other combinations of enzymes and buffers might work in some cases, the protocol was optimized with the reagents described, and we cannot guarantee success with other reagents.
27. Be sure to use BsaI (NEB # R0535), and NOT BsaI-HF, the high fidelity version of the enzyme. BsaI-HF does not work well in the T4 DNA ligase buffer, and its use is a very common reason for Golden Gate reaction #1 to fail.
28. The protocol described in Cermak et al. [18] uses Quick Ligase for the Golden Gate reactions, but we have found that normal T4 Ligase from NEB (M0202) is adequate.
29. It is extremely critical to use fresh T4 DNA ligase buffer. The Golden Gate reaction #1 with ten repeats is the most sensitive step in the protocol. Fusing ten DNA fragments with a plasmid backbone requires all reagents to be working with 100 % efficiency. T4 DNA ligase buffer is the reagent that is most likely to go bad, and using sub-optimal ligase buffer is the most common reason for the Golden Gate reaction #1 to fail. The ligase buffer can tolerate only a few freeze/thaw cycles. It can be aliquoted into ~25 μl aliquots to avoid excessive freeze-thawing. Always thaw ligase buffer on ice, and before using, try to detect the sulfuric smell of dithiothreitol (DTT). If absent, discard or add fresh DTT to the concentration of 10 mM (or 1 mM in the final reaction). Bad ligase buffer results in very few or no white colonies, or white colonies with incomplete arrays of repeats. We have had inquiries from a large number of people about the failure of Golden Gate reaction #1, and replacing the ligase buffer has solved many of these problems. Although the Golden Gate reaction #2 is more efficient, as only 3–4 DNA fragments are being fused into the vector backbone, the quality of T4 DNA ligase buffer is equally important for this step. In this case, the Esp3I enzyme requires DTT for its activity, and so the same precautions should be taken as described above. Bad ligase buffer with insufficient DTT will result in no cleavage by Esp3I and subsequently thousands of blue colonies will appear, representing the uncut expression vector. If no or very few white colonies are found, this is often the problem.
30. Golden Gate reactions #1 and #2 were originally optimized for 20 μl reaction volumes, but 10 μl reactions are reliably effective (same concentrations of reagents). If a particular cloning reaction is somewhat difficult (fails more than once), it may be useful to use a 20 μl reaction volume.
31. The Plasmid-Safe nuclease manual says you should inactivate the enzyme by heating the reaction to 70 °C for 30 min, but in our experience with bacterial transformation, inactivation is not necessary.

32. Golden Gate reaction with five or fewer repeats (when pFUS_B5 or lower is used) are very efficient, and plating 100 µl of the transformation reaction is sufficient to get a high number of colonies. The same is true for Golden Gate reaction #2, if 10 restriction/ligation cycles are used.
33. We have found that for most high-copy-number plasmids, IPTG is not necessary for the blue-white screening in DH5α *E. coli*.
34. Screening three colonies is usually sufficient to identify one clone with a correctly assembled TALE repeat array. If a specific reaction is inefficient and only a few white colonies are obtained (i.e., the ratio of blue to white colonies is >1), we recommend screening up to ten white colonies.
35. The same primers are used to screen pFUS_A30A and pFUS_A30B transformants.
36. If screening higher number of colonies (e.g., from several TAL arrays) the screening/inoculation procedure can be simplified as follows. To start the bacterial cultures on the same day as colony PCR screening, bacterial cells are stored in 50 µl of liquid LB medium with no antibiotics. First, prepare two PCR plates. In the first plate, fill each well with 25 µl of the colony PCR master mix and in the other, using the same layout, fill each well with 50 µl of liquid LB. Lift a bacterial colony from the LB agar plate using a pipette tip, then wash in the PCR mix by pipetting up and down several times with the tip; then, put the tip in the corresponding well of the plate filled with liquid LB. After all colonies are picked, all tips can be washed in the liquid LB in the PCR plate by pipetting the medium up and down several times and then discard the tip. This can be done with a multichannel pipettor. Seal the plate and store at 4 °C. After correct clones are identified by PCR and gel electrophoresis, the overnight cultures can be started by inoculating the 50 µl of LB stock from each correct clone into 3 ml of liquid LB supplemented with the appropriate antibiotic.
37. Restriction enzyme BspEI cuts only the repeat units containing the RVD HD. However, note that the HD repeats in the first, 11th, 21st, and the last position (positions in which pHD1 or pLR-HD modules are used) do not contain the BspEI site. Combine about ~600 ng of the plasmid DNA with 2 µl of the NEBuffer 3.1 and 0.5 µl of BspEI and add ddH_2O to 20 µl. Incubate the reaction at 37 °C for 1 h. On the gel, you will get a pattern resembling your TALEN sequence and HD repeats position.
38. For TALENs with >21 RVDs, three ten-repeat arrays will be fused: ten repeats in pFUS_A30A, ten repeats in pFUSA30B, and ten or fewer repeats in a pFUS_B.

39. Vectors pTAL1 and pTAL2 do not contain the FokI catalytic domain and are not for the production of TALENs, but rather for TALE assembly. TALEs made in pTAL1 include the TALE transcription activation domain and can be used as transcriptional activators. pTAL2 was derived from pTAL1 but lacks the stop codon and is designed for fusions of TALEs to other protein domains of user choice. Both vectors are Gateway compatible. For TALENs, use one of the vectors pTAL3, pTAL4, pZHY500, or pZHY501. The only difference between pTAL3 and 4 and pZHY500 and 501 is in the selectable marker gene for propagation in yeast. All of the TALEN vectors in the kit are yeast expression vectors that can be used directly to test TALEN activity in vivo in yeast. For this purpose, the pCP5b reporter plasmid is included in the kit. It is very similar to the published pCP5 used for the single strand annealing assay in yeast [40], but differs in the origin of replication and the antibiotic resistance gene. To use TALENs in organisms other than yeast, they must be sub-cloned into an appropriate expression vector. Final expression vectors that are optimized for expression in many systems, including mammalian cells, zebrafish, and *Drosophila*, can be directly used in the place of the yeast expression vectors in Golden Gate reaction #2 and are available from Addgene (http://www.addgene.org/TALeffector/goldengate/add-ons/). The Voytas lab is also planning to include ready-for-use expression vectors for plants. There are two different FokI architectures found in various TALEN expression vectors. The wild-type, "homodimeric" FokI is able to dimerize with itself. Consequently, any given TALEN monomer can not only pair with its desired partner, but also with another monomer of itself if such target sequence is available. This increases the number of potential sites for off-target cleavage. To increase specificity, several "heterodimeric" FokI architectures were developed in which an active dimer is formed only by two different forms of the FokI monomers. Monomers cannot create active pairs with themselves, thus decreasing the number of possible off-targets and increasing specificity when used in TALENs. This gain in specificity is often offset by decreased activity compared with the homodimeric FokI, although heterodimeric architectures with increased activity have been described [41]. All vectors in the Golden Gate kit have homodimeric FokI nucleases, whereas most of the add-on expression vectors use a heterodimeric architecture.

40. Other groups have reported the use of the restriction enzyme BsmBI (New England Biolabs) in the place of Esp3I in Golden Gate reaction #2. BsmBI and Esp3I share the same recognition site and DNA cleavage profile; however, BsmBI has an optimal

activity temperature of 55 °C, which is not compatible with T4 DNA ligase. Despite this, other groups report using BsmBI with good success, digesting at the non-optimal temperature for BsmBI of 37 °C. However, we still recommend using Esp3I as the preferred enzyme.

41. With these cycling conditions, the reaction will in most cases work very well, and you will get hundreds of white colonies. To save time, we found that one cycle of 37 °C, 10 min + 16 °C, 15 min + 37 °C, 15 min + 80 °C, 5 min is sufficient to get tens of white colonies with correct inserts.

42. TAL_Seq_5-1 and TAL_R3 primers were designed to sequence longer TALENs. They anneal next to the repeat array, whereas the TAL_F1 and TAL_R2 primers anneal further away. With the TAL_Seq_5-1 and TAL_R3 primers, the bases encoding the first and last few amino acids of the first TAL repeat will not be sequenced, as they are too close to the primers. The first RVD is far enough to be sequenced well with TAL_Seq_5-1 primer. To sequence the last RVD; however, TAL_R2 should be used. TAL_Seq_5-1 and TAL_R3 primers allow us to sequence through TALENs up to 20 RVDs long. For longer TALENs, pFUS vectors should be sequenced to determine the identity of the RVDs. We currently do not have a way to sequence the middle of a long, completed array in an expression vector.

43. Although rare, recombination may occur between TALE array repeats due to their repetitive nature. Even though the recombination may occur with a very low frequency in each TALEN plasmid prep, no gross evidence of recombination has been seen in more than 1 in 100 plasmid preps of TALEN plasmids [J. Keith Joung, personal communication]. In addition, random mutations in the RVD coding triplets may have detrimental effects on TALEN activity. Some users have observed that the recombination-mediated mutation rate can be decreased to undetectable levels by growing *E. coli* at 30 °C or by using less rich LB medium. Nevertheless, sequencing of the complete TALEN repeat arrays is important to rule out the possibility of mutations. Sequencing TALENs (or any TAL array) can be challenging due to their repetitive nature (*see* also **Note 42**). We very seldom identify any mutations within the repeats. After sequencing several hundred TALE arrays, we have only seen two or three problems within the repeat arrays. Most problems occur when the user adds the wrong repeat-containing plasmid. We primarily use two companies for sequencing. ACGT, Inc has given us the longest sequence reads, but we also use Functional Biosciences. We do not have any connection to these companies except as users (websites: www.acgtinc.com; functionalbio.com).

44. An online tool developed by the Bao lab at Georgia Tech (http://baolab.bme.gatech.edu/Research/BioinformaticTools/assembleTALSequences.html) can be used to assemble TALEN array sequences in pTAL1-4 backbones and most of the add-on final expression vectors available from Addgene. The DNA sequence of the full length target site or the RVD array of the TALEN can be used as input. The outputs are sequences of all intermediate plasmids (pFUS) needed for the full-length TALEN assembly as well as the sequence of the final expression vector. The sequences can be downloaded in GenBank format and thus opened as annotated maps in preferred DNA analysis software.
45. TALEN fragments released using XbaI will not include the SV40 NLS. There is another XbaI site between the TALE DNA binding domain and FokI of pTAL3 and 4. However, this site is methylated and will not be cut by XbaI unless the plasmid is grown in a methylation deficient *E. coli* strain.

Acknowledgment

We thank Rebecca Greenstein for critical reading of the manuscript and helpful comments.

References

1. Gaj T, Gersbach CA, Barbas CF 3rd (2013) ZFN, TALEN, and CRISPR/Cas-based methods for genome engineering. Trends Biotechnol 31:397–405
2. Christian M, Cermak T, Doyle EL, Schmidt C, Zhang F, Hummel A, Bogdanove AJ, Voytas DF (2010) Targeting DNA double-strand breaks with TAL effector nucleases. Genetics 186:757–761
3. Li T, Huang S, Jiang WZ, Wright D, Spalding MH, Weeks DP, Yang B (2011) TAL nucleases (TALNs): hybrid proteins composed of TAL effectors and FokI DNA-cleavage domain. Nucleic Acids Res 39:359–372
4. Cong L, Ran FA, Cox D, Lin S, Barretto R, Habib N, Hsu PD, Wu X, Jiang W, Marraffini L, Zhang F (2013) Multiplex genome engineering using CRISPR/Cas systems. Science 339:819–823
5. Mali P, Yang L, Esvelt KM, Aac J, Guell M, DiCarlo JE, Norville JE, Church GM (2013) RNA-guided human genome engineering via Cas9. Science 339:823–826
6. Bogdanove AJ, Schornack S, Lahaye T (2010) TAL effectors: finding plant genes for disease and defense. Curr Opin Plant Biol 13: 394–401
7. Boch J et al (2009) Breaking the code of DNA binding specificity of TAL-type III effectors. Science 326:1509–1512
8. Moscou MJ, Bogdanove AJ (2009) A simple cipher governs DNA recognition by TAL effectors. Science 326:1501
9. Mak ANS, Bradley P, Cernadas RA, Bogdanove AJ, Stoddard BL (2012) The crystal structure of TAL effector PthXo1 bound to its DNA target. Science 335:716–719
10. Deng D, Yan C, Pan X, Mahfouz M, Wang J, Zhu J, Shi Y, Yan N (2012) Structural basis for sequence-specific recognition of DNA by TAL effectors. Science 335:720–723
11. Christian ML, Demorest ZL, Starker CG, Osborn MJ, Nyquist MD, Zhang Y, Carlson DF, Bradley P, Bogdanove AJ, Voytas DF (2012) Targeting G with TAL effectors: a comparison of activities of TALENs constructed with NN and NK repeat variable Di-residues. PLoS One 7:e45383
12. Miller JC, Tan S, Qiao G, Barlow KA, Wang J, Xia DF et al (2011) A TALE nuclease

architecture for efficient genome editing. Nat Biotechnol 29:143–148

13. Mussolino C, Morbitzer R, Lutge F, Dannemann N, Lahaye T, Cathomen T (2011) A novel TALE nuclease scaffold enables high genome editing activity in combination with low toxicity. Nucleic Acids Res 39: 9283–9293
14. Carlson DF, Tan W, Lillico SG, Stverakova D, Proudfoot C, Christian M, Voytas DF, Long CR, Whitelaw CBA, Fahrenkrug SC (2012) Efficient TALEN-mediated gene knockout in livestock. Proc Natl Acad Sci U S A 109: 17382–17387
15. Zhang Y, Zhang F, Li X, Baller JA, Qi Y, Starker CG, Bogdanove AJ, Voytas DF (2013) Transcription activator-like effector nucleases enable efficient plant genome engineering. Plant Physiol 161:20–27
16. Engler C, Kandzia R, Marillonnet S (2008) A one pot, one step, precision cloning method with high throughput capability. PLoS One 3:e3647
17. Engler C, Gruetzner R, Kandzia R, Marillonnet S (2009) Golden gate shuffling: a one-pot DNA shuffling method based on type IIs restriction enzymes. PLoS One 4:e5553
18. Cermak T et al (2011) Efficient design and assembly of custom TALEN and other TAL effector based constructs for DNA targeting. Nucleic Acids Res 39:e82
19. Morbitzer R, Elsaesser J, Hausner J, Lahaye T (2011) Assembly of custom TALE-type DNA binding domains by modular cloning. Nucleic Acids Res 39:5790–5799
20. Geissler R, Scholze H, Hahn S, Streubel J, Bonas U, Behrens SE, Boch J (2011) Transcriptional activators of human genes with programmable DNA-specificity. PLoS One 6:e19509
21. Weber E, Gruetzner R, Werner S, Engler C, Marillonnet S (2011) Assembly of designer TAL effectors by Golden Gate cloning. PLoS One 6:e19722
22. Li L, Piatek MJ, Atef A, Piatek A, Wibowo A, Fang X, Sabir JSM, Zhu JK, Mahfouz MM (2012) Rapid and highly efficient construction of TALE-based transcriptional regulators and nucleases for genome modification. Plant Mol Biol 78:407–416
23. Zhang Z, Zhang S, Huang X, Orwig KE, Sheng Y (2013) Rapid assembly of customized TALENs into multiple delivery systems. PLoS One 8:e80281
24. Liang J, Chao R, Abil Z, Bao Z, Zhao H (2013) FairyTALE: a high-throughput TAL effector synthesis platform. ACS Syn Biol 3:67. doi:10.1021/sb400109p
25. Li T, Huang S, Zhao X, Wright DA, Carpenter S, Spalding MH, Weeks DP, Yang B (2011) Modularly assembled designer TAL effector nucleases for targeted gene knockout and gene replacement in eukaryotes. Nucleic Acids Res 39:6315–6325
26. Zhang F, Cong L, Lodato S, Kosuri S, Church GM, Arlotta P (2011) Efficient construction of sequence-specific TAL effectors for modulating mammalian transcription. Nat Biotechnol 29:149–153
27. Sanjana NE, Cong L, Zhou Y, Cunniff MM, Feng G, Zhang F (2012) A transcription activator-like effector toolbox for genome engineering. Nat Protoc 7:171–192
28. Uhde-Stone C, Gor N, Chin T, Huang J, Lu B (2013) A do-it-yourself protocol for simple transcription activator-like effector assembly. Biol Proced Onl 15:3
29. Yang J, Yuan P, Wen D, Sheng Y, Zhu S, Yu Y, Gao X, Wei W (2013) ULtiMATE system for rapid assembly of customized TAL effectors. PLoS One 8:e75649
30. Schmid-Burgk JL, Schmidt T, Kaiser V, Höning K, Hornung V (2013) A ligation-independent cloning technique for high-throughput assembly of transcription activator–like effector genes. Nat Biotech 31:76–82
31. Reyon D, Tsai SQ, Khayter C, Foden JA, Sander JD, Joung JK (2012) FLASH assembly of TALENs for high-throughput genome editing. Nat Biotech 30:460–465
32. Briggs AW, Rios X, Chari R, Yang L, Zhang F, Mali P, Church GM (2012) Iterative capped assembly: rapid and scalable synthesis of repeat-module DNA such as TAL effectors from individual monomers. Nucleic Acids Res 40:e117
33. Doyle EL, Booher NJ, Standage DS, Voytas DF, Brendel VP, VanDyk JK, Bogdanove AJ (2012) TAL effector-nucleotide targeted (TALE-NT) 2.0: tools for TAL effector design and target prediction. Nucleic Acids Res 40: 117–122
34. Meckler JF, Bhakta MS, Kim MS, Ovadia R, Habrian CH, Zytovich A, Yu A, Lockwood SH, Morbitzer R, Elsäesser J, Lahaye T, Segal DJ, Baldwin EP (2013) Quantitative analysis of TALE-DNA interactions suggests polarity effects. Nucleic Acids Res 41:4118–4128
35. Morbitzer R, Römer P, Boch J, Lahaye T (2010) Regulation of selected genome loci using de novo-engineered transcription activator-like effector (TALE)-type transcrip-

tion factors. Proc Natl Acad Sci U S A 107: 21617–21622

36. Streubel J, Blücher C, Landgraf A, Boch J (2012) TAL effector RVD specificities and efficiencies. Nat Biotech 30:593–595
37. Huang P, Xiao A, Zhou M, Zhu Z, Lin S, Zhang B (2011) Heritable gene targeting in zebrafish using customized TALENs. Nat Biotech 29:699–700
38. Deng D, Yin P, Yan C, Pan X, Gong X, Qi S, Xie T, Mahfouz M, Zhu JK, Yan N, Shi Y (2012) Recognition of methylated DNA by TAL effectors. Cell Res 22:1502–1504
39. Valton J, Dupuy A, Daboussi F, Thomas S, Maréchal A, Macmaster R, Melliand K, Juillerat A, Duchateau P (2012) Overcoming transcription activator-like effector (TALE) DNA binding domain sensitivity to cytosine methylation. J Biol Chem 287:38427–38432
40. Zhang F, Maeder ML, Unger-Wallace E, Hoshaw JP, Reyon D, Christian M, Li X, Pierick CJ, Dobbs D, Peterson T, Joung JK, Voytas DF (2010) High frequency targeted mutagenesis in *Arabidopsis thaliana* using zinc finger nucleases. Proc Natl Acad Sci U S A 107:12028–12033
41. Doyon Y, Vo TD, Mendel MC, Greenberg SG, Wang J, Xia DF, Miller JC, Urnov FD, Gregory PD, Holmes MC (2011) Enhancing zinc-finger-nuclease activity with improved obligate heterodimeric architectures. Nat Methods 8:74–79

Chapter 8

Ligation-Independent Cloning (LIC) Assembly of TALEN Genes

Jonathan L. Schmid-Burgk, Tobias Schmidt, and Veit Hornung

Abstract

Modular DNA binding protein architectures hold the promise of wide application in functional genomic studies. Functionalization of DNA binding proteins, e.g. using the *FokI* nuclease domain, provides a potent tool to induce DNA double strand breaks at user-defined genomic loci. In this regard, TAL (transcription activator-like) effector proteins, secreted by bacteria of the Xanthomonas family, provide the highest degree of modularity in their DNA binding mode. However, the assembly of large and highly repetitive TALE protein coding genes can be challenging. We describe a ligation-independent cloning (LIC) based method to allow high-throughput assembly of TALE nuclease genes at high fidelity and low effort and cost.

Key words TALE nucleases, LIC assembly, Genome editing

1 Introduction

Deepened understanding of the DNA binding modes of natural meganucleases, zinc finger proteins, and TAL-effector proteins has enabled the engineering of novel DNA binding proteins with near-arbitrary sequence specificity [1–3]. TALE proteins derived from Xanthomonas bacteria provide a fully modular DNA binding architecture as these proteins consist of an array of 15–30 protein domains, each of which is 34-amino acids in length and determines binding to one DNA base through its repeat variable di-residue (RVD). Every repeat unit binds it target base in a largely context-independent manner, thus allowing the rearrangement of the individual repeat units to obtain a TALE protein with user-defined sequence specificity. Fusing synthetic DNA binding proteins to nuclease domains, e.g., derived from the type-IIS endonuclease *FokI*, enables cutting of eukaryotic genomes within cells [4–7]. This has successfully been employed to induce mutagenesis in mouse, rat, zebrafish, insect, and human genomes [8–14].

The challenge of exploiting the genome engineering potential of the TALE nuclease technology for large-scale functional genomic

Shondra M. Pruett-Miller (ed.), *Chromosomal Mutagenesis*, Methods in Molecular Biology, vol. 1239,
DOI 10.1007/978-1-4939-1862-1_8, © Springer Science+Business Media New York 2015

studies mainly lies within the capability of synthesizing the large, repetitive TALE repeat array in a streamlined fashion. To this end, several approaches have been proposed [15–19]. Recently, we have described a ligation-independent cloning (LIC) based assembly method for manual or semi-automated high-throughput assembly of TALE nuclease genes [20]. LIC based DNA assembly [21, 22] constitutes a high fidelity method to synthesize TALE repeat arrays with a minimal number of pipetting steps. One major advantage of the LIC approach is the fact that no PCR and no ligation steps are required in the course of the assembly process. Consequently, correctly sized assembly fragments do not contain any de novo mutations, which abolishes the need of sequence verification of the final assembly product. Furthermore, the method can easily be performed on lab automation systems and only requires standard reagents present in every molecular biology lab. In the following, we provide a detailed protocol of a two-step hierarchical cloning scheme. In a first assembly reaction three LIC ready 2-mer fragments (each coding for two RVD domains) are assembled to obtain a 6-mer fragment (six RVD domains). In a second assembly step three LIC ready 6-mer fragments give rise to an 18-mer RVD array in an expression-ready TALEN backbone construct (Fig. 1).

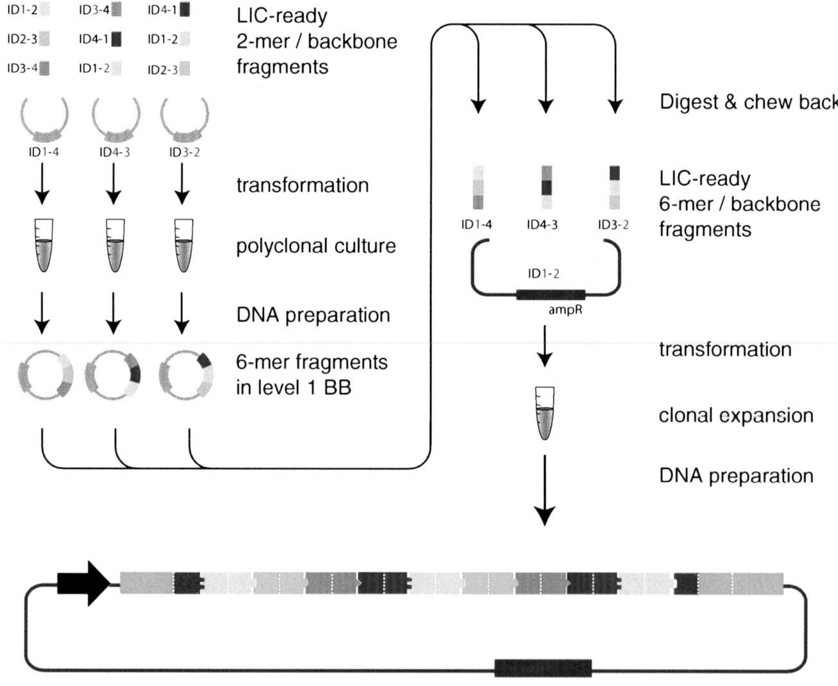

Fig. 1 Scheme for LIC based TALEN assembly. Starting from a library of 64 LIC-ready DNA fragments each encoding two TALE repeat units, three 6-mer intermediates are assembled using LIC. The resulting plasmids are amplified in bacteria, are purified, and made LIC-ready by digestion and incubation with T4 DNA polymerase. Three 6-mer fragments are then combined together with a backbone DNA fragment in order to give rise to an 18.5-RVD TALE nuclease plasmid ready for transfection into mammalian cells

2 Materials

2.1 Oligonucleotide Primers for Sequencing

Primer	Sequence (5′–3′)
6-mer seq	CGGCCTTTTTACGGTTCCTG
18.5-mer seq 1	GCTGCTGAAGATCGCCAAGAG
18.5-mer seq 2	CCAGGTGGTCGTTGGTCA

2.2 Enzymes

1. 10 U/μl Mva1269I.
2. PstI FastDigest.
3. KpnI FastDigest.
4. 3 U/μl T4 DNA polymerase.

2.3 Buffers and Solutions

1. 10× FastDigest buffer (Thermo Scientific Fermentas, Pittsburg, PA, USA).
2. 10× buffer R (Thermo Scientific Fermentas).
3. 10× NEB2 buffer.
4. 10 g/l BSA solution.
5. dATP and dTTP solution 100 mM.
6. LB medium.
7. LB-AGAR.
8. 100 mg/ml Ampicillin stock solution.
9. 30 mg/ml Kanamycin stock solution.
10. Digestion master mix A: 0.3 μl 10× Buffer R, 0.1 μl Mva1269I, 2.6 μl H_2O.
11. Digestion master mix B: 6 μl 10× buffer FastDigest, 2 μl KpnI FastDigest, 2 μl PstI FastDigest, 30 μl H_2O.
12. Digestion master mix C: 6 μl 10× buffer FastDigest, 2 μl KpnI FastDigest, 32 μl H_2O.
13. Chew-back mix A: 1 μl 10× NEB2 buffer, 0.1 μl (10 g/l) BSA, 0.1 μl 100 mM dATP (3 U/μl) 0.33 μl T4 DNA polymerase, 5.47 μl H_2O.
14. Chew-back mix T: 1 μl 10× NEB2 buffer, 0.1 μl (10 g/l) BSA, 0.1 μl 100 mM dTTP, 0.33 μl (3 U/μl) T4 DNA polymerase, 5.47 μl H_2O.

2.4 Plasmids

1. 2-mer fragment library (TALEN Kit #1000000023, Addgene, Cambridge, MA, USA).
2. Level 1 backbone plasmids (TALEN Kit #1000000023, Addgene).
3. Level 2 backbone plasmids (TALEN Kit #1000000023, Addgene).

2.5 Bacteria and Kits

1. Chemically competent DH10B bacterial cells (competence >10^7 CFU/μg) (Life Technologies, Carlsbad, CA, USA) (*see* **Note 1**).
2. Silica-based plasmid purification kit.
3. Agarose gel purification kit.

2.6 Gel Electrophoresis

1. UltraPure Agarose.
2. Agarose gel running chamber and power supply.
3. 25 mg/ml Ethidium Bromide stock solution.
4. 10× Agarose gel loading buffer (2.5 % Ficoll 400, 11 mM EDTA, 3.3 mM Tris–HCl, 0.017 % SDS, 0.015 % Bromophenol Blue, pH 8.0 @ 25 °C).
5. GeneRuler DNA Ladder Mix.
6. UV illumination and camera system.

2.7 Software and Devices

1. TALE design homepage http://www.imm-bonn.de/research-facilities/genome-engineering-platform.
2. PCR thermal cycler.
3. Spectrophotometer.
4. Heating block.

3 Methods

Carry out all procedures on ice unless otherwise specified.

3.1 Calculation of Assembly Scheme for Given TALEN Target Sequences

1. Launch a web browser and enter the URL http://www.imm-bonn.de/research-facilities/genome-engineering-platform/185-rvd-talens.html
2. Enter the target sites as they appear on one strand of the target genome comprising left binding site, spacer, and right binding site separated by spaces.
3. Click "START" and print out the resulting pipetting scheme.

3.2 Preparation of Master Mixes and LIC Ready 2-mer Fragments and Level 1 Backbones

1. Prepare digestion master mixes A, and B on ice (*see* Subheading 2).
2. Prepare chew-back mixes A and T on ice (*see* Subheading 2).
3. Add 3 μl digestion master mix A to 7 μl 2-mer plasmid (100 ng/μl, eluted in 1× Buffer R) and incubate at 37 °C for 6 min.
4. Add 40 μl digestion master mix B to 20 μl level 1 backbone (1 μg/μl, eluted in water) and incubate at 37 °C for 60 min.
5. Gel-purify digested level 1 backbones according to manufacturer's instructions (typical concentration after gel purification: 50 ng/μl).

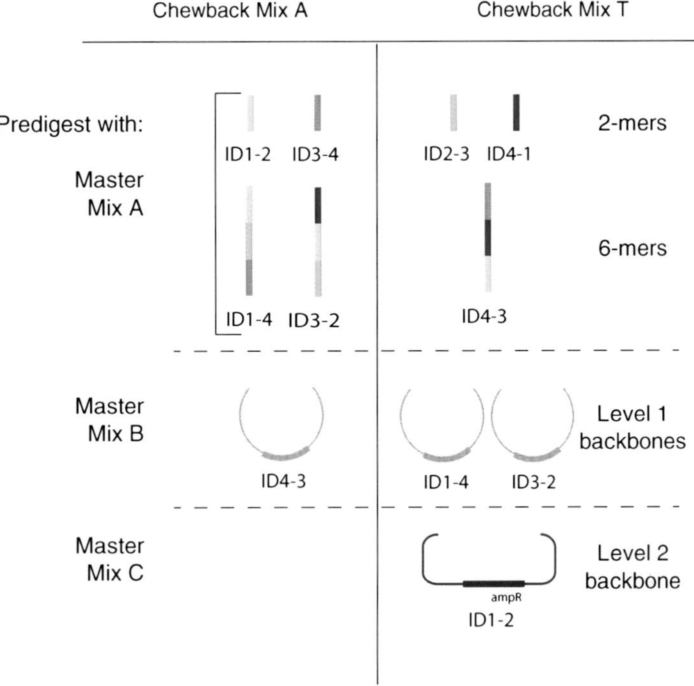

Fig. 2 Overview of all assembly components and the respective digestion mixes. Depicted are the single components for TALEN assembly and the digestion master mix and chewback mix that are used for the preparation of the 2-mer, 6-mer, and backbone fragments

6. Add 3 μl of the digested and purified DNA solution of above prepared 2-mers and level 1 backbones to 7 μl of chew back master mix containing the appropriate STOP dNTP (Fig. 2).
   - *2-mer ID 1-2* and *ID 3-4/Level 1 Backbone ID 4-3*: *Chew back mix A*.
   - *2-mer ID 2-3* and *ID 4-1/Level 1 Backbone ID 1-4* and *ID 3-2*: *Chew back mix T*.
7. Incubate at 27 °C for 5 min and inactivate at 75 °C for 20 min (*see* **Note 2**).
8. 2-mers and backbones are then diluted 20-fold in 1× NEB2 (*see* **Notes 3** and **4**).

3.3 6-mer Assembly Reaction

1. Prepare a mix of 2.5 μl of three appropriate 2-mer fragments and 2.5 μl of the appropriate chewed level 1 backbone to a total volume of 10 μl.
2. Incubate the mixture at 55 °C for 30 min followed by 25 °C for 3 h.

3.4 Transformation 6-mer Assembly

1. Thaw a 10 μl aliquot of chemically competent DH10B *E.* coli on ice for 5 min.
2. Add 2 μl of the assembly reaction to the bacteria, mix gently and incubate on ice for 5 min.
3. Heat shock at 42 °C for 45 s.
4. Incubate on ice for 2 min.
5. Add 100 μl LB medium and incubate at 37 °C for 1 h, shaking at 900 rpm (*see* **Note 5**).
6. Prepare LB medium supplemented with kanamycin (30 μg/ml) and add 1 ml to the transformation mix.
7. Incubate at 37 °C for 16 h at 900 rpm (*see* **Note 5**).

3.5 Plasmid Purification

A standard silica column based plasmid purification kit is recommended. Follow the manufacturer's instructions and finally elute the samples in 100 μl 1× Buffer R (For quality control measures refer to **Note 6**).

3.6 Preparation of Master Mix C and LIC Ready 6-mer Fragments and Level 2 Backbones

1. Prepare digestion master mixes A and C on ice (*see* Subheading 2).
2. Prepare chew-back mixes A and T on ice (*see* Subheading 2).
3. Add 3 μl digestion master mix A to 7 μl 6-mer plasmid (100 ng/μl, eluted in 1× Buffer R) and incubate at 37 °C for 60 min.
4. Add 40 μl digestion master mix C to 20 μl level 2 backbone (1 μg/μl, eluted in water) and incubate at 37 °C for 60 min.
5. Gel-purify digested level 2 backbones according to manufacturer's instructions (recommended concentration after gel purification 50 ng/μl). 6-mer digestions should not be purified either by a gel or a column.
6. Add 3 μl of the digested and purified DNA solution of above prepared 6-mers and level 2 backbone to 7 μl of chewback master mix containing the appropriate STOP dNTP (Fig. 2).
   - *6-mer ID 1-4 and ID 3-2 Chew back mix A.*
   - *6-mer ID 4-3/level 2 backbone ID1-2 Chew back mix T.*
7. Incubate at 27 °C for 5 min and inactivate at 75 °C for 20 min (*see* **Note 2**).
8. 6-mers and backbones are then diluted 20-fold in 1× NEB2 (*see* **Notes 3** and **4**).

3.7 18-mer Assembly Reaction

1. Prepare a mix of 2.5 μl of three appropriate 6-mer fragments and 2.5 μl of the appropriate chewed level 2 backbone to a total volume of 10 μl.
2. Incubate the mixture at 55 °C for 30 min followed by 25 °C for 3 h.

3.8 Transformation 6-mer Assembly

1. Thaw a 10 μl aliquot of chemically competent DH10B *E.* coli on ice for 5 min.
2. Add 2 μl of the assembly reaction to the bacteria, mix gently, and incubate on ice for 5 min.
3. Heat shock at 42 °C for 45 s.
4. Incubate in ice for 2 min.
5. Add 100 μl LB medium and incubate at 37 °C for 30 min, shaking at 900 rpm (*see* **Note 5**).
6. Streak out the entire transformation mix on LB agar plate supplemented with ampicillin (100 μg/ml).
7. Incubate the plate at 37 °C for 16 h.
8. Pick 1–2 single colonies an inoculate 1,000 μl LB medium supplemented with ampicillin (100 μg/ml) (*see* **Note** 7).
9. Let bacteria grow at 37 °C for 16 h.

3.9 Plasmid Purification

A standard silica column based plasmid purification kit is recommended. Follow the manufacturer's instructions and finally elute the samples in 100 μl H_2O (for quality control measures refer to **Note 6**).

4 Notes

1. We usually prepare 10 μl stocks of chemo-competent DH10B bacteria that we then use for the LIC assembly process. We recommend using the DH10B strain from Life Technologies, as in our hands other bacterial strains (e.g., Stable-3) do not perform properly in the LIC process. You will find a detailed description of preparing chemo-competent bacteria in the "Promega Subcloning Notebook" at www.promega.com.
2. Make sure that the reaction is not gradually heated up between incubation and inactivation. Otherwise, double stranded DNA will slowly melt and be digested nonspecifically by the T4 DNA polymerase. After incubation at 27 °C put the samples on ice and wait until the cycler heated up to 75 °C. Then put the samples back in the cycler.
3. These dilutions can be stored at 4 °C for further use. It is recommended to generate a pre-digested and pre-chewed fragment library for subsequent assembly reactions. Such a library should typically contain each 2-mer fragment in a quantity allowing to be used in 80 assembly reactions.
4. In case of polyclonal cultures remaining sterile or the number of colonies after transformation being low, it is possible to boost cloning efficiency by omitting the dilution step at the end of Subheadings 3.3 and 3.7 and by scaling up the transformation reactions fivefold.

5. Usually we use 2 ml tubes for liquid bacterial cultures and shake them at 900 rpm. Any other system to propagate bacteria should work just as well.
6. Correct 6-mer assemblies yield a 0.6 kb fragment after Mva1269I digestion (backbone 2.6 kB). Correct 18-mer assemblies can be test-digested by XhoI/XbaI resulting in a 3.2 kb 18-mer and a 5.3 kb backbone fragment. Optionally, a sequencing reaction can be performed to validate the assembly reaction. The sequencing primer "6-mer seq" will sequence 6-mer fragments from the 5′ end. The sequencing primer "18.5-mer seq 1" and "18.5-mer seq 2" will sequence the RVD array in final 18.5-mer constructs from the 5′ end and the 3′ end, respectively. Of note, LIC assembly does not require any PCR steps. Thus test digestions are sufficient to confirm correct assemblies.
7. The fidelity of the 18-mer assembly is typically around 80 %. We recommend picking 2 colonies to be sure that one colony bears the correct fragment size (combined fidelity around 96 %).

Acknowledgement

J.L.S.-B. is supported by a scholarship of the Studienstiftung des Deutschen Volkes. V.H. is member of the excellence cluster ImmunoSensation and supported by grants from the German Research Foundation (SFB704 and SFB670) and the European Research Council (ERC-2009-StG 243046).

References

1. Boch J, Scholze H, Schornack S, Landgraf A, Hahn S, Kay S, Lahaye T, Nickstadt A, Bonas U (2009) Breaking the code of DNA binding specificity of TAL-type III effectors. Science 326(5959):1509 1512. doi:10.1126/science.1178811
2. Moscou MJ, Bogdanove AJ (2009) A simple cipher governs DNA recognition by TAL effectors. Science 326(5959):1501. doi:10.1126/science.1178817
3. Boch J, Bonas U (2010) Xanthomonas AvrBs3 family-type III effectors: discovery and function. Annu Rev Phytopathol 48:419–436. doi:10.1146/annurev-phyto-080508-081936
4. Carroll D (2011) Genome engineering with zinc-finger nucleases. Genetics 188(4):773–782. doi:10.1534/genetics.111.131433
5. Li T, Huang S, Zhao X, Wright DA, Carpenter S, Spalding MH, Weeks DP, Yang B (2011) Modularly assembled designer TAL effector nucleases for targeted gene knockout and gene replacement in eukaryotes. Nucleic Acids Res 39(14):6315–6325. doi:10.1093/nar/gkr188
6. Miller JC, Tan S, Qiao G, Barlow KA, Wang J, Xia DF, Meng X, Paschon DE, Leung E, Hinkley SJ, Dulay GP, Hua KL, Ankoudinova I, Cost GJ, Urnov FD, Zhang HS, Holmes MC, Zhang L, Gregory PD, Rebar EJ (2011) A TALE nuclease architecture for efficient genome editing. Nat Biotechnol 29(2):143–148. doi:10.1038/nbt.1755
7. Christian M, Cermak T, Doyle EL, Schmidt C, Zhang F, Hummel A, Bogdanove AJ, Voytas DF (2010) Targeting DNA double-strand breaks with TAL effector nucleases. Genetics 186(2):757–761. doi:10.1534/genetics.110.120717

8. Watanabe T, Ochiai H, Sakuma T, Horch HW, Hamaguchi N, Nakamura T, Bando T, Ohuchi H, Yamamoto T, Noji S, Mito T (2012) Non-transgenic genome modifications in a hemimetabolous insect using zinc-finger and TAL effector nucleases. Nat Commun 3:1017. doi:10.1038/ncomms2020
9. Liu J, Li C, Yu Z, Huang P, Wu H, Wei C, Zhu N, Shen Y, Chen Y, Zhang B, Deng WM, Jiao R (2012) Efficient and specific modifications of the Drosophila genome by means of an easy TALEN strategy. J Genet Genomics 39(5):209–215. doi:10.1016/j.jgg.2012.04.003
10. Huang P, Xiao A, Zhou M, Zhu Z, Lin S, Zhang B (2011) Heritable gene targeting in zebrafish using customized TALENs. Nat Biotechnol 29(8):699–700. doi:10.1038/nbt.1939
11. Sander JD, Cade L, Khayter C, Reyon D, Peterson RT, Joung JK, Yeh JR (2011) Targeted gene disruption in somatic zebrafish cells using engineered TALENs. Nat Biotechnol 29(8):697–698. doi:10.1038/nbt.1934
12. Tesson L, Usal C, Menoret S, Leung E, Niles BJ, Remy S, Santiago Y, Vincent AI, Meng X, Zhang L, Gregory PD, Anegon I, Cost GJ (2011) Knockout rats generated by embryo microinjection of TALENs. Nat Biotechnol 29(8):695–696. doi:10.1038/nbt.1940
13. Hockemeyer D, Wang H, Kiani S, Lai CS, Gao Q, Cassady JP, Cost GJ, Zhang L, Santiago Y, Miller JC, Zeitler B, Cherone JM, Meng X, Hinkley SJ, Rebar EJ, Gregory PD, Urnov FD, Jaenisch R (2011) Genetic engineering of human pluripotent cells using TALE nucleases. Nat Biotechnol 29(8):731–734. doi:10.1038/nbt.1927
14. Wood AJ, Lo TW, Zeitler B, Pickle CS, Ralston EJ, Lee AH, Amora R, Miller JC, Leung E, Meng X, Zhang L, Rebar EJ, Gregory PD, Urnov FD, Meyer BJ (2011) Targeted genome editing across species using ZFNs and TALENs. Science 333(6040):307. doi:10.1126/science.1207773
15. Reyon D, Tsai SQ, Khayter C, Foden JA, Sander JD, Joung JK (2012) FLASH assembly of TALENs for high-throughput genome editing. Nat Biotechnol 30(5):460–465. doi:10.1038/nbt.2170
16. Weber E, Gruetzner R, Werner S, Engler C, Marillonnet S (2011) Assembly of designer TAL effectors by Golden Gate cloning. PLoS One 6(5):e19722. doi:10.1371/journal.pone.0019722
17. Cermak T, Doyle EL, Christian M, Wang L, Zhang Y, Schmidt C, Baller JA, Somia NV, Bogdanove AJ, Voytas DF (2011) Efficient design and assembly of custom TALEN and other TAL effector-based constructs for DNA targeting. Nucleic Acids Res 39(12):e82. doi:10.1093/nar/gkr218
18. Morbitzer R, Elsaesser J, Hausner J, Lahaye T (2011) Assembly of custom TALE-type DNA binding domains by modular cloning. Nucleic Acids Res 39(13):5790–5799. doi:10.1093/nar/gkr151
19. Kim HJ, Lee HJ, Kim H, Cho SW, Kim JS (2009) Targeted genome editing in human cells with zinc finger nucleases constructed via modular assembly. Genome Res 19(7):1279–1288. doi:10.1101/gr.089417.108
20. Schmid-Burgk JL, Schmidt T, Kaiser V, Honing K, Hornung V (2013) A ligation-independent cloning technique for high-throughput assembly of transcription activator-like effector genes. Nat Biotechnol 31(1):76–81. doi:10.1038/nbt.2460
21. Schmid-Burgk JL, Xie Z, Frank S, Virreira Winter S, Mitschka S, Kolanus W, Murray A, Benenson Y (2012) Rapid hierarchical assembly of medium-size DNA cassettes. Nucleic Acids Res 40(12):e92. doi:10.1093/nar/gks236
22. Aslanidis C, de Jong PJ (1990) Ligation-independent cloning of PCR products (LIC-PCR). Nucleic Acids Res 18(20):6069–6074

Chapter 9

Assembly and Characterization of megaTALs for Hyperspecific Genome Engineering Applications

Sandrine Boissel and Andrew M. Scharenberg

Abstract

Rare-cleaving nucleases have emerged as valuable tools for creating targeted genomic modification for both therapeutic and research applications. MegaTALs are novel monomeric nucleases composed of a site-specific meganuclease cleavage head with additional affinity and specificity provided by a TAL effector DNA binding domain. This fusion product facilitates the transformation of meganucleases into hyperspecific and highly active genome engineering tools that are amenable to multiplexing and compatible with multiple cellular delivery methods. In this chapter, we describe the process of assembling a megaTAL from a meganuclease, as well as a method for characterization of nuclease cleavage activity in vivo using a fluorescence reporter assay.

Key words megaTAL, Meganuclease, Homing endonuclease, TAL effector, Genome engineering, Gene therapy, Traffic Light Reporter assay

1 Introduction

Genome engineering provides a means for achieving a wide range of therapeutic and technological goals, including correcting genomic disorders, eradicating infectious diseases, and repurposing microorganisms for sustainable resource production. While traditional methods of genetic manipulation using semi-random viral integration are functionally limited to applications requiring only gene supplementation and raise safety concerns due to their capacity for promoting cancer formation through proto-oncogene activation [1–4], site-specific nucleases have emerged as alternative reagents for safely producing a variety of targeted genomic alterations. The most common application of site specific nuclease technology involves generating an enzyme to cleave an endogenous gene of interest in order to correct or alter the genetic information in its native environment. This helps ensure normal regulation of the targeted gene as well as those neighboring it, avoids potential

Shondra M. Pruett-Miller (ed.), *Chromosomal Mutagenesis*, Methods in Molecular Biology, vol. 1239, DOI 10.1007/978-1-4939-1862-1_9,

issues arising from copy number variation and allows for the treatment of disorders arising from dominant negative mutations. Additionally, nucleases form double-strand breaks at their DNA target sites, which can be repaired by the homologous recombination or nonhomologous end joining DNA repair pathways to create either targeted correction or knockout events, respectively, in both coding and noncoding DNA [5–10].

Although site-specific nucleases present a number of advantages over previous genetic manipulation strategies, a significant challenge faced by the field today is the ability to develop a reagent that can be efficiently delivered to cells, while balancing the need for generating highly active reagents in a cost- and time-efficient manner with that of developing highly specific nucleases that yield minimal off-target cleavage. To this end, a number of natural and hybrid nuclease platforms have been characterized, each exhibiting distinct properties that provides both advantages and disadvantages towards these competing goals. For applications such as gene therapy, which involve manipulations of primary cells and where great importance is placed on target specificity, nuclease architectures that promote reduced off-target cleavage while simultaneously facilitating delivery to many cell types are desired.

There are currently four commonly used site-specific nuclease platforms: meganucleases (also known as homing endonucleases, HEs), zinc-finger nucleases (ZFNs), transcription activator-like effector nucleases (TALENs) and clustered, regularly interspaced, short palindromic repeat (CRISPR) RNA-guided nucleases. Meganucleases are small, single-chain proteins with coupled binding and cleavage activity, making them easy to deliver to many cells types and amenable to simultaneous targeting of more than one DNA site using concurrent delivery of different meganuclease variants. They also exhibit low potential for sequence-independent cleavage as evidenced by their low toxicity in vivo [11–13]. However, their compact structures and integrated activities make meganucleases notoriously difficult to engineer toward new target specificities. The design of meganucleases towards novel targets commonly results with significantly reduced DNA binding affinity [14–16]. ZFNs and TALENs, on the other hand, possess separate DNA binding and cleavage domains, reducing the effort required to reengineer their target specificity, particularly for the latter [17–20]. However, the uncoupled binding and cleavage activities of these artificial enzymes leads to a higher potential for off-target cleavage. In addition, because the Fok-I nuclease is only active as a dimer, two separate nuclease halves are required to target a single DNA sequence. The resulting large construct sizes for both ZFNs and TALENs make cellular delivery difficult and severely limits the ability to target multiple loci in tandem [21–25]. Lastly, CRISPR nucleases achieve site specificity via RNA guide molecules which are complementary to the target DNA sequence, eliminating any need for any protein engineering and thus making this platform the simplest to engineer towards any given locus [26, 27]. The high off-target cleavage rates observed with CRISPR nucleases, however, offsets the utility of these enzymes for applications requiring high levels of specificity [28, 29].

To address the limitations of these currently available nuclease platforms, we have developed a novel hybrid fusion which combines the advantages of meganuclease size and specificity with the target-specific affinity and ease of engineering afforded by TAL effector DNA-binding domains [30]. The resulting megaTAL nuclease requires reduced engineering efforts, yet allows the generation of single-chain, hyperspecific enzymes compatible with both viral vectorization and multiplexing [31, 32]. The combination of these desirable properties in a single nuclease platform holds significant promise to extend the impact of genome engineering for any application requiring hyperspecificity.

Here we describe a step-by-step procedure for the assembly of a megaTAL from a native or engineered meganuclease. We also detail a protocol for characterizing nuclease activity and comparing relative levels of both mutagenic nonhomologous end-joining (mutNHEJ) and homologous recombination (HR) using the Traffic Light Reporter (TLR) assay [33, 34].

2 Materials

2.1 megaTAL Cloning

2.1.1 Plasmid Map Generation and Cloning Design

1. TALEN and megaTAL plasmid map generator "TALEN_map_generator_prompt" available from Addgene at http://www.addgene.org/50627/files/.
2. Python version 2.7—can be downloaded and installed from www.python.org/download.

2.1.2 TAL Effector Golden Gate Cloning: Step 1

1. Golden Gate TALEN and TAL effector Kit 2.0 (Addgene, Cambridge, MA, USA) DNA preps of all of the necessary plasmids diluted to a final concentration of 150 ng/μl.
2. Kit add-on plasmids pFUSC2-10 (Addgene) DNA prep diluted to a final concentration of 75 ng/μl.
3. PCR thermal cycler.
4. 0.2 ml PCR tubes.
5. T4 DNA ligase.
6. BsaI restriction enzyme.
7. Plasmid-Safe DNase.
8. ATP.
9. Chemically competent bacteria.

10. Selection plates: Luria–Bertani agar plates with 50 μg/ml spectinomycin.
11. Blue-white Select Screening Reagent (Sigma, St. Louis, MO, USA).
12. Luria–Bertani medium with 50 μg/ml spectinomycin.
13. Miniprep kit.
14. PCR/sequencing primers pCR8_F1 (TTG ATG CCT GGC AGT TCC CT) and pCR8_R1 (CGA ACC GAA CAG GCT TAT GT).

2.1.3 TAL Effector Golden Gate Cloning: Step 2

1. Golden Gate TALEN and TAL effector Kit 2.0 (Addgene) DNA preps of all of the necessary plasmids diluted to a final concentration of 150 ng/μl.
2. Kit add-on plasmid pCVL TAL(Nd154C63) SFFV Xba-Sal linker (Addgene) DNA prep diluted to a final concentration of 75 ng/μl.
3. PCR thermal cycler.
4. 0.2 ml PCR tubes.
5. T4 DNA ligase.
6. BsmBI restriction enzyme.
7. Chemically competent bacteria.
8. Selection plates: Luria–Bertani agar plates with 100 μg/ml carbenicillin.
9. Blue-white Select Screening Reagent (Sigma).
10. Luria–Bertani medium with 100 μg/ml carbenicillin.
11. Miniprep kit.
12. PCR/sequencing primers TAL_F1 (TTG GCG TCG GCA AAC AGT GG) and TAL_R2 (GGC GAC GAG GTG GTC GTT GG).

2.1.4 Cloning the Meganuclease into the megaTAL Vector

1. PCR primers for amplifying the meganuclease ORF, as described in **step 1**, Subheading 3.2.4.
2. High-fidelity polymerase.
3. PCR thermal cycler.
4. 0.2 ml PCR tubes.
5. PCR cleanup kit.
6. XbaI and SalI restriction enzymes (or appropriate Type IIS restriction enzyme—*see* **Note 9**).
7. Calf-intestinal or shrimp alkaline phosphatase.
8. T4 DNA ligase.
9. Chemically competent cells.

10. Selection plates: Luria–Bertani agar plates with 100 μg/ml carbenicillin.
11. Luria–Bertani medium with 100 μg/ml carbenicillin.
12. Miniprep kit.
13. PCR/sequencing primers TALC_FP (GCA GTG AAA AAG GGA TTG C) and WPRE_RP (TGC CCC ACC ATT TTG TTC).

2.1.5 Cloning the Meganuclease into the Control Vector

1. Kit add-on plasmid pCVL SFFV Sbf-Sal linker (Addgene) DNA prep diluted to a final concentration of 75 ng/μl.
2. PCR primers for amplifying the meganuclease ORF, as described in **step 1**, Subheading 3.2.5.
3. High-fidelity polymerase.
4. PCR thermal cycler.
5. 0.2 ml PCR tubes.
6. PCR cleanup kit.
7. SbfI and SalI restriction enzymes (or appropriate Type IIS restriction enzyme—*see* **Note 12**).
8. Calf-intestinal or shrimp alkaline phosphatase.
9. T4 DNA ligase.
10. Chemically competent bacteria.
11. Selection plates: Luria–Bertani agar plates with 100 μg/ml carbenicillin.
12. Luria–Bertani medium with 100 μg/ml carbenicillin.
13. Miniprep kit.
14. PCR/sequencing primers BFP_FP (CAG ATA CTG CGA CCT CCC) and WPRE_RP (TGC CCC ACC ATT TTG TTC).

2.2 Traffic Light Reporter Cell Line Derivation

2.2.1 Traffic Light Reporter Plasmid Cloning

1. Traffic Light Reporter plasmid pCVL TLR 3.1 (Sce target) iRFP puro (Addgene) DNA prep.
2. TLR top and bottom insert, as described in **step 1**, Subheading 3.3.1.
3. T4 Polynucleotide Kinase (PNK).
4. PCR thermal cycler.
5. 0.2 ml PCR tubes.
6. SbfI and SpeI restriction enzymes.
7. Calf-intestinal or shrimp alkaline phosphatase.
8. PCR cleanup kit.
9. T4 DNA ligase.
10. Chemically competent bacteria.

11. Selection plates: Luria–Bertani agar plates with 100 μg/ml carbenicillin.
12. Luria–Bertani medium with 100 μg/ml carbenicillin.
13. Miniprep kit.
14. PCR/sequencing primer eGFP_RP (AGA TGG TGC GCT CCT GGA C).

2.2.2 Traffic Light Reporter Lentiviral Production

1. Tissue culture facilities.
2. HEK 293T cells.
3. Culture media: Dulbecco's Modified Eagle's Medium with 4.5 g/l glucose and sodium pyruvate, without L-glutamine, supplemented with 10 % fetal bovine serum, 1 % L-glutamine, 1 % pen-strep, and 1 % 1 mM HEPES.
4. Phosphate buffered saline (PBS).
5. Trypsin.
6. Hemacytometer or other cell counting device.
7. 6-well tissue culture plate.
8. psPAX2 (Addgene) DNA prep.
9. pMD2.G (Addgene) DNA prep.
10. Diluent: 10 mM HEPES, 150 mM NaCl, pH 7.05 (sterile filter, store at room temp and protect from light).
11. Polyethyleneimine (PEI) solution 1 mg/ml (sterile).
12. Transfection media: Dulbecco's Modified Eagle's Medium without phenol red, supplemented with 4 % fetal bovine serum and 1 % L-glutamine.

2.2.3 Traffic Light Reporter Cell Line Derivation

1. Tissue culture facilities.
2. HEK 293T cells.
3. Culture media: Dulbecco's Modified Eagle's Medium with 4.5 g/l glucose and sodium pyruvate, without L-glutamine, supplemented with 10 % fetal bovine serum, 1 % L-glutamine, 1 % pen-strep, and 1 % 1 mM HEPES.
4. Phosphate buffered saline (PBS).
5. Trypsin.
6. Hemacytometer or other cell counting device.
7. 24-well tissue culture plate.
8. Polybrene.
9. 6-well tissue culture plates.
10. Flow cytometer, such as the BD LSR, with proper filters (*see* **Note 20**).
11. Tubes for flow cytometer.

12. Puromycin.
13. 10 cm tissue culture dishes.
14. Fluorescence-activated cell sorter (Optional—*see* **Note 24**).

2.3 Testing megaTALs Using the Traffic Light Reporter Assay

1. Tissue culture facilities.
2. Culture media: Dulbecco's Modified Eagle's Medium with 4.5 g/l glucose and sodium pyruvate, without L-glutamine, supplemented with 10 % fetal bovine serum, 1 % L-glutamine, 1 % pen-strep, and 1 % 1 mM HEPES.
3. Phosphate buffered saline (PBS).
4. Trypsin.
5. Hemacytometer or other cell counting device.
6. 24-well tissue culture plate.
7. Donor plasmid pCVL SFFV d14GFP Donor (Addgene) DNA prep.
8. pCVL SFFV Sbf-Sal linker plasmid (Addgene) DNA prep.
9. X-tremeGENE 9 (Roche, Basel, Switzerland).
10. Serum-free media: Dulbecco's Modified Eagle's Medium with 4.5 g/l glucose and sodium pyruvate.
11. Sterile tubes.
12. Flow cytometer, such as the BD LSR, with proper filters (*see* **Note 21**).
13. Tubes for flow cytometer.
14. FlowJo software (Tree Star, Inc., Ashland, OR, USA).

3 Methods

This protocol describes methods for the assembly of a megaTAL (mT) from a native or engineered meganuclease (mn) variant and subsequent characterization of enzyme activity using an in vivo reporter cell assay. For details on how to engineer meganucleases with specificity towards a custom DNA target site, *see* Chapter 6 by Stoddard and colleagues (*see* **Note 1**).

3.1 TAL Effector Target Site Selection

Identify the genomic sequence upstream of your meganuclease target site. Locate a thymine base approximately 14–19 bp upstream of the start of your meganuclease target (assuming a mn target of 20 bp in length, *see* **Note 2**). The base immediately following this thymine will be the first nucleotide of the TAL effector target and the full target will extend to the nucleotide positioned 8 bp upstream of the meganuclease target site (Fig. 1). For a more detailed description of factors that affect target site selection, please *see* **Note 3**.

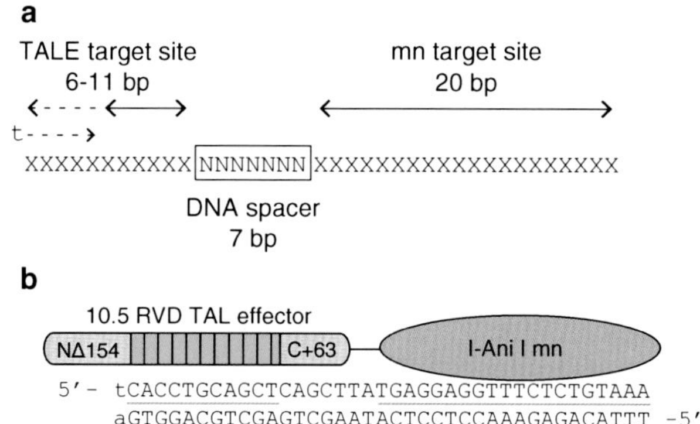

Fig. 1 Selection of a TAL effector binding site upstream of a meganuclease target site for megaTAL design. (**a**) The meganuclease target site should be assumed to be 20 bp in length, centering on the central four bases and site of DNA DSB formation. The TAL effector target should be 6–11 bp in length, positioned immediately following a thymine base, and should be followed by a 7 bp DNA spacer just 5 of the meganuclease target. (**b**) Schematic representation of a megaTAL and example DNA target

3.2 megaTAL Cloning

The following protocol was adapted from instructions provided with the Golden Gate TALEN and TAL effector Kit 2.0 [35].

3.2.1 Plasmid Map Generation and Cloning Design

We have written a program provided in the "TALEN_map_generator_prompt" folder that prompts the user for information in order to generate the plasmid maps for each megaTAL and meganuclease cloning step as a genbank file and outputs the design steps needed to clone each plasmid (*see* **Notes 4** and **5**).

1. Download and unzip the "TALEN_map_generator_prompt" folder from on the Addgene website at http://www.addgene.org/50627/files/ and move the folder to the desired location on your computer.
2. Open a command-line interface on your computer:

   Windows: open Command Prompt

   Program Files → Accessories → Command Prompt

   Mac: open Terminal

   Applications → Utilities → Terminal
3. In the command line of the command-line interface window, navigate to the location of the downloaded folder "TALEN_map_generator_prompt". For example, if the folder is located on the Desktop, type:

   Windows:

   ```
   cd Desktop\TALEN_map_generator_prompt
   ```

Mac:

```
cd Desktop/TALEN_map_generator_prompt
```

and hit enter.

4. In the command line type:

```
python make_plasmids.py
```

and hit enter.

5. The program will guide you through a series of questions regarding the desired TAL effector target site and name, the date, preferred repeat variable diresidue (RVD) usage (*see* **Note 6**) and meganuclease name and DNA sequence. Plasmids maps for the megaTAL, meganuclease, and the intermediate cloning step into a pFUSC vector will be output into the program folder. Plasmids needed for each cloning step (Subheadings 3.2.2–3.2.5) will be displayed in the command window and should be noted.

3.2.2 TAL Effector Golden Gate Cloning: Step 1

The first step of the Golden gate protocol clones all but the last RVD into the pFUSC vector (*see* **Note 7**).

1. Using the plasmids identified from Subheading 3.2.1, mix the following reaction in a 0.2 ml PCR tube:

10× T4 DNA ligase buffer	1 μl
BsaI	0.5 μl
T4 DNA ligase	0.5 μl
RVD plasmids (150 ng/μl stock)	0.5 μl each
pFUSC plasmid (75 ng/μl stock)	0.5 μl
Water	to 10 μl

2. Run the reaction with the following thermal cycler program:

   10 cycles: 37 °C/5 min, 16 °C/10 min.

   37 °C/10 min.

   65 °C/20 min.

3. Treat the reaction with the following mixture for 1 h at 37 °C:

Plasmid-Safe DNase	1 μl
ATP (25 mM)	0.4 μl

4. Thaw the blue-white screening reagent and spread 40 μl onto a dried spectinomycin plate.

5. Transform 5 μl of the reaction (no purification necessary) into competent bacterial cells and plate onto the prepared spectinomycin plates.
6. The following day, pick several white colonies (*see* **Note 8**) into LB + spectinomycin cultures for plasmid minipreps.
7. Isolate the plasmid DNA from each culture using a miniprep kit.
8. Sequence the plasmid DNA preps using primers pCR8_F1 and pCR8_R1 (*see* **Note 9**) to identify a clone containing the desired insert. A correct insert will match the sequence between "cut 1" and "cut 2" on the generated plasmid map.

3.2.3 TAL Effector Golden Gate Cloning: Step 2

The second step of the Golden Gate protocol clones the RVD sequence from the pFUSC vector and the last RVD into the megaTAL plasmid.

1. Using the plasmids identified from Subheading 3.2.1 for megaTAL cloning **step 2**, mix the following reaction in a 0.2 ml PCR tube:

10× T4 DNA ligase buffer	1 μl
BsmBI	0.5 μl
T4 DNA ligase	0.5 μl
pFUSC with RVDs from **step 8** of Subheading 3.2.2	75 ng
pLR plasmid (150 ng/μl stock)	0.5 μl
pCVL TAL(Nd154C63) SFFV Xba-Sal linker (75 ng/μl stock)	0.5 μl
Water	to 10 μl

2. Run the reaction with the following thermal cycler program:

   10 cycles: 37 °C/5 min, 16 °C/10 min.

   37 °C/10 min.

   65 °C/20 min.
3. Thaw the blue-white screening reagent and spread 40 μl onto a dried carbenicillin plate.
4. Transform 5 μl of the reaction (no purification necessary) into competent bacterial cells and plate onto the prepared carbenicillin plates.
5. The following day, pick several white colonies (*see* **Note 8**) into LB + carbenicillin cultures for plasmid miniprep.
6. Isolate the plasmid DNA from each culture using a miniprep kit.
7. Sequence the plasmid preps using primers TAL_F1 and TAL_R2 (*see* **Note 9**) to identify a clone containing the desired insert. A correct insert will match the sequence between "cut A" and "cut B" on the map generated for cloning **step 2/3** from Subheading 3.2.1 (*see* **Note 10**).

3.2.4 Cloning the Meganuclease into the megaTAL Vector

The last step of the cloning procedure clones the meganuclease ORF downstream of the TAL effector in the megaTAL vector (*see* **Note 11**).

1. Design and order primers to amplify the meganuclease ORF:

   FP with *XbaI* site: XXX*TCTAGA*GTGGGAGGAAGC$(N)_{10+}$

   RP with *SalI* site: XXX*GTCGAC*TCA$(N)_{10+}$

   where X represents any nucleotide and N represents the meganuclease sequence, excluding start and stop codons and making sure that no XbaI or SalI sites are present within the amplicon (*see* **Note 12**).
2. Set up a 50 μl PCR reaction to amplify the meganuclease ORF with the forward and reverse primers designed above, according to the manufacturer specifications for the chosen polymerase.
3. Clean up the PCR reaction and elute the purified DNA in 40 μl water.
4. Digest the purified PCR amplicon for 1 h at 37 °C (or appropriate temperature given manufacturer's specifications):

Purified PCR DNA	32 μl
XbaI	2 μl
SalI	2 μl
10× restriction enzyme buffer	4 μl

5. During this time, digest the pCVL TAL(Nd154C63) SFFV Xba-Sal linker vector with inserted RVDs (from **step 7**, Subheading 3.2.3) for 1 h at 37 °C (or appropriate temperature given manufacturer's specifications):

Plasmid DNA	~1 μg
XbaI	2 μl
SalI	2 μl
Phosphatase	1 μl
10× restriction enzyme buffer	4 μl
water	to 40 μl

6. Clean up both insert and vector digest reactions using a PCR cleanup kit.
7. Combine 20 ng of digested PCR product with 50 ng of digested plasmid in a 10 μl ligation reaction (*see* **Note 13**).
8. Transform 5 μl of the ligation reaction into competent cells and plate onto carbenicillin plates.

9. The following day, pick several colonies into LB + carbenicillin cultures for plasmid minipreps.
10. Isolate plasmid DNA using a miniprep kit.
11. Sequence the plasmid preps using primers TALC_FP and WPRE_RP to identify a clone containing the desired insert. A correct insert will match the sequence between the XbaI and SalI sites on the map generated for cloning **step 2/3** from Subheading 3.2.1 (*see* **Note 14**).

3.2.5 Cloning the Stand-Alone Meganuclease into the Control Vector

This step can be used to clone the stand-alone meganuclease into the same backbone vector as the megaTAL in order to provide direct comparison between the meganuclease and megaTAL activity in the reporter assay.

1. Design and order primers to amplify the meganuclease ORF:

   FP1 with NLS: CCACCTAAGAAGAAACGCAAAGTC$(N)_{10+}$

   FP2 with SbfI-HA-partial NLS: XXX*CCTGCA GG*TATCCAT ATGATGTCCCAGATTATGCGCCACCTAAGA AGAAACG

   RP with *SalI* site: XXX*GTCGAC*TCA$(N)_{10+}$

   where X represents any nucleotide and N represents the meganuclease ORF sequence, excluding start and stop codons and making sure that no SbfI or SalI sites are present within the amplicon (*see* **Note 15**).
2. Set up a 50 μl PCR reaction to amplify the meganuclease ORF with the FP1 and RP primers designed above, according to the manufacturer specifications for the polymerase.
3. Set up a second 50 μl PCR reaction to amplify 1 μl of the product from the previous step with the FP2 and RP above, according to the manufacturer specifications for the polymerase.
4. Clean up the PCR reaction and elute the DNA product in 40 μl water.
5. Digest the PCR amplicon for 1 h at 37 °C (or appropriate temperature given manufacturer's specifications):

purified PCR DNA	32 μl
SbfI	2 μl
SalI	2 μl
10× restriction enzyme buffer	4 μl

6. During this time, digest the pCVL SFFV Sbf-Sal linker vector for 1 h at 37 °C (or appropriate temperature given manufacturer's specifications):

Plasmid DNA	~1 μg
SbfI	2 μl
SalI	2 μl
Phosphatase	1 μl
10× restriction enzyme buffer	4 μl
Water	to 40 μl

7. Clean up both vector and insert digest reactions using a PCR cleanup kit.
8. Combine 20 ng of digested PCR product with 50 ng of digested plasmid in a 10 μl ligation reaction (*see* **Note 12**).
9. Transform 5 μl of the ligation reaction into competent bacterial cells and plate onto carbenicillin plates.
10. The following day, pick several colonies into LB + carbenicillin cultures for plasmid minipreps.
11. Isolate plasmid DNA using a miniprep kit.
12. Sequence the plasmid DNA preps using primers BFP_FP and WPRE_RP to identify a clone containing the desired insert. A correct insert will match the sequence between the SbfI and SalI sites on the map generated for the control meganuclease cloning from Subheading 3.2.1 (*see* **Note 14**).

3.3 Traffic Light Reporter Cell Line Derivation

This protocol has been adapted from methods published by Certo et al., using a modified Traffic Light Reporter plasmid, for the generation of cell lines with an integrated copy of the fluorescent reporter embedded into the genomic DNA [33].

3.3.1 Traffic Light Reporter Plasmid Cloning

1. Order DNA oligos containing the megaTAL binding site to be tested with SbfI/SpeI overhangs as indicated:

SbfI-target-SpeI top	GG$(N)_{3X+1}$A
SpeI-target-SbfI bottom	CTAGT$((N)_{3X+1})_{RC}$CCTGCA

Where N indicates the desired sequence to be tested, including the TAL effector binding site and preceding thymine (*see* **Note 16**), spacer DNA and meganuclease binding site and $(N)_{RC}$ indicates its reverse complement (*see* **Note 17**). The length of the TLR insert must be $3X+1$ nucleotides long to maintain the correct coding frame of the reporter and should include the thymine base immediately upstream of the TAL effector binding site Ensure that there are no stop codons in the +1 frame of the gene (+2 frame of the top oligo) upstream of base 12 of the meganuclease target or in the +3 frame of the gene

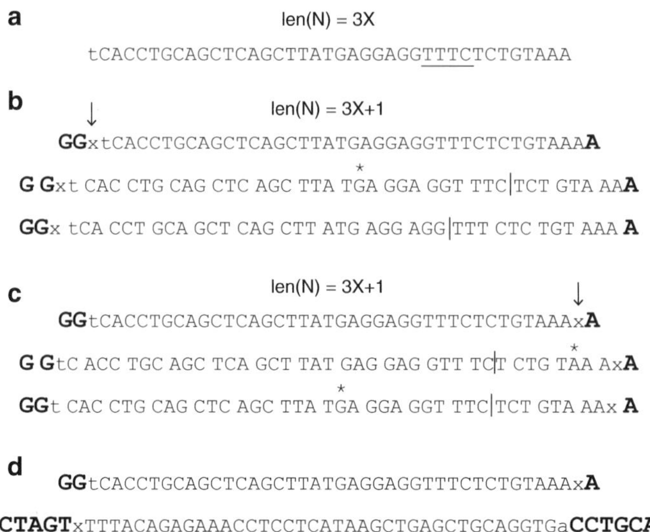

Fig. 2 Traffic Light Reporter target insert design. (**a**) An example megaTAL binding site (shown in Fig. 1b), including the thymine base upstream of the TAL effector binding site. The central four bases of the meganuclease at which DSB formation occurs are underlined. The length of the full target site is 39 bp long, or 3*X* long and so an additional nucleotide must be added to the insert to achieve a length of 3*X*+1 and maintain the correct reading frame of the TLR. This may be done by including an extra base (*X*, indicated by *arrow*) upstream (**b**) or downstream (**c**) of the megaTAL sequence. (**b**) Inserting a single nucleotide upstream of the megaTAL target (*top*) will place a stop codon (indicated by *) in the +1 reading frame of the TLR (+2 reading frame of the insert, *middle*) and thus will affect readout of mutNHEJ events. No stop codons are present in the +3 reading frame of the TLR (+1 reading frame of the insert, *bottom*). (**c**) Inserting a single nucleotide downstream of the megaTAL target (*top*) will place stop codons in both the +1 (*middle*) and +3 (*bottom*) reading frames of the TLR. However, because the stop codon in the +1 reading frame occurs downstream of base 12 of the meganuclease target (or downstream of the central four bases, as indicated by the *vertical line*) it will not affect readout of the mCherry gene after a +3→+1 frameshift mutation occurs. Similarly, the stop codon in the +3 reading frame occurs upstream of base 8 of the meganuclease target (upstream of the central four bases, as indicated by the *vertical line*) will not affect readout of mutNHEJ. (**d**) *Top* and *bottom* oligos for the TLR insert from the sequence identified in (**c**)

(+1 frame of the top oligo) downstream of base 8 of the meganuclease target (*see* **Note 18**, Fig. 2). Extra nucleotides can be added upstream and downstream of the megaTAL site to move stop codons out of these frames, as long as the final insert length totals 3*X*+1 nucleotides.

2. Set up the following reaction in a 0.2 ml PCR tube:

10× restriction enzyme buffer	1 μl
Top oligo (10 μM)	0.5 μl
Bottom oligo (10 μM)	0.5 μl
T4 PNK	0.5 μl
Water	7.5 μl

3. Anneal the oligos by running the reaction on a thermal cycler with the following program:

   37 °C/30 min.

   95 °C/3 min.

   14 cycles: 94 °C/5 s, decrease by 1 °C/cycle.

   80 °C/10 min.

4. During this time, digest the pCVL TLR 3.1 (Sce target) iRFP puro plasmid for 1 h at 37 °C (or appropriate temperature given manufacturer's specifications):

pCVL TLR 3.1 (Sce target) iRFP puro	~1 µg
SpeI	2 µl
SbfI	2 µl
Phosphatase	1 µl
10× restriction enzyme buffer	4 µl
Water	to 40 µl

5. Clean up the digested vector using a PCR cleanup kit.
6. Dilute the annealed oligo mix 1:100 in water and combine 1 µl of the diluted mix with 50 ng digest TLR plasmid in a 10 µl ligation reaction (*see* **Note 13**).
7. Transform 5 µl of the ligation reaction in competent bacterial cells and plate onto carbenicillin plates.
8. The following day, pick several colonies into LB + carbenicillin cultures for plasmid minipreps.
9. Isolate plasmid DNA with a miniprep kit.
10. Sequence the plasmid preps using primer eGFP_RP to identify a clone containing the correct insert matching the annealed oligo sequence.

3.3.2 Traffic Light Reporter Lentiviral Production

All steps must be done using sterile reagents and technique in a tissue culture hood. This protocol details the steps for producing lentivirus for a single TLR plasmid. For multiple plasmids (if different targets are to be tested), plate one well (during **step 1** below) for each plasmid generated from Subheading 3.3.1 and adjust the remaining steps accordingly.

Day 0

1. Plate 8×10^5 293T cells in 2 ml total volume of culture media into a single well of a 6-well culture plate.

Day 1

2. Combine 1.44 µg TLR vector with insert from Subheading 3.3.1 with 0.72 µg psPAX2 and 0.36 µg pMD2.G in 115 µl diluent.

3. Add 10 μl of PEI to the DNA mix and incubate for 10–15 min at room temperature.
4. Add the PEI/DNA mix dropwise to the plated 293T cells and incubate overnight.

Day 2

5. Warm PBS and transfection media in a 37 °C water bath.
6. Wash cells with 1 ml PBS, then add 2 ml of transfection media and incubate overnight (*see* **Note 19**).

Day 3

7. Draw supernatant off of the cells and discard the culture plate. Viral supernatant can be used immediately in Subheading 3.3.3, stored for 2 weeks at 4 °C, or stored indefinitely at −80 °C (*see* **Note 20**).

3.3.3 Traffic Light Reporter Cell Line Derivation

All steps must be performed using sterile reagents and technique in a tissue culture hood. This protocol details the steps for generating a cell line from a single TLR lentiviral prep. For multiple preps (if different targets are to be tested), plate 6 wells (during **step 1** below) for each prep generated from Subheading 3.3.2 and adjust the remaining steps accordingly.

Day 1

1. Plate 2×10^5 293T cells/well in 0.5 ml of culture media containing 4 μg/ml polybrene into seven wells of a 24-well culture plate (six wells will be treated with lentivirus, one will serve as a control) (*see* **Note 21**).
2. Add 0.5, 1, 2, 4, 8, and 16 μl viral supernatant from Subheading 3.3.2 to one well each and incubate the cells overnight.

Day 2

3. Passage cells to 6-well plates in 2 ml of culture media and incubate for ~48 h.

Day 4

4. Aspirate media and resuspend cells in 1 ml of fresh culture media (*see* **Note 22**). Transfer 100 μl of the resuspended cells to FACS tubes.
5. Read samples from the FACS tubes on a flow cytometer and analyze data to determine the percent of live cells expressing iRFP for each sample (Fig. 3), using the control treatment as an iRFP-control. Record FSC-A, FSC-H, SSC-A, and iRFP (*see* **Note 23**).
6. Identify the sample where iRFP expression is closest to, but below 10 % of the total cell population (*see* **Note 24**) and add puromycin to a final concentration of 1 μg/ml to the corresponding well to select for cells with an integrated copy of the TLR.

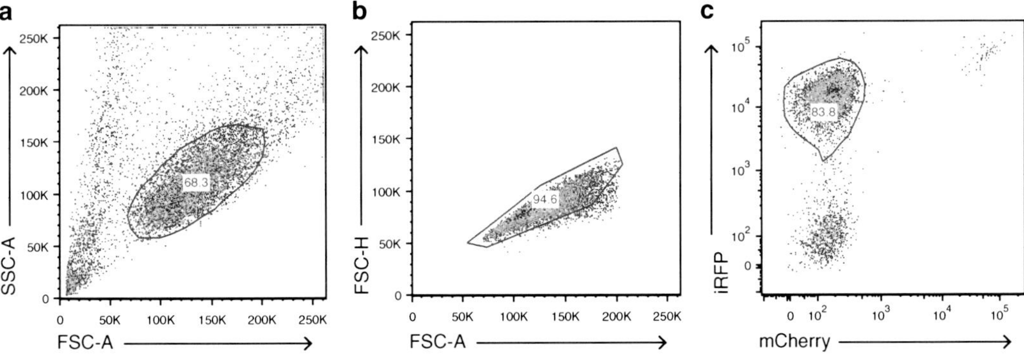

Fig. 3 Traffic Light Reporter cell line sorting for an iRFP+/mCherry– population. (**a**) Gate on live cells using the FSC-A and SSC-A parameters, then (**b**) singlets using FSC-A and FSC-H. (**c**) Finally, sort a minimum of 500,000 cells from the subpopulation of cells expressing iRFP, but not mCherry

Days 5+

7. Continue to grow and passage selected cell lines as needed, adding fresh puromycin one or more additional times (*see* **Note 25**).
8. *Optional*: Once the cell line has been expanded to confluency in a 10 cm dish, sort for iRFP+/mCherry– cells on a FACS. Expand, freeze, and passage sorted cells as needed (*see* **Note 26**, Fig. 3).

3.4 Testing megaTALs Using the Traffic Light Reporter Assay

The traffic light reporter assay provides a means to compare nuclease activity based on levels of mutagenic NHEJ (in one of three potential frameshifts) and homologous recombination, as readout by mCherry and GFP expression, respectively, in nuclease-treated TLR cells (*see* **Note 27**). All steps must be performed using sterile reagents and technique in a tissue culture hood.

Day 0

1. Plate $1–2 \times 10^5$ TLR cells generated from Subheading 3.3.3 in 0.5 ml of culture media/well into 24-well culture plates (*see* **Note 28**). Include one well to serve as an untreated control one to serve as a BFP control.

Day 1

2. Add 0.5 μg nuclease plasmid (megaTAL from Subheading 3.2.4 or meganuclease from Subheading 3.2.5) and 0.5 μg donor plasmid to separate tubes for each nuclease treatment to be tested (*see* **Note 29**). For the BFP control, 0.5 μg of the pCVL SFFV Sbf-Sal linker plasmid can be used in place of the nuclease plasmid.
3. Dilute 1 μl of X-tremeGENE 9 in 50 μl serum-free medium for every well to be transfected (make an excess to be sure to have

enough diluted reagent for all treatments) and incubate 5 min at room temperature (*see* **Note 30**).

4. Add 50 μl diluted X-tremeGENE 9 to each tube containing DNA and incubate for 15 min at room temperature.
5. Add X-tremeGENE 9/DNA mixtures to plated cells.

Day 2

6. Passage treated cells to 6-well plates in 2 ml of culture media.

Day 4

7. Aspirate media and resuspend cells in 1 ml of fresh culture media. Transfer resuspended cells to FACS tubes.
8. Run treated cells on a flow cytometer (or FACS, *see* **Note 27**). Record FSC-A, FSC-H, SSC-A, iRFP, BFP, mCherry, and GFP (*see* **Note 23**).
9. Analyze data using FlowJo software, gating on live cells→ singlets→iRFP positive→BFP positive cells and looking at percent mCherry positive and GFP positive populations within this gate (Fig. 4).

4 Notes

1. MegaTALs made from low and high affinity meganucleases have both shown increases in cleavage activity; however, the highest level of specificity can be achieved using a catalytically active, low affinity meganuclease that exhibits little or no cleavage on its own (due to lack of DNA binding affinity). No observable effect on cleavage activity has been observed with megaTALs made from meganucleases that exhibit catalytic defects (low k_{cat}), when compared with the activity of the stand-alone meganuclease.
2. The length of the meganuclease target site may vary slightly depending on the family of enzymes chosen. For example, the I-AniI meganuclease shows no nucleotide preference outside of a 20 bp target centered on the central four bases where double strand break formation occurs, while members of the I-OnuI family show specificity for a 22 bp target (www.homingendonuclease.net) [36, 37]. In this protocol, we assume the meganuclease target is 20 bp long, so any bases upstream of this 20 bp target should be considered part of the DNA spacer. For the I-OnuI meganuclease, the first base of the 22 bp target would be part of the 7 bp long DNA spacer. For meganucleases with structures differing significantly from I-AniI, *see* **Note 3** for tips on choosing the megaTAL DNA spacer length.

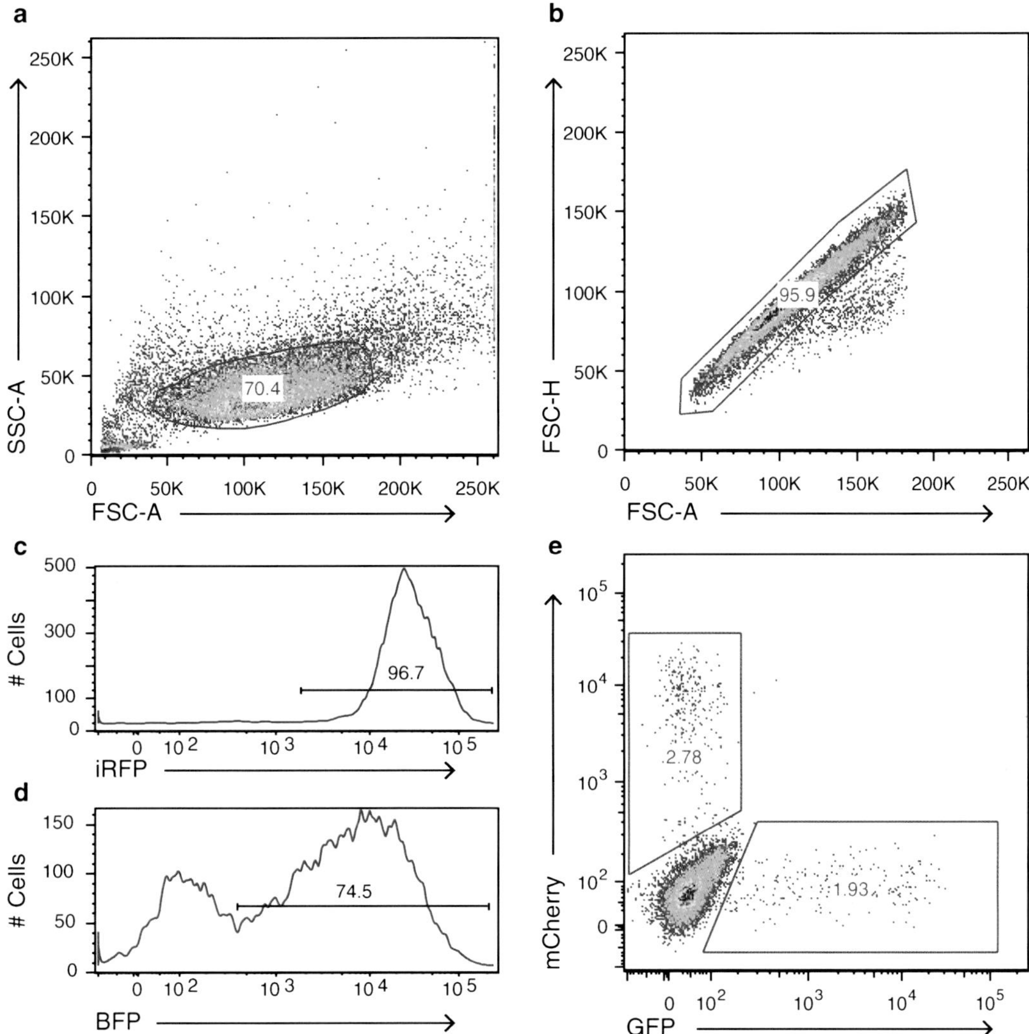

Fig. 4 Flow cytometry and analysis of a Traffic Light Reporter experiment. Gating should be done on live (**a**—FSC-A vs. SSC-A) singlets (**b**—FSC-A vs. FSC-H), expressing both iRFP (**c**) and BFP (**d**). Within this subpopulation, cells expressing mCherry provide a readout of mutNHEJ events, while cells expressing GFP provide a readout of homologous recombination (**e**)

3. Factors that affect TAL effector target site selection: *Thymine at position 0*: It has been observed that TAL effectors exhibit a preference for a thymine base at the nucleotide immediately preceding the first base of the TALE target (position 0) [38–40]. The closely related RipTALs show a preference for a guanine at position 0 and thus can be used in place of the TAL if a thymine cannot be located at an appropriate positions [41]. The cloning vector encoding the RipTAL domain can be obtained through Addgene and used in place of the pCVL TAL(Nd154C63) SFFV Xba-Sal linker cloning vector (Addgene).

Mutations to the N-terminus of TAL effectors that eliminate nucleotide preference at the 0th position of the DNA target have been identified [42]; however, a megaTAL cloning vector containing these changes has not yet been developed. *Sequence of the TAL effector target site*: different DNA-binding affinities have been observed for the various TAL effector RVD-nucleotide pairs, with the HD RVD providing the highest affinity (for cytosine), followed by NN (guanine), NI (adenine), and NG (thymine). TAL effector target sequences rich in cytosine and guanine bases, particularly at either end of their target sequence, may result in megaTALs with higher activity [30, 43]. *megaTAL DNA spacer*: Maximal cleavage activity was obtained using a 7 bp DNA spacer separating the TAL effector and meganuclease target sites when assessed using megaTALs made from variants of the I-AniI meganuclease [30]. This ideal spacer length may vary for megaTALs made from other meganuclease scaffolds, particularly those with little structural similarity to I-AniI. For megaTALs made with variants derived from alternate meganuclease scaffolds, it may be useful to build a number of megaTALs with varying DNA spacer lengths, or test a single megaTAL for activity against targets with increasing DNA spacer length using the Traffic Light Reporter assay described in Subheadings 3.3 and 3.4. *Length of the TAL effector target site*: We have tested the effect of the number of TAL effector RVD units (and thus the length of the TALE target site) on megaTAL activity and found that a 5.5–7.5 RVD TALE effector was sufficient to achieve maximal activity with the lowest affinity meganuclease tested [30]. The length of the TAL effector DNA target suggested in Subheading 3.1 (6–11 bp long) is recommended in order to obtain high genomic specificity and maximal cleavage activity with even a low affinity meganuclease.

4. The program provided in the "TALEN_map_generator_prompt" folder prompts the user for information in order to generate a cloning scheme and plasmid maps. Alternatively, the program provided in "TALEN_map_generator" folder (http://www.addgene.org/50627/files/) allows the user to input multiple target sites using a text file. More information and detailed instructions for both program versions are available in the "tutorial.txt" file within the folder.

5. Plasmid maps and a cloning scheme may also be generated by hand: *TAL effector cloning step 1* (Subheading 3.2.2)—Determine which backbone vector (pFUSC1-10) is needed by subtracting the length of the TALE target site by one. For example, if the TAL effector binding site is 9 bp long, pFUSC8 will be used for cloning. Determine which RVD plasmids are needed to clone into your pFUSC vector based on your TALE

target sequence (excluding the last base of the target)—pNI for A, pHD for C, pNN for G, and pNG for T. The RVD plasmids are numbered such that they correspond to the position of the nucleotide within the TAL effector target, with the plasmid for the first base being labelled as pXX1, etc. For example, if the TAL effector binding site is "ACGTCA", vectors pNI1, pHD2, pNN3, pNG4, and pHD5 will be used for cloning. *TAL effector cloning step 2* (Subheading 3.2.3)—Determine the plasmid needed for the last RVD based on the last nucleotide of the TALE binding site—pLR-NI for A, pLR-HD for C, pLR-NN for G, and pLR-NG for T. For example, for the TALE binding site "ACGTCA" would require the pLR-NI plasmid.

6. Repeat variable diresidue (RVD) refers to the two amino acids (at positions 12 and 13 of the TAL effector repeat unit) that confer specificity for a single TAL effector repeat to a single nucleotide on the DNA target. Different protein residue pairs at the RVD positions have been shown to contribute to nucleotide specificity and affinity to varying degrees [44–47].
7. The pFUSC vectors and protocol have been adapted from the Voytas lab Golden Gate TALEN and TAL effector Kit 2.0 (Addgene) to generate megaTALs with TAL effectors 2.5–10.5 RVDs long. MegaTALs with TAL effectors longer than 10.5 RVDs can be generated by following the original kit protocol and using vectors pFUSA and pFUSB1-10 in place of pFUSC1-10 as needed.
8. For blue-white screening on plates containing X-gal/IPTG, colonies do not appear turn blue until they reach a certain size. Allow sufficient time (16+ h) for the color change to occur.
9. Alternatively, if cloning efficiency appears to be low, a PCR screen can be used to identify colonies with the correct insert size (as described in the Golden Gate TALEN and TAL effector Kit 2.0 protocol) prior to growing cultures and sequencing miniprepped DNA.
10. For TAL effectors made with pFUSA/B in place of pFUSC, sequencing may not cover the whole length of the insert. Colonies for which sequencing indicates RVD insertion can be used for subsequent steps, regardless of having only partial sequencing of the insert.
11. The meganuclease can be cloned into the pCVL TAL(Nd154C63) SFFV Xba-Sal linker vector prior to the TAL effector portion. If multiple megaTALs are to be built using the same meganuclease it would useful to clone the meganuclease ORF into the vector first.
12. Alternatively, if XbaI or SalI sites are present in the meganuclease ORF and cannot be removed, a type IIS restriction enzyme that generates a 4 base 5′ overhang, such as BsaI or BsmBI, with

no restriction sites present in the ORF can be used instead. For example, the BsaI restriction enzyme can be used to generate the appropriate overhangs using the FP X*GGTCTC*XCTAGGT GGGAGGAAGC$(N)_{10+}$ and RP X*GGTCTC*XTCGATCA$(N)_{10+}$ to amplify the meganuclease ORF. Use 2 μl BsaI (or other type IIS restriction enzyme) and 2 μl water in place of XbaI and SalI in **step 4**, Subheading 3.2.4.

13. It is helpful to set up a control ligation with only 50 ng of "open" vector to ensure that the vector digestion and phosphatase treatment has gone to completion.

14. If cloning efficiency appears to be low, or high background was obtained for the control ligation reaction, a PCR screen using the sequencing primers can be helpful to identify colonies with the correct insert size prior to growing cultures and sequencing plasmid DNA preps.

15. As described in **Note 8**, if SalI restriction sites are present in the meganuclease ORF and cannot easily be removed, type IIS restriction enzymes that generate 4 base 5′ overhangs can be used in place of SalI. No type IIS restriction enzyme has been identified that can leave a 4 base 3′ overhang compatible with a SbfI digested end. The restriction enzyme NsiI can be used in place of SbfI in **step 1** of Subheading 3.2.5 if no NsiI restriction site are present in the meganuclease ORF, using an alternate primer NsiI-HA-partial NLS XXXATGCAT XXX*ATGCAT* CTATCCATATGATGTCCCAGATTATGCG CCACCTAAGAAGAAACG in place of FP2. This change will result in a single codon/amino acid change to the plasmid insert, but the resulting protein product should not be affected. Use the appropriate enzymes in **step 5** of Subheading 3.2.5 in place of SbfI and/or SalI.

16. As discussed in **Note 3**, TAL effectors show higher activity against target DNA sequences preceded by a thymine base and thus this base should be included in the TLR insert to achieve maximum cleavage activity. Alternatively, if a megaTAL has been assembled using the RipTAL cloning vector, include a guanine base upstream of the TAL effector DNA target.

17. The level of DNA repair readout may vary across different cell line derivations for the same TLR reporter construct (same target site insert). If megaTAL activity is to be measured and compared across different TLR cell lines it is important to have a means to control for such variations in activity. The stand-alone meganuclease may serve this purpose. However, if the stand-alone meganuclease is not expected to show activity on its own, a second meganuclease target (i.e., for I-SceI or Y2 I-AniI, [48, 49]) may be included in the TLR insert alongside the megaTAL target for this purpose.

18. Mutagenic NHEJ is measured by DNA insertions or deletions causing a frameshift mutation to place the mCherry gene in the +1 reading frame from the +3 reading frame. Stop codons within the TLR insert that prevent proper mCherry expression will affect readout of mutNHEJ. Because indels occur at the 4 bp overhang generated by the meganuclease at the center of its target site, any stop codons in the +1 reading frame of the TLR ORF (+2 reading frame of the insert) upstream or +3 reading frame of the TLR ORF (+1 reading frame of the insert) downstream of the center of this target can prevent mCherry expression after a +3 → +1 frameshift mutation. If stop codons that affect mutNHEJ readout cannot be removed, nuclease activity can only be measured using readout of homologous recombination.
19. HEK 293T cells are loosely adherent, therefore, take special care to avoid disturbing the cells when adding media and PBS by slowly pipetting solutions along the wall of the plate.
20. Freeze-thaw cycles will decrease the titer of the viral stock.
21. Different cell lines may be used in place of 293T cells if desired. However, this may require modifications to the transfection protocol in Subheading 3.4 to achieve equivalent levels of nuclease transfection and expression.
22. Because 293T cells are loosely adherent, the cells can be resuspended using gentle pipetting without trypsinization.
23. Flow cytometer and FACS filters will vary by instrument. Determine which filters should be used for visualization of iRFP, BFP, mCherry, and GFP fluorophores.
24. iRFP expression within less than 10 % of the total cell population suggests single copy integration of the TLR locus into the cell genome.
25. Once the wild-type cell population (without an integrated TLR) has been selected against, puromycin selection is no longer needed.
26. Some copies of the TLR become improperly integrated into the cell genome during lentiviral integration, resulting in expression of mCherry that should otherwise not occur. To more accurately compare levels of nuclease activity, these mCherry-positive cells should be sorted out. However, typically less than 0.5 % of untreated TLR cells express mCherry and so for rapid characterization of nuclease activity, this step can be omitted.
27. The traffic light reporter assay typically under-reports levels of cleavage activity and repair events (both mutNHEJ and HR) and therefore should not be used as a quantitative readout of nuclease activity. This assay can however provide a qualitative

means of comparing different nucleases across a single cell line. For a quantitative measure of nuclease activity, genomic DNA may be isolated from treated cells sorted for BFP expression at Day 4 of the protocol (Subheading 3.4) and subjected to high-throughput sequencing analysis of the TLR locus. The megaTAL and meganuclease plasmids all encode BFP-T2A-nuclease so that BFP expression may serve as a marker of nuclease expression.

28. Transfection with X-tremeGENE 9 works optimally on highly confluent cells (~70–90 % confluent).

29. If your designed megaTAL is to be used for gene knockout rather than repair, the donor plasmid can be omitted from the experiment; however, this change should be reflected by adding half as much X-tremeGENE 9 transfection reagent during **step 3**, Subheading 3.4 (*see* **Note 30**).

30. The ratio of XtremeGENE 9 to DNA should be 2 μl: 1 μg in 50 μl serum-free media—any changes to final DNA concentrations should compensated by an equivalent change in the volume of the transfection reagent added.

References

1. Hacein-Bey-Abina S, Garrigue A, Wang GP et al (2008) Insertional oncogenesis in 4 patients after retrovirus-mediated gene therapy of SCID-X1. J Clin Invest 118:3132–3142. doi:10.1172/JCI35700
2. Bushman F, Lewinski M, Ciuffi A et al (2005) Genome-wide analysis of retroviral DNA integration. Nat Rev Microbiol 3:848–858. doi:10.1038/nrmicro1263
3. Baum C, Kustikova O, Modlich U et al (2006) Mutagenesis and oncogenesis by chromosomal insertion of gene transfer vectors. Hum Gene Ther 17:253–263. doi:10.1089/hum.2006.17.253
4. Nowrouzi A, Glimm H, Von Kalle C, Schmidt M (2011) Retroviral vectors: post entry events and genomic alterations. Viruses 3:429–455. doi:10.3390/v3050429
5. (2012) Method of the Year 2011. Nature Methods 9:1–1. doi: 10.1038/nmeth.1852
6. Baker M (2012) Gene-editing nucleases. Nat Methods 9:23–26. doi:10.1038/nmeth.1807
7. McMahon MA, Rahdar M, Porteus M (2012) Gene editing: not just for translation anymore. Nat Methods 9:28–31. doi:10.1038/nmeth.1811
8. Schiffer JT, Aubert M, Weber ND et al (2012) Targeted DNA mutagenesis for the cure of chronic viral infections. J Virol 86:8920–8936. doi:10.1128/JVI.00052-12
9. Lieber MR (2010) The mechanism of double-strand DNA break repair by the nonhomologous DNA end-joining pathway. Annu Rev Biochem 79:181–211. doi:10.1146/annurev.biochem.052308.093131
10. Shrivastav M, Haro LPD, Nickoloff JA (2008) Regulation of DNA double-strand break repair pathway choice. Cell Res 18:134–147. doi:10.1038/cr.2007.111
11. Paques F, Duchateau P (2007) Meganucleases and DNA double-strand break-induced recombination: perspectives for gene therapy. Curr Gene Ther 7:49–66. doi:10.2174/156652307779940216
12. Stoddard BL (2011) Homing endonucleases: from microbial genetic invaders to reagents for targeted DNA modification. Structure 19:7–15. doi:10.1016/j.str.2010.12.003
13. Mussolino C, Morbitzer R, Lutge F et al (2011) A novel TALE nuclease scaffold enables high genome editing activity in combination with low toxicity. Nucleic Acids Res 39:9283–9293. doi:10.1093/nar/gkr597
14. Ashworth J, Taylor GK, Havranek JJ et al (2010) Computational reprogramming of homing endonuclease specificity at multiple adjacent base pairs. Nucleic Acids Res 38:5601–5608
15. Takeuchi R, Lambert AR, Mak AN-S et al (2011) Tapping natural reservoirs of homing

endonucleases for targeted gene modification. Proc Natl Acad Sci U S A 108:13077–13082. doi:10.1073/pnas.1107719108
16. Szeto MD, Boissel SJS, Baker D, Thyme SB (2011) Mining endonuclease cleavage determinants in genomic sequence data. J Biol Chem 286:32617–32627. doi:10.1074/jbc.M111.259572
17. Kim YG, Cha J, Chandrasegaran S (1996) Hybrid restriction enzymes: zinc finger fusions to Fok I cleavage domain. Proc Natl Acad Sci U S A 93:1156–1160
18. Klug A (2010) The discovery of zinc fingers and their applications in gene regulation and genome manipulation. Annu Rev Biochem 79:213–231. doi:10.1146/annurev-biochem-010909-095056
19. Christian M, Cermak T, Doyle EL et al (2010) Targeting DNA double-strand breaks with TAL effector nucleases. Genetics 186:757–761. doi:10.1534/genetics.110.120717
20. Li T, Huang S, Jiang WZ et al (2011) TAL nucleases (TALNs): hybrid proteins composed of TAL effectors and FokI DNA-cleavage domain. Nucleic Acids Res 39:359–372. doi:10.1093/nar/gkq704
21. Cornu TI, Thibodeau-Beganny S, Guhl E et al (2008) DNA-binding specificity is a major determinant of the activity and toxicity of zinc-finger nucleases. Mol Ther 16:352–358. doi:10.1038/sj.mt.6300357
22. Gabriel R, Lombardo A, Arens A et al (2011) An unbiased genome-wide analysis of zinc-finger nuclease specificity. Nat Biotechnol 29:816–823. doi:10.1038/nbt.1948
23. Pattanayak V, Ramirez CL, Joung JK, Liu DR (2011) Revealing off-target cleavage specificities of zinc-finger nucleases by in vitro selection. Nat Methods 8:765–770. doi:10.1038/nmeth.1670
24. Söllü C, Pars K, Cornu TI et al (2010) Autonomous zinc-finger nuclease pairs for targeted chromosomal deletion. Nucleic Acids Res 38:8269–8276. doi:10.1093/nar/gkq720
25. Holkers M, Maggio I, Liu J et al (2012) Differential integrity of TALE nuclease genes following adenoviral and lentiviral vector gene transfer into human cells. Nucleic Acids Res 41:e63. doi:10.1093/nar/gks1446
26. Jinek M, Chylinski K, Fonfara I et al (2012) A programmable dual-RNA–guided DNA endonuclease in adaptive bacterial immunity. Science 337:816–821. doi:10.1126/science.1225829
27. Mali P, Yang L, Esvelt KM et al (2013) RNA-guided human genome engineering via Cas9. Science 339:823. doi:10.1126/science.1232033
28. Fu Y, Foden JA, Khayter C et al (2013) High-frequency off-target mutagenesis induced by CRISPR-Cas nucleases in human cells. Nat Biotechnol 31:822. doi:10.1038/nbt.2623
29. Cradick TJ, Fine EJ, Antico CJ, Bao G (2013) CRISPR/Cas9 systems targeting β-globin and CCR5 genes have substantial off-target activity. Nucleic Acids Res 41:9584. doi:10.1093/nar/gkt714
30. Boissel S, Jarjour J, Astrakhan A et al (2014) megaTALs: a rare-cleaving nuclease architecture for therapeutic genome engineering. Nucleic Acids Res 42:2591. doi:10.1093/nar/gkt1224
31. Takeuchi R. Choi M, Stoddard BL (2013) Efficient engineering of multiple meganucleases and MegaTALs using bioinformatics and in vitro compartmentalization (in press)
32. Wang Y, Khan I, Boissel S, Jarjour J, Pangallo J, Thyme S, Baker D, Scharenberg A, Rawlings D (2014) Progressive engineering of a homing endonuclease genome editing reagent for the murine X-linked immunodeficiency locus. Nucleic Acids Res 42:6463
33. Certo MT, Ryu BY, Annis JE et al (2011) Tracking genome engineering outcome at individual DNA breakpoints. Nat Methods 8:671–676. doi:10.1038/nmeth.1648
34. Kuhar R, Gwiazda KS, Humbert O et al (2013) Novel fluorescent genome editing reporters for monitoring DNA repair pathway utilization at endonuclease-induced breaks. Nucleic Acids Res 42:e4. doi:10.1093/nar/gkt872
35. Cermak T, Doyle EL, Christian M et al (2011) Efficient design and assembly of custom TALEN and other TAL effector-based constructs for DNA targeting. Nucleic Acids Res 39:e82. doi:10.1093/nar/gkr218
36. Scalley-Kim M, McConnell-Smith A, Stoddard BL (2007) Coevolution of a homing endonuclease and its host target sequence. J Mol Biol 372:1305–1319. doi:10.1016/j.jmb.2007.07.052
37. Sethuraman J, Majer A, Friedrich NC et al (2009) Genes within genes: multiple LAGLIDADG homing endonucleases target the ribosomal protein S3 gene encoded within an rnl group I intron of Ophiostoma and related taxa. Mol Biol Evol 26:2299–2315. doi:10.1093/molbev/msp145
38. Römer P, Recht S, Strauß T et al (2010) Promoter elements of rice susceptibility genes are bound and activated by specific TAL effectors from the bacterial blight pathogen, Xanthomonas oryzae pv. oryzae. New Phytol 187:1048–1057
39. Scholze H, Boch J (2010) TAL effector-DNA specificity. Virulence 1:428–432. doi:10.4161/viru.1.5.12863
40. Mak AN-S, Bradley P, Cernadas RA et al (2012) The crystal structure of TAL effector

PthXo1 bound to its DNA target. Science 335:716–719. doi:10.1126/science.1216211

41. De Lange O, Schreiber T, Schandry N et al (2013) Breaking the DNA-binding code of Ralstonia solanacearum TAL effectors provides new possibilities to generate plant resistance genes against bacterial wilt disease. New Phytol 199:773–786. doi:10.1111/nph.12324
42. Lamb BM, Mercer AC, Barbas CF 3rd (2013) Directed evolution of the TALE N-terminal domain for recognition of all 5′ bases. Nucleic Acids Res 41:9779–9785. doi:10.1093/nar/gkt754
43. Meckler JF, Bhakta MS, Kim M-S et al (2013) Quantitative analysis of TALE–DNA interactions suggests polarity effects. Nucleic Acids Res 41:4118–4128. doi:10.1093/nar/gkt085
44. Moscou MJ, Bogdanove AJ (2009) A simple cipher governs DNA recognition by TAL effectors. Science 326:1501. doi:10.1126/science.1178817
45. Boch J, Scholze H, Schornack S et al (2009) Breaking the code of DNA binding specificity of TAL-type III effectors. Science 326:1509–1512. doi:10.1126/science.1178811
46. Christian ML, Demorest ZL, Starker CG et al (2012) Targeting G with TAL effectors: a comparison of activities of TALENs constructed with NN and NK repeat variable di-residues. PLoS One 7:e45383. doi:10.1371/journal.pone.0045383
47. Streubel J, Blücher C, Landgraf A, Boch J (2012) TAL effector RVD specificities and efficiencies. Nat Biotechnol 30:593–595. doi:10.1038/nbt.2304
48. Colleaux L, D'Auriol L, Betermier M et al (1986) Universal code equivalent of a yeast mitochondrial intron reading frame is expressed into E. coli as a specific double strand endonuclease. Cell 44:521–533. doi:10.1016/0092-8674(86)90262-X
49. Takeuchi R, Certo M, Caprara MG et al (2009) Optimization of in vivo activity of a bifunctional homing endonuclease and maturase reverses evolutionary degradation. Nucleic Acids Res 37:877–890. doi:10.1093/nar/gkn1007

Chapter 10

Genome Engineering Using CRISPR-Cas9 System

Le Cong and Feng Zhang

Abstract

The Clustered Regularly Interspaced Short Palindromic Repeats (CRISPR)-Cas9 system is an adaptive immune system that exists in a variety of microbes. It could be engineered to function in eukaryotic cells as a fast, low-cost, efficient, and scalable tool for manipulating genomic sequences. In this chapter, detailed protocols are described for harnessing the CRISPR-Cas9 system from *Streptococcus pyogenes* to enable RNA-guided genome engineering applications in mammalian cells. We present all relevant methods including the initial site selection, molecular cloning, delivery of guide RNAs (gRNAs) and Cas9 into mammalian cells, verification of target cleavage, and assays for detecting genomic modification including indels and homologous recombination. These tools provide researchers with new instruments that accelerate both forward and reverse genetics efforts.

Key words Genome engineering, Cas9, CRISPR, CRISPR-Cas9, PAM, Guide RNA, sgRNA, DNA cleavage, Mutagenesis, Homologous recombination, *Streptococcus pyogenes*

1 Introduction

1.1 General Principle of Genome Engineering with Designer Nucleases

Genome engineering is a highly desirable technology in biological research and biomedical applications, allowing us to change genomic sequences at will to control the fundamental genetic information within a cell. It has been implemented in creating model cell lines or organisms for the study of gene functions or genetics of human diseases, and in gene therapy to correct deleterious genomic changes or confer beneficial ones. Genome engineering using engineered nucleases has been one of the most designable and scalable paths to achieve precise editing of genomic sequences. The basic principle for this process could be summarized into a two-step process (Fig. 1). The first step is the introduction of targeted DNA cleavage, in the form of typically double strand breaks (DSBs) or possibly single strand nicks (SSNs), through the activity of nucleases or nickases, respectively. The second step involves DNA repair process activated by the targeted cleavage, carried out normally by the endogenous DNA damage repair

Shondra M. Pruett-Miller (ed.), *Chromosomal Mutagenesis*, Methods in Molecular Biology, vol. 1239,
DOI 10.1007/978-1-4939-1862-1_10, © Springer Science+Business Media New York 2015

machinery within the cells. In this step, two major pathways could be employed, achieving different types of genome modification. One pathway is to repair the cleaved site via nonhomologous end joining (NHEJ) pathway, which applies only to the DSBs as SSNs are thought to be unable to induce NHEJ pathway. During NHEJ-mediated repair, insertion/deletion (indels) will be engineered into the target site. The other pathway requires the artificial supply of a repair template with chosen sequence alterations into the cell in addition to the DSBs or SSNs, so that homology-directed repair (HDR) pathway would be activated. The latter pathway requires homologous recombination (HR) between chromosomal DNA and the supplied foreign DNA stimulated by the local DNA cleavage, resulting in virtually any type of desired editing through replacement of native sequence by the designed HR template.

Overall, one of the most essential rate-limiting steps in this type of genome engineering technology is the targeted cleavage of endogenous genome. Hence, effort that centers on the development of better designer DNA-cleaving enzymes has been a focus of this field. In the past decades, genome editing tools based on sequence-specific nucleases and DNA-binding proteins such as zinc-finger nucleases (ZFNs) [1–4], meganucleases [5], and transcription activator like effectors (TALEs) [6–11], among others, have opened up the possibility of performing precise perturbation of genome at single-nucleotide resolution. Recently, the emergence of a new type of genome engineering technology based on the Clustered Regularly Interspaced Short Palindromic Repeats (CRISPR) system, a microbial adaptive immune system, has transformed the field because of its easiness to design, rapidity to implement, low-cost, and scalability in eukaryotic cells.

The CRISPR locus in microbes typically consists of a set of noncoding RNA elements and enzymes that harbors the ability to recognize and cleave foreign nucleic acids based on their sequence signature. Three known CRISPR systems, types I, II, and III, have been described so far [12–15]. Among these different CRISPR systems, type II CRISPR systems usually only require a single protein, called Cas9, to perform the target cleavage. Recent years, the type II CRISPR-Cas9 systems have been studied by several groups in their native microbial domain to elucidate their molecular organizations and functions [16–18]. These studies demonstrated that Cas9 is a RNA-guided nuclease that is capable of binding to a target DNA and introduce a double strand break in a sequence-specific manner, and its specificity is determined by sequence encoded within the RNA components.

Following these breakthroughs, this new family of RNA-guided nucleases has been successfully developed into a multiplex genome engineering system that is simple to design, efficient, and cost-effective for mammalian genome editing purposes [19, 20]. Application of this system could enable introduction of different

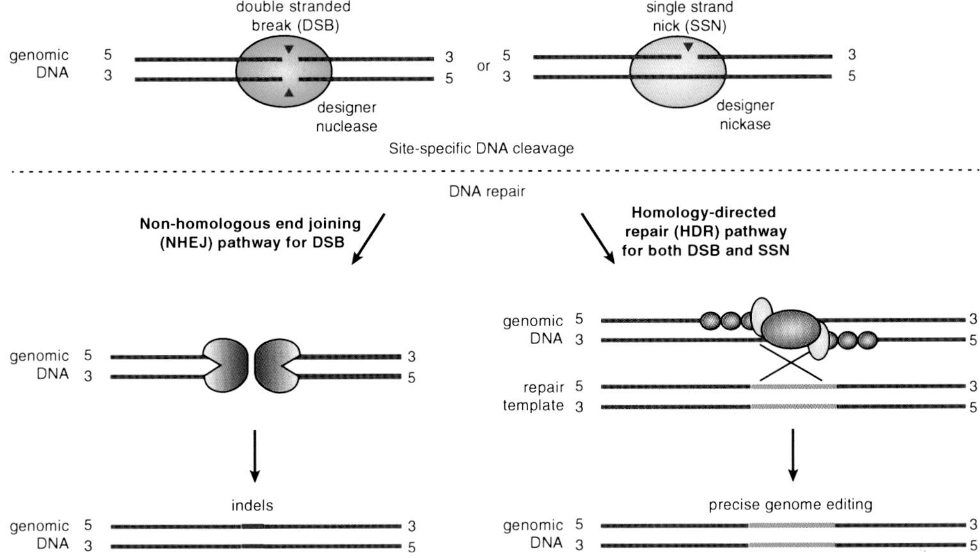

Fig. 1 Principle for genome engineering using designer nucleases or nickases. DNA cleavage induced by designer nucleases or DNA nicks created by designer nickases can result in double strand break (DSB), single strand nick (SSN), or double nicks (DNs) (*top*). These site-specific DNA cleavages can be repaired via two different pathways (*bottom*). In the error-prone NHEJ pathway for DSB or DNs, the ends of the breaks are processed by endogenous DNA repair machinery and rejoined, which can result in random indels at the site of junction. The other pathway is the HDR pathway for both DSB, SSN, and DNs, where a repair template is supplied, allowing precise editing of the endogenous genome sequences. Parts of this figure are adapted from [41]

type of modifications, such as mutation, insertion, deletion, or even larger chromosomal changes, through NHEJ or HDR pathways, in many different cell types and organisms [19–27].

1.2 Genome Engineering with CRISPR-Cas9 System

The required components for CRISPR-Cas9 mediated DNA cleavage are the Cas9 protein and its bound RNA component that guide the Cas9 to target sequence. The RNA component consists of a RNA duplex formed by partial pairing between the CRISPR RNA (crRNA), which includes the guide sequence that bound to target DNA through Watson–Crick base-pairing to specify the site of cleavage, and the trans-activating crRNA (tracrRNA), which facilitate the maturation of crRNA and the loading of crRNA onto Cas9 (Fig. 2a) [16–18]. Our previous work demonstrated that by harnessing the well-understood CRISPR-Cas9 system from *Streptococcus pyogenes* SF370, we could develop a system that is sufficient for carrying out site-specific DNA cleavage in mammalian cells [19]. A simplified version of this design is achieved by fusion of the mature form of crRNA and tracrRNA into a chimeric single guide RNA (sgRNA), which demonstrated even superior efficiency compared with the split design [19, 28, 20]. Hereafter we center on this two-component design, which enhances the convenience of using the CRISPR-Cas9 system and multiplexing of genome targeting applications.

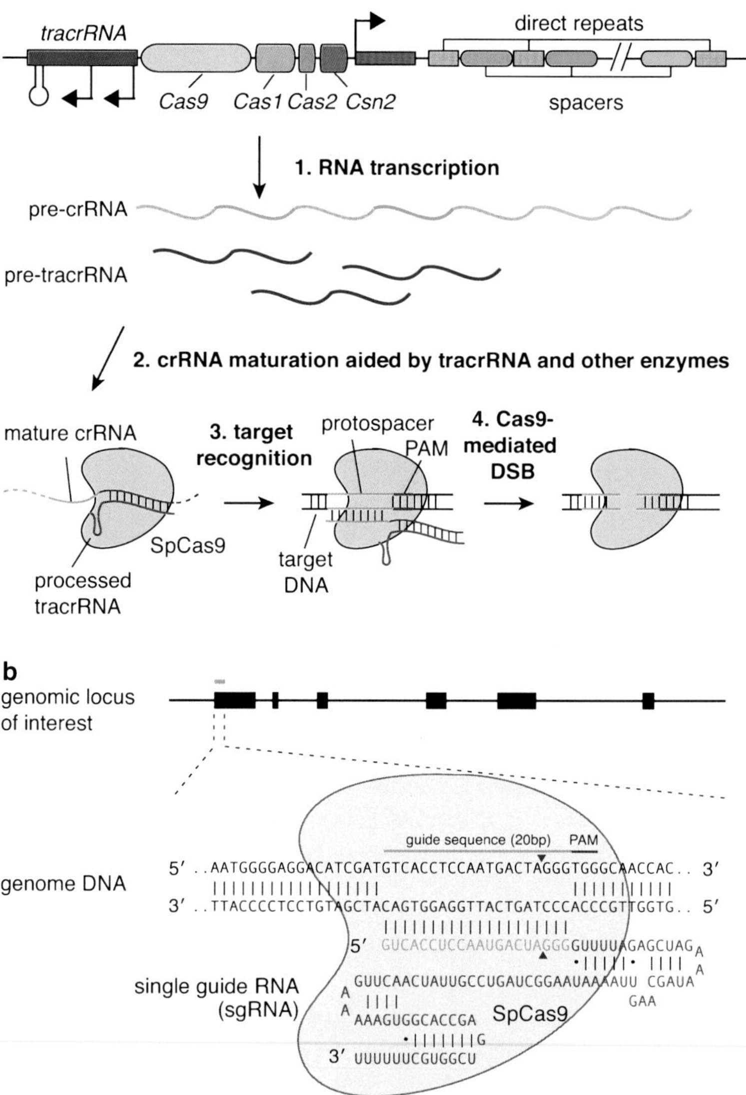

Fig. 2 Schematic of the type II CRISPR-mediated DNA double-strand break and the design of chimeric sgRNA. (**a**) The type II CRISPR locus from *Streptococcus pyogenes* SF370 contains a cluster of four genes, Cas9, Cas1, Cas2, and Csn1, as well as two noncoding RNA elements, tracrRNA and a characteristic array of repetitive sequences (direct repeats) interspaced by short stretches of non-repetitive sequences (spacers, 30 bp each). Each spacer is typically derived from foreign genetic material (protospacer), and directs the specificity of CRISPR-mediated nucleic acid cleavage. In the target nucleic acid, each protospacer is associated with a protospacer adjacent motif (PAM) whose recognition is specific to individual CRISPR systems. The Type II CRISPR system carries out targeted DNA double-strand break (DSB) in sequential steps. First, the pre-crRNA array and tracrRNA are transcribed from the CRISPR locus. Second, tracrRNA hybridizes to the direct repeats of pre-crRNA and associates with Cas9 as a duplex, which mediates the processing of the pre-crRNA into mature crRNAs containing individual, truncated spacer sequences.

The guide sequence within sgRNA has a length of 20 bp and is the exact complementary sequence of the target site within the genome, sometimes referred to as "protospacer" following the convention of microbial CRISPR research (Fig. 2a). In choosing the target site, it is important to note the requirement of having the "NGG" trinucleotide motif, called protospacer adjacent motif (PAM), right next to the protospacer target on the 3′ end. The Cas9 cleavage site is located within the protospacer and positioned at around 3 bp upstream the 5′ end of the PAM, indicating a potential anchor role of the PAM for Cas9-mediated DNA cleavage (Fig. 2a, b). The PAM sequence depends on the Cas9 protein employed in the CRISPR-Cas9 system, and hence, the "NGG" PAM sequence applies specifically to the *Streptococcus pyogenes* Cas9 (SpCas9) discussed in this protocol.

The Cas9 enzyme typically contains two nuclease domains: the RuvC-like (N-terminal RNase H fold) domain and the HNH (McrA-like) domain, each responsible for cleavage of one strand of the duplex DNA molecule. Therefore, there are two different versions of Cas9 that can be used for genome modification, the wild type Cas9 that induces DSBs to the target site, and the nickase version of Cas9 (Cas9n) bearing a deactivating mutation in one of the two nuclease domains, which nicks the target DNA on one strand. For SpCas9 that we described in this protocol, the two derived nickases are SpCas9n (D10A) and SpCas9n (H840A), with mutation at one of the two catalytic residues of the nuclease domain, D10 and H840, respectively. Whereas the wild type SpCas9 is highly efficient in executing NHEJ- and HDR-mediated genome engineering, its off-target effects due to the induction of DSBs at mismatched targets within the genome cannot be neglected [28–30]. The NHEJ-incompetent SpCas9n, in particular in the form of a double-nicking design where a pair of sgRNAs cooperates two adjacent nickings at target genomic locus, might be superior in terms of editing accuracy as it maintains the ability to achieve comparable genome engineering capacity and efficiency while reducing the likelihood of off-target NHEJ events [31, 32]. Furthermore, if the catalytic residues within the two domains are

Fig. 2 (continued) Third, the mature crRNA:tracrRNA duplex directs Cas9 to the DNA target consisting of the protospacer and the requisite PAM via heteroduplex formation between the spacer region of the crRNA and the protospacer DNA. Finally, Cas9 mediates cleavage of target DNA upstream of PAM to create a DSB within the protospacer. (**b**) Design of chimeric sgRNA for genome cleavage. The SpCas9 nuclease (*yellow*) binds to genomic DNA directed by the sgRNA consisting of a 20-nt guide sequence (*blue*) and a chimeric RNA scaffold (*red*). The guide sequence base-pairs with protospacer target (*blue*), located upstream of a requisite "NGG" called protospacer adjacent motif (PAM; colored in *pink*). The SpCas9 mediates a DSB around 3 bp upstream to the 5′ end of the PAM (*red triangles*). Parts of this figure are adapted from [19, 41]

simultaneously mutated, the deactivated non-cleaving version of SpCas9 (dSpCas9) can serve as a generic RNA-guided DNA-binding protein, priming additional applications in mammalian cells such as targeted transcriptional modulation [33–37]. More recently, one study showed that by shortening the guide length within the sgRNA from 20 to 17 bp, the genome targeting specificity of wild type SpCas9 protein could be improved [38]. Shortly afterwards, two independent pieces of work demonstrated the utilization of Cas9-FokI fusion protein as a designer dimer nuclease for improved Cas9 specificity [39, 40]. All these work on SpCas9-based system and earlier demonstration that orthogonal Cas9 proteins from different microbial species can be adapted for mammalian genome targeting [19], implicate the potential of optimizing the guide RNA and the Cas9 protein as valid strategies to address the off-target issue of CRISPR-Cas9 system, raising the possibility that an efficient and specific CRISPR-Cas9 system will be realized for powerful and precision genome engineering.

Here we describe a detailed protocol for applying the CRISPR-Cas9 system from *Streptococcus pyogenes* for genome engineering purposes in mammalian cells. The protocol includes steps arranged in order of actual experiments: selection of targets and cloning of targeting constructs (Subheading 3.1), delivery into mammalian cells (Subheading 3.2), assay for genomic cleavage (Subheading 3.3), implementation of homologous recombination using CRISPR-Cas system (Subheading 3.4), and the verification and quantification of HR efficiency (Subheading 3.5). There are still numerous challenges related to efficiency of homologous recombination, off-target effects, and multiplexed targeting before we could deploy CRISPR-Cas9 system to achieve robust, efficient, and accurate genome engineering in all type of biological systems, especially for in vivo settings. Nonetheless, the general principle and design guidelines for SpCas9 in this chapter should be relevant for these future efforts as well.

2 Materials

2.1 Molecular Biology Reagents

1. Backbone plasmids: pX330 (U6-chimeric guide RNA + CBh-SpCas9 backbone, Addgene ID 42230), pX335 (U6-chimeric guide RNA + CBh-SpCas9n backbone, Addgene ID 42335). These and additional backbone plasmids mentioned in the protocol are all available from Addgene, at www.addgene.org.
2. DNA oligos (Standard de-salted) for cloning of chosen guides. For oligos longer than 60 nt, the ultramer oligo (Integrated DNA Technology, Coralville, IA, USA) is recommended.
3. PCR reagents: Herculase II fusion polymerase (Agilent, Santa Clara, CA, USA). Other high-fidelity polymerases, such as Kapa HiFi (Kapa Biosystems, Wilmington, MA, USA) can also be used for PCR in this protocol.

4. Cloning enzymes: FastDigest *Bbs*I, FastDigest *Age*I, FastAP (Thermo Scientific/Fermentas, Pittsburg, PA, USA), Plasmid-Safe exonuclease (Epicentre, Madison, WI, USA), T7 DNA ligase.
5. 10× T4 DNA ligase buffer.
6. 10× Tango buffer (Thermo Scientific/Fermentas).
7. DTT (DL-Dithiothreitol; Cleland's reagent).
8. 10 mM Adenosine 5′-triphosphate.
9. Ultrapure water (RNAse/DNAse-free).
10. Tris-EDTA (TE) buffer.
11. *E. coli* competent cells. Strains such as Stbl3 (Life Technologies, Carlsbad, CA, USA) or other recombination-deficient strain is recommended to avoid recombination of repetitive elements in the plasmids.
12. Luria Broth (LB) media.
13. Ampicillin.
14. Spin miniprep kit.
15. Plasmid Midi/Maxi Kit.
16. Standard gel electrophoresis reagents and apparatus.

2.2 Cell Culture and Processing Reagents

1. Cell line: human embryonic kidney (HEK) 293FT cell line.
2. Dulbecco's Modified Eagle's Medium (DMEM).
3. 10 % fetal bovine serum (FBS).
4. GlutaMAX solution (Life Technologies).
5. 100× Pen-Strep: 100 U/μL penicillin, 100 μg/μL streptomycin.
6. OptiMEM medium (Life Technologies).
7. TrypLE™ Express Enzyme (Life Technologies).
8. Transfection reagent: Lipofectamine 2000 (Life Technologies) can be used for HEK 293FT or Neuro-2a cell lines with this protocol. Other transfection reagents can also be used depending on the cell type and after trial/optimization.
9. 24-well tissue culture plates.
10. pCMV-EGFP plasmid (Addgene, Cambridge, MA, USA).
11. pUC19 plasmid (NEB).

2.3 Genome Modification and Homologous Recombination Assay Reagents

1. 1× PBS.
2. QuickExtract™ DNA extraction kit (Epicentre).
3. SURVEYOR Mutation Detection Kit (Transgenomic, Omaha, NE, USA).
4. 4–20 % Novex TBE polyacrylamide gels and accompanying gel-running reagents (Life Technologies).
5. PCR primer for amplification of genomic DNA used in SURVEYOR assay and HR assay.

6. PCR reagents: Herculase II fusion polymerase (Agilent). Other high-fidelity polymerases, such as Kapa HiFi (Kapa Biosystems) can also be used for PCR in this protocol.
7. 10× Taq PCR buffer.
8. Novex Hi-Density TBE sample buffer (Life Technologies).
9. 10,000× SYBR Gold nucleic acid gel stain (Life Technologies).
10. PCR purification kit.
11. Gel extraction kit.
12. Restriction enzyme for HR assay depends on the design of the HR template.

3 Methods

Ultrapure water is used for all the steps below. The setup of the experiments is carried out at room temperature unless otherwise stated.

3.1 Design and Cloning of Genome Engineering Constructs

Below is a step-wise protocol for cloning and verification of targeting vectors. Prior to the cloning of the constructs for genome engineering, the target sites should be selected as described in Subheading 4 (*see* **Notes 1–3**).

1. Prepare sgRNA oligo duplex.
   Resuspend the oligos (*see* **Note 2**) to a final concentration of 100 μM with TE buffer. Mix regents listed below to prepare the reaction mixture for each pair of oligos to phosphorylate and anneal them into ligation-ready duplex DNA fragments.

1 μl	forward oligo (100 μM)
1 μl	reverse oligo (100 μM)
1 μl	10× T4 Ligation Buffer
6 μl	ddH_2O
1 μl	T4 PNK (NEB)
10 μl	total volume

   Place the mixture in a thermocycler using the following parameters to phosphorylate and anneal the oligos.

37 °C	30 min
95 °C	5 min and then ramp down to 25 °C at 5 °C/min
4 °C	hold until ready to proceed

   Add 2 μl of annealed oligos into 398 μl of water to dilute the products 200-fold.

2. Cloning of targeting constructs.
   Select the appropriate cloning backbone vector to insert the annealed oligos (*see* **Notes 2** and **4**). Mix reagents listed below to set up the ligation reaction. Mix the same reagents without adding the sgRNA oligo duplex to prepare a negative control reaction.

X μl	backbone plasmid (total amount 100 ng)
2 μl	sgRNA oligo duplex (200-fold diluted)
2 μl	10× Tango buffer
1 μl	DTT
1 μl	ATP
1 μl	FastDigest *Bbs*I
0.5 μl	T7 ligase
Y μl	ddH_2O
20 μl	total

   Incubate the ligation reaction in a thermocycler following the parameters below.

37 °C	5 min
23 °C	5 min
Cycle the previous two steps for 6 cycles (total run time 1 h)	
4 °C	hold until ready to proceed

3. (Optional but highly recommended) Plasmid-Safe treatment.
   Add Plasmid-Safe buffer, ATP, and the ATP-dependent Plasmid-Safe exonuclease as listed below to treat the ligation reaction and the negative control to increase the efficiency of cloning by degrading unligated DNA fragments in the product mixture.

11 μl	ligation reaction from previous step
1.5 μl	10× Plasmid-Safe Buffer
1.5 μl	10 mM ATP
1 μl	ATP-dependent Plasmid-Safe exonuclease
15 μl	total

   Incubate the reaction mixture in a thermocycler following the parameters below.

37 °C	30 min
70 °C	30 min
4 °C	hold until ready to proceed

4. Transformation.
   Add 2 μl of each reaction from previous step (including the negative control ligation reaction) into chosen competent cells, such as Stbl3, transform following the manufacturer's protocol.
   Plate the transformed cells on Ampicillin selection LB agar plates or other type of plates depending on the selection marker of the backbone vector.
5. Plasmid preparation.
   Check the plate for presence of bacterial colonies after overnight incubation. Usually there should be no or less than five colonies on the negative control plate, whereas over tens to hundreds of colonies will grow on the cloning plate, indicating high cloning efficiency.
   Pick two to three colonies from each agar plate and set up a small volume (typically 5 ml) LB culture for miniprep. Incubate and shake under 37 °C overnight, prepare plasmid DNA using a spin miniprep kit according to manufacturer's protocol.
6. (Optional) Digest the plasmid DNA to verify the insertion of the oligos.
   Digest the miniprep DNA with diagnostic restriction enzymes, BbsI and AgeI, if using the vector backbone PX330/PX335. Run the digested product on a 1 % agarose gel to visualize the band pattern of the digested product.
   When a lot of cloning is done at the same time, it is possible to screen for correct insertion of the target sequence oligos by this digestion because a successful insertion will destroy the BbsI sites. After double digestion, clones with insertion of annealed sgRNA oligos will show only linearized plasmid, whereas clones without insertion will yield two fragments, with sizes of ~980 and ~7,520 bp in the case of pX330 when running and visualizing on an agarose gel.
7. Sequencing verification of positive clones.
   Sequence the clones with a forward primer that binds the human U6 promoter to verify the successful insertion of guide sequences. The verified clones can be then prepared using a Midi or Maxi plasmid extraction kit for downstream experiments.

3.2 Mammalian Cell Culture and Transfection

As a general protocol, the steps below use HEK 293FT cells as an exemplar system to demonstrate CRISPR-Cas9 genome engineering. Additional suggestions can be found in Subheading 4 (*see* **Note 5**).

1. HEK 293FT cell maintenance.
   Maintain the HEK cells in DMEM medium supplemented with 10 % FBS (D10 medium) in an incubator at 37 °C temperature and supplemented with 5 % CO_2 as recommended by the manufacturer. Feed the cells every other day, and passage the cells so they never reach over 75 % confluence.

2. Preparing cells for transfection.
   At 16–24 h before transfection, dislodge and disassociate the cells by trypsinization. Plate cells into 24-well plates containing 500 μl antibiotic free medium at a density of 250,000 cells per mL of culture medium, or 125,000 cells per well. Scale up or down the culture volume proportionally to the plate format based on the number of cells needed in the experiment. At the time of transfection, the cells should be at around 75–85 % confluence.

3. Transfection.
   Mix a total of up to 600 ng DNA containing the targeting constructs in a microcentrifuge tube for each well of transfection on a 24-well plate. For transfecting plasmid derived from pX330 or pX335, where a guide sequence has been cloned into the backbone vector, directly use 600 ng of the prepared plasmid in the transfection mixture. When transfecting more than one construct for alternative designs, e.g., the double nickase method, or when multiplex genome cleavage is desired, mix equal molar ratio of all constructs to a total of 600 ng in the transfection mixture (*see* **Notes 2** and **4**). Add 50 μl OptiMEM medium OptiMEM medium to the DNA mixture. Mix well and spin down.

   Dilute Lipofectamine 2000 transfection reagent by adding 1.5 μl of the reagent into 50 μl of OptiMEM, Mix well and incubate at room temperature for 5 min. Within 15 min of diluting the transfection reagent, add all the diluted Lipofectamine 2000 reagent into the DNA-OptiMEM mixture prepared earlier. Mix well and spin down. Incubate the mixture for another 20 min to allow the formation of DNA–Lipofectamine complex. Add the final complex directly onto the culture medium of each well on the 24-well plate. (Not required but highly recommended) Include a transfection control to monitor transfection efficiency. Use a control plasmid expressing a fluorescent protein, such as pCMV-EGFP, and transfect this plasmid following same protocol as above into the cell. Transfect another well of cells using the SpCas9 backbone vector, e.g. pX330 or pX335, as a negative control for downstream processing and validation of assay.

   Replace the medium with warmed fresh medium around 12–24 h post transfection. Maintain the cells for another 48–72 h to allow sufficient time for genome engineering mediated by CRISPR-Cas9 system.

3.3 Genome Cleavage Analysis

1. Genomic DNA extraction.
   Extract genomic DNA from transfected cells using the QuickExtract DNA extraction kit following the manufacturer's recommended protocol. Briefly, disassociate cells from the plate, harvest by spin down the cell suspension at 250 × *g* for

5 min. Wash cell pellet with 500 μl of PBS and then resuspended in QuickExtract solution. We typically use 50 μl for one well in a 24-well plate (scale up and down accordingly). Vortex the suspension and incubate at 65 °C for 15 min, 68 °C for 15 min, and 98 °C for 10 min.

2. SURVEYOR assay to detect genomic cleavage.
   Use SURVEYOR assay to detect genomic cleavage after extraction of genomic DNA (*see* **Note 6**), following the stepwise instructions provided in the SURVEYOR Mutation Detection Kit manual. A brief description of the steps is listed below.
   (a) Amplify extracted genomic DNA with a pair of primers designed for the target region of interest using a high-fidelity enzyme such as Herculase II fusion polymerase of Kapa Hifi Hotstart polymerase. Typically an amplicon size of less than 1,000 bp is preferred as shorter amplicon gives more specific amplification products.
   (b) Visualize the PCR product on an agarose gel to check the specificity of the amplification. It is important to have very specific amplification of genomic region to yield accurate SURVEYOR assay results as nonspecific bands will interfere with the interpretation of gel electrophoresis analysis.
   (c) Purify and quantify the PCR products, set up a denaturing/re-annealing reaction by mixing up to 400 ng of PCR products with water and re-annealing buffer (we typically use the PCR reaction buffer and add to a final concentration of 1×, refer to the SURVEYOR assay manual for more information).
   (d) Run the reaction in a thermocycler with following parameters.

95 °C	5 min and then ramp down to 85 °C at −2 °C/s
85 °C	1 min and then ramp down to 25 °C at −0.1 °C/s
4 °C	hold until ready to proceed

   (e) Digest the re-annealed products with SURVEYOR enzyme kit at 42 °C for 1 h as recommended by the manufacturer's protocol.
   (f) Visualize the digested product using gel electrophoresis.
   For visualization of SURVEYOR assay results, we recommended loading of the Surveyor Nuclease digestion products with Polyacrylamide gel electrophoresis (PAGE) method as it gives better solution compared with agarose gels.

Quantification of the assay results and the method to convert it to an estimation of the frequency of indels generated by CRISPR-Cas system in the population of cells are described in Subheading 4 (*see* **Note 7**).

3.4 Implementation of Homologous Recombination (HR) Using CRISPR-Cas System

The guideline and considerations for designing a HR experiment to use CRISPR-Cas system to precisely modify the genomic sequence of interest by inserting, deleting, or replacing part of the genome are described in Subheading 4 (*see* **Note 8**). Briefly, following design and cloning of the HR template, perform HR experiment following steps below.

1. HEK 293FT cell maintenance and the preparation of cells for transfection.
   This part is same as the corresponding steps in Subheading 3.2. Briefly, plate cells into 24-well plates and make sure at the time of transfection, the cells should be at around 75–85 % confluence.
2. Transfection.
   Mix a total of up to 800 ng DNA containing the targeting constructs and the HR template vector (or single-stranded DNA oligos, *see* **Note 9**) in a microcentrifuge tube for each well of transfection on a 24-well plate. Generally, apply a molar ratio of 1:3–5:1 for the targeting vector and HR template vector. (Optional but recommended) Titrate different molar ratio between the targeting vector and HR template vector to test the optimal condition for the HR experiment. Add 50 μl OptiMEM medium to the DNA mixture. Mix well and spin down.

   Dilute Lipofectamine 2000 transfection reagent by adding 2 μl of the reagent into 50 μl of OptiMEM, Mix well and incubate at room temperature for 5 min. Within 15 min of diluting the transfection reagent, add all the diluted Lipofectamine 2000 reagent into the DNA-OptiMEM mixture prepared earlier. Mix well and spin down. Incubate the mixture for another 20 min to allow the formation of DNA–Lipofectamine complex. Add the final complex directly onto the culture medium of each well on the 24-well plate.

   Transfect another well of cells using the SpCas9 backbone vector, e.g., pX330 or pX335, together with the same amount of HR template vector as a negative control.

   Replace the medium with warmed fresh medium around 12–24 h post transfection. Maintain the cells for another 72 h to allow sufficient time for HR mediated by CRISPR-Cas9 system and the template.

3.5 Verification and Quantification of HR Efficiency

To verify the homologous recombination between the HR template and the endogenous genome, restriction fragment length polymorphism (RFLP) assay for HR can be applied.

1. Genomic DNA extraction.
   Extract the genomic DNA using the same extraction protocol using the QuickExtract DNA extraction kit as in the SURVEYOR assay (Subheading 3.3).
2. Target region amplification.
   Amplify the genomic region of interest by a HR testing primer set where the two primers bind outside the homology region to avoid false positive results given by amplification of the residue HR template.
3. Perform RFLP digestion.
   Run the resulting PCR product on an agarose gel to check for specificity of amplification, as nonspecific PCR products will interfere with the assay and prevent accurate quantification of HR efficiency. In many cases, several pairs of HR testing primer sets should be screened to obtain robust, specific amplicons.

   Purify the PCR amplification product by standard PCR purification, or in the case where clean PCR product cannot be obtained, gel extract the desired amplicon following separation of PCR product on an agarose gel.

   Digest the purified products with the appropriate enzyme corresponding to the design of the HR template (*see* **Note 8**), and visualize on an agarose gel or PAGE gel. The latter usually gives better resolution and is highly recommended. The efficiency of HR in the population of cells assayed can be estimated by the following formula:

$$\text{HR percentage }(\%) = (m + n / m + n + p) \times 100$$

   Here the number "m" and "n" indicate the relative quantity of bands from digested genomic PCR products, whereas the "p" equals the relative quantity of undigested products.
4. Perform additional Sanger and next-generation DNA sequencing to verify the presence of desired engineered sequences within the genome. Briefly, clone the genomic PCR product into a sequencing vector, TOPO-TA, or other blunt-end cloning method, and perform Sanger sequencing to detect recombined genomic amplicons. Alternatively for higher throughput, subject the genomic PCR products to next-generation sequencing.

4 Notes

1. Identification and selection of target genomic site. The two primary rules for identifying a target site for the SpCas9 system are: (1) finding the "NGG" PAM sequence which is required for SpCas9 targeting, and (2) picking a sequence of 20 bp in length upstream of the PAM to its 5′ end as the guide sequence. Following these two guidelines, multiple potential target sites

can be usually found within the genomic region of interest (Fig. 2). Additionally, when using U6 promoter to express sgRNA (such as pX330, pX335), we suggest adding the G (not replacing but add one more base) because the human U6 promoter requires a "G" at the transcription start site to have highest level of expression (Fig. 2). While we do notice that sometimes the sgRNA will still work without the extra "G", it is generally better to have this additional base. In the case where the guide sequence starts with a base "G", this addition can be omitted. In our open-source online resources website (http://crispr.genome-engineering.org), we provide the most up-to-date information for using the CRISPR-Cas9 system for genome engineering, focusing on the SpCas9 system. Additionally, we also developed an online tool for the selection of SpCas9 targets for different organisms including human, mouse, zebrafish, *C. elegans*, etc. This tool can greatly facilitate and simplify the process of performing target selection in batch (http://crispr.mit.edu/). Because the efficiency of different targets could vary considerably depending on the guide sequences, we highly recommend testing multiple target sites for each gene or region of interest and selecting the most effective target (*see* **Note 1**). In the case of double nickase design, we recommend individually testing each target with the wild type SpCas9 system to assess the cleavage efficiency of individual guides and then combine the most efficient pair of guides with opposite directionality and appropriate spacing for the genome engineering application.

2. Design of oligos for inserting guide sequences into backbone vectors. The cloning vectors we use for typical SpCas9 genome targeting are pX330 for the wild type SpCas9 and pX335 for the nickase version SpCas9n. Both vectors are mammalian dual-expression vectors, which enables the co-expression of SpCas9 protein driven by the potent constitutive promoter CBh and sgRNA driven by the RNA Pol III human U6 promoter in mammalian cells (Fig. 3). CBh promoter is a hybrid promoter derived from the CAG promoter, which have been validated to support strong expression of transgene in multiple cell types/lines, including HEK 293FT, mouse Neuro-2a, mouse Hepa1-6, HepG2, HeLa, human ESCs, and mouse ESCs. To clone custom guide sequences into these backbone vectors, a pair of oligos encoding the guide sequences can be ordered with the appropriate overhangs (Fig. 3), then annealed to form a clone-ready duplex DNA fragment. The vector can be then digested using BbsI, and a pair of annealed oligos can be cloned into the backbone to express the corresponding sgRNA (Fig. 3). The oligos are designed based on the target site sequence selected in previous section. A common confusion sometimes in cloning the guide into backbone vector is to

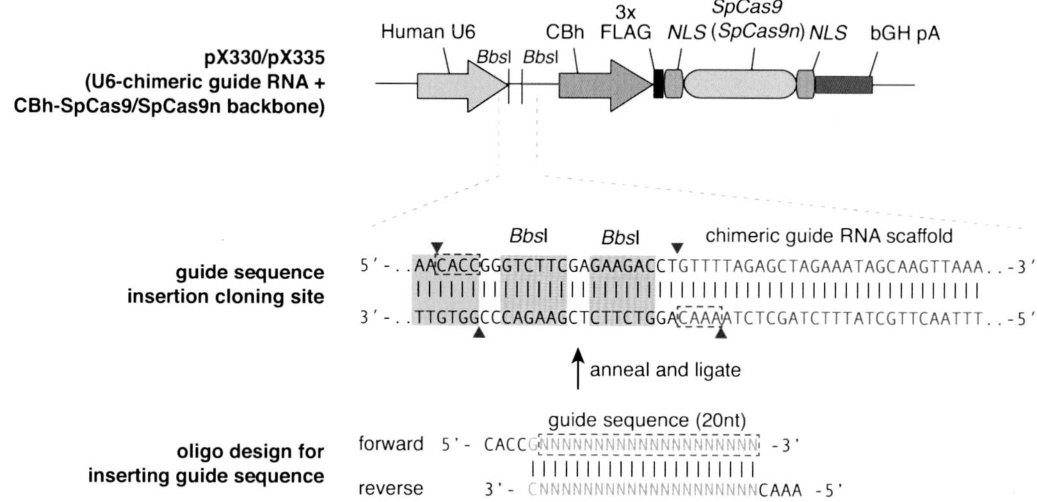

Fig. 3 Schematics for the cloning backbone vectors pX330/pX335 with oligo design for inserting guide sequences. The pX330 and pX335 vectors contains dual-expression cassettes for both the SaCas9 protein and the sgRNA. Digestion of the backbone with *BbsI* Type II restriction sites (*blue*) generates the complementary cloning overhangs to the annealed oligos (*purple boxed*). Note that a G–C base pair is added at the 5′ end of the guide sequence for optimal U6 transcription. The oligos contain overhangs for ligation into the overhangs of *BbsI* sites. The top and bottom strand orientations is exactly identical to those of the genomic target but exclude the "NGG" PAM. Parts of this figure are adapted from [19, 41]

include the "NGG" PAM sequence in the guide sequence. Hence, it is important to check that only the 20 bp sequence to the 5′ end of the PAM is being used for designing the oligo sequences to order. An alternative way of designing oligos for directly amplifying a PCR fragment that contains the U6 promoter driving a sgRNA could also be employed for testing the guide sequences, which simplifies the test by avoiding the need of cloning, but might be less efficient than using the cloned vector plasmids (*see* **Note 4**, [41]).

3. Screening of multiple guides. For most applications, we screen for at least three guide sequences within the target genomic region in an effort to find the most efficient ones. This is because while CRISPR-Cas9 system works very efficiently, the actual cleavage efficiency could be affected by the sequence of the guide, the accessibility of local chromatin, the activity of the endogenous DNA repair pathways, and other guide-specific or cell-type-specific factors. Hence, to ensure that a valid guide sequence is obtained, this screening process is highly recommended. Following the same logic, a guide sequence that has been verified in one cell type will not necessarily work to the same efficiency in another cell type or condition. Hence, additional optimization or re-screening of new guide sequences might be required when moving from one experimental sys-

tem to another. This same situation is also applicable to the HR experiment where the HDR efficiency can very considerably among different types of cells or tissues.

4. Additional strategy for screening guides and backbones for different applications. We have also developed another way of quickly screening guide sequences with amplified PCR products. In this design, two primers are used to amplify the U6 RNA-expression promoter, where the forward primer binds to the 5′ beginning of U6 promoter, and the reverse primer binds to the 3′ end of the U6 promoter. Because the reverse primer also contains a long extension that can add on the guide sequence and the chimeric sgRNA scaffold, the amplified PCR product contains all necessary elements for expressing a sgRNA containing the guide specified in the reverse primer. Hence, the screening of guide sequences can be done by co-transfecting this PCR product with a backbone vector expressing the SaCas9 protein. Because many application of CRISPR-Cas9 genome engineering involve cell lines that might be difficult to work with, e.g., cell lines that are hard to transfect, we developed additional backbone vectors to facilitate selection and screening for transfected cells. These vectors contain the fluorescent maker protein, GFP, or the selectable puromycin resistance gene, linked to the expressing of SaCas9 via a 2A peptide linker. These constructs will enable fluorescence activated cell sorting (FACS) or the selection of transfected population, which can further improve the overall efficiency of genome engineering particularly in the case of HR applications. Additional details on these designs and backbones can be found in our recent publication [41].

5. Cell line choice for validation of guide design. Functional validation of targeting constructs bearing the designed guides can be carried out in relevant cell lines, e.g., HEK 293FT, K562, Hela for human genome engineering, or Neuro-2a, Hepa1-6 for mouse. This process takes advantage of some favorable experimental properties of these lines, such as robust and easy maintenance, efficient transfection, etc., before embarking on complicated procedures in other mammalian systems. Nonetheless, achieving best results for each experiment might require additional optimization (*see* **Note 2**). Moreover, due to the genetic and epigenetic differences between cell types or subjects of study, results obtained from one cell type might not necessarily correspond to those from another cell type of the same species (*see* **Note 2**).

6. Mechanism of SURVEYOR nuclease assay. Following the delivery of SpCas9 and the sgRNAs into mammalian cells, the induced genomic cleavage could be assayed by the SURVEYOR assay, which could detect modification of genomic DNA within

a population of cells. This assay works when a certain portion of the cells will be modified by SpCas9 so that their genomic sequence at target site is different from the un-modified population. Hence, in the assay, it is possible to amplify region of interest from genomic DNA via PCR, then through a denaturing and re-annealing process to form mismatched DNA. This mismatched DNA can then be recognized by the SURVEYOR nuclease and cleaved for visualization on analytical gels. To quantify the efficiency of genomic cleavage, one can then assess the percentage of cleaved products as a surrogate for the percentage of indels generated within the target genomic region.

7. Analysis of SURVEYOR assays results. To calculate the genome cleavage efficiency of a tested target, quantify the band intensity of SURVEYOR assay products visualized by PAGE using the following formula:

   Indel percentage (%) $= (1 - \sqrt{(1-x)}) \times 100$, where $x = (a+b)/(a+b+c)$

   In this formula, the number "*a*" and "*b*" represent the relative quantities of the cleaved bands, while "*c*" equals to the relative quantity of the non-cut full-length PCR product.

   Other methodology of detecting the genomic cleavage can also be applied. One such method is to clone the SURVEYOR PCR products into a sequencing vector, e.g., pUC19, and transformed into *E. coli*. These individual clones can be then sequenced via Sanger sequencing to reveal the identity of genome modifications. Additionally, the percentage of modified clones can also be used as a measurement for the efficiency of genome engineering. Alternatively, the PCR products could also be sequenced in a more high-throughput way with next-generation sequencing.

8. Design and synthesis of repair template for HR experiment. For introducing a precise genomic modification into the genome, the HDR pathway can be employed. This is achieved by co-transfecting SpCas9 constructs (derived from pX330 or pX335) bearing guide sequences with a HR template in the target cell line. After recombination, modifications such as point mutation, small and large insertions/deletions, or other type of chromosomal changes could be engineered into the endogenous genome. A few considerations for the choice of guide:

   (a) Typically, a screening for the most efficient guide sequence is performed first. We recommend picking several (three to six) targets within the genomic region of interest following protocols listed earlier. Tests are then performed to assay the cleavage efficiency of each of these guides. Then, the actual HR experiments can be carried out with the most efficient guides (also see additional considerations in **Notes 3** and **4**).

(b) For maximize the efficiency of HR, it is recommended that the cleavage site of the guide is as close to the junction of the homology arm, i.e., the size at which genome modifications are introduced, as possible. Usually this distance should be less than 100 bp, ideally less than 10 bp.

(c) To minimize the off-target cleavage, the double nickase design can be used. In this case, multiple guide sequences can be first tested individually, and typically the combination of highest cutting guide designs with appropriate directionality will yield highest cleavage when used in the paired fashion, thus giving best results in HR experiment.

9. The HR template is essentially the desired sequence that needs to be present in the engineered genome, flanked by two homology arms bearing the same sequence as the reference genome. Below are considerations for the choice of HR template:

(a) It is usually advised to insert a testable marker in the HR template to facilitate the assay for successful HR events. For example, a restriction site could be inserted to allow RFLP assay. Alternatively, the insertion of fluorescent proteins or selectable drug-resistance genes such as puromycin-resistance cassette can also be used.

(b) For introducing single-point mutation the best HR template for transfection is usually single-stranded DNA (ssDNA) oligos. For ssDNA oligo design, we typically use around 50–90 bp homology arms on each side and introduce your mutation/modification in between the two arms. When ordering long oligos, ultramer oligo (IDT) is recommended.

(c) For introducing larger genomic modification, plasmid DNA vector can be used because of the length limit of ssDNA oligos. When designing a plasmid-based HR template, a minimum of 800 bp homology arms on each side is recommended.

(d) If you have intact "protospacer + PAM" sequence within the HR template, it can lead to the HR template being degraded by Cas9. Hence, it is recommended to make silent mutations to destroy the sgRNA-binding site, or avoid putting in the full target site in the HR template by choosing target sites that span the site of modification. For making silent mutations, one good option is to mutate the PAM "NGG" within the HR template, as the PAM is required for cleavage. For example, change the "NGG" to "NGT" or "NGC", in addition to mutations in the spacer itself, could usually prevent degradation of donor plasmid.

References

1. Porteus MH, Baltimore D (2003) Chimeric nucleases stimulate gene targeting in human cells. Science 300(5620):763. doi:10.1126/science.1078395, 300/5620/763 [pii]
2. Miller JC, Holmes MC, Wang J, Guschin DY, Lee YL, Rupniewski I, Beausejour CM, Waite AJ, Wang NS, Kim KA, Gregory PD, Pabo CO, Rebar EJ (2007) An improved zinc-finger nuclease architecture for highly specific genome editing. Nat Biotechnol 25(7):778–785. doi:10.1038/nbt1319
3. Sander JD, Dahlborg EJ, Goodwin MJ, Cade L, Zhang F, Cifuentes D, Curtin SJ, Blackburn JS, Thibodeau-Beganny S, Qi Y, Pierick CJ, Hoffman E, Maeder ML, Khayter C, Reyon D, Dobbs D, Langenau DM, Stupar RM, Giraldez AJ, Voytas DF, Peterson RT, Yeh JR, Joung JK (2011) Selection-free zinc-finger-nuclease engineering by context-dependent assembly (CoDA). Nat Methods 8(1):67–69. doi:10.1038/nmeth.1542
4. Wood AJ, Lo TW, Zeitler B, Pickle CS, Ralston EJ, Lee AH, Amora R, Miller JC, Leung E, Meng X, Zhang L, Rebar EJ, Gregory PD, Urnov FD, Meyer BJ (2011) Targeted genome editing across species using ZFNs and TALENs. Science 333(6040):307. doi:10.1126/science.1207773
5. Stoddard BL (2005) Homing endonuclease structure and function. Q Rev Biophys 38(1):49–95. doi:10.1017/S0033583505004063
6. Boch J, Scholze H, Schornack S, Landgraf A, Hahn S, Kay S, Lahaye T, Nickstadt A, Bonas U (2009) Breaking the code of DNA binding specificity of TAL-type III effectors. Science 326(5959):1509–1512. doi:10.1126/science.1178811, 1178811 [pii]
7. Moscou MJ, Bogdanove AJ (2009) A simple cipher governs DNA recognition by TAL effectors. Science 326(5959):1501. doi:10.1126/science.1178817, 1178817 [pii]
8. Zhang F, Cong L, Lodato S, Kosuri S, Church GM, Arlotta P (2011) Efficient construction of sequence-specific TAL effectors for modulating mammalian transcription. Nat Biotechnol 29(2):149–153. doi:10.1038/nbt.1775
9. Miller JC, Tan S, Qiao G, Barlow KA, Wang J, Xia DF, Meng X, Paschon DE, Leung E, Hinkley SJ, Dulay GP, Hua KL, Ankoudinova I, Cost GJ, Urnov FD, Zhang HS, Holmes MC, Zhang L, Gregory PD, Rebar EJ (2011) A TALE nuclease architecture for efficient genome editing. Nat Biotechnol 29(2):143–148. doi:10.1038/nbt.1755
10. Christian M, Cermak T, Doyle EL, Schmidt C, Zhang F, Hummel A, Bogdanove AJ, Voytas DF (2010) Targeting DNA double-strand breaks with TAL effector nucleases. Genetics 186(2):757–761. doi:10.1534/genetics.110.120717, genetics.110.120717 [pii]
11. Reyon D, Tsai SQ, Khayter C, Foden JA, Sander JD, Joung JK (2012) FLASH assembly of TALENs for high-throughput genome editing. Nat Biotechnol 30(5):460–465. doi:10.1038/nbt.2170
12. Deveau H, Garneau JE, Moineau S (2010) CRISPR/Cas system and its role in phage-bacteria interactions. Annu Rev Microbiol 64:475–493. doi:10.1146/annurev.micro.112408.134123
13. Horvath P, Barrangou R (2010) CRISPR/Cas, the immune system of bacteria and archaea. Science 327(5962):167–170. doi:10.1126/science.1179555
14. Makarova KS, Haft DH, Barrangou R, Brouns SJ, Charpentier E, Horvath P, Moineau S, Mojica FJ, Wolf YI, Yakunin AF, van der Oost J, Koonin EV (2011) Evolution and classification of the CRISPR-Cas systems. Nat Rev Microbiol 9(6):467–477. doi:10.1038/nrmicro2577
15. Bhaya D, Davison M, Barrangou R (2011) CRISPR-Cas systems in bacteria and archaea: versatile small RNAs for adaptive defense and regulation. Annu Rev Genet 45:273–297. doi:10.1146/annurev-genet-110410-132430
16. Deltcheva E, Chylinski K, Sharma CM, Gonzales K, Chao Y, Pirzada ZA, Eckert MR, Vogel J, Charpentier E (2011) CRISPR RNA maturation by trans-encoded small RNA and host factor RNase III. Nature 471(7340):602–607. doi:10.1038/nature09886
17. Jinek M, Chylinski K, Fonfara I, Hauer M, Doudna JA, Charpentier E (2012) A programmable dual-RNA-guided DNA endonuclease in adaptive bacterial immunity. Science 337(6096):816–821. doi:10.1126/science.1225829
18. Garneau JE, Dupuis ME, Villion M, Romero DA, Barrangou R, Boyaval P, Fremaux C, Horvath P, Magadan AH, Moineau S (2010) The CRISPR/Cas bacterial immune system cleaves bacteriophage and plasmid DNA. Nature 468(7320):67–71. doi:10.1038/nature09523
19. Cong L, Ran FA, Cox D, Lin S, Barretto R, Habib N, Hsu PD, Wu X, Jiang W, Marraffini LA, Zhang F (2013) Multiplex genome engineering using CRISPR/Cas systems. Science 339(6121):819–823. doi:10.1126/science.1231143
20. Mali P, Yang L, Esvelt KM, Aach J, Guell M, DiCarlo JE, Norville JE, Church GM (2013) RNA-guided human genome engineering via Cas9. Science 339(6121):823–826. doi:10.1126/science.1232033
21. Jinek M, East A, Cheng A, Lin S, Ma E, Doudna J (2013) RNA-programmed genome

editing in human cells. Elife 2:e00471. doi:10.7554/eLife.00471

22. Hwang WY, Fu Y, Reyon D, Maeder ML, Tsai SQ, Sander JD, Peterson RT, Yeh JR, Joung JK (2013) Efficient genome editing in zebrafish using a CRISPR-Cas system. Nat Biotechnol 31(3):227–229. doi:10.1038/nbt.2501
23. Wang H, Yang H, Shivalila CS, Dawlaty MM, Cheng AW, Zhang F, Jaenisch R (2013) One-step generation of mice carrying mutations in multiple genes by CRISPR/Cas-mediated genome engineering. Cell 153(4):910–918. doi:10.1016/j.cell.2013.04.025
24. Yang H, Wang H, Shivalila CS, Cheng AW, Shi L, Jaenisch R (2013) One-step generation of mice carrying reporter and conditional alleles by CRISPR/Cas-mediated genome engineering. Cell 154(6):1370–1379. doi:10.1016/j.cell.2013.08.022
25. DiCarlo JE, Norville JE, Mali P, Rios X, Aach J, Church GM (2013) Genome engineering in Saccharomyces cerevisiae using CRISPR-Cas systems. Nucleic Acids Res 41(7):4336–4343. doi:10.1093/nar/gkt135
26. Jiang W, Bikard D, Cox D, Zhang F, Marraffini LA (2013) RNA-guided editing of bacterial genomes using CRISPR-Cas systems. Nat Biotechnol 31(3):233–239. doi:10.1038/nbt.2508
27. Cho SW, Kim S, Kim JM, Kim JS (2013) Targeted genome engineering in human cells with the Cas9 RNA-guided endonuclease. Nat Biotechnol 31(3):230–232. doi:10.1038/nbt.2507
28. Hsu PD, Scott DA, Weinstein JA, Ran FA, Konermann S, Agarwala V, Li Y, Fine EJ, Wu X, Shalem O, Cradick TJ, Marraffini LA, Bao G, Zhang F (2013) DNA targeting specificity of RNA-guided Cas9 nucleases. Nat Biotechnol 31(9):827–832. doi:10.1038/nbt.2647
29. Fu Y, Foden JA, Khayter C, Maeder ML, Reyon D, Joung JK, Sander JD (2013) High-frequency off-target mutagenesis induced by CRISPR-Cas nucleases in human cells. Nat Biotechnol 31(9):822–826. doi:10.1038/nbt.2623
30. Pattanayak V, Lin S, Guilinger JP, Ma E, Doudna JA, Liu DR (2013) High-throughput profiling of off-target DNA cleavage reveals RNA-programmed Cas9 nuclease specificity. Nat Biotechnol 31(9):839–843. doi:10.1038/nbt.2673
31. Mali P, Aach J, Stranges PB, Esvelt KM, Moosburner M, Kosuri S, Yang L, Church GM (2013) CAS9 transcriptional activators for target specificity screening and paired nickases for cooperative genome engineering. Nat Biotechnol 31(9):833–838. doi:10.1038/nbt.2675
32. Ran FA, Hsu PD, Lin CY, Gootenberg JS, Konermann S, Trevino AE, Scott DA, Inoue A, Matoba S, Zhang Y, Zhang F (2013) Double nicking by RNA-guided CRISPR Cas9 for enhanced genome editing specificity. Cell 154(6):1380–1389. doi:10.1016/j.cell.2013.08.021
33. Cheng AW, Wang H, Yang H, Shi L, Katz Y, Theunissen TW, Rangarajan S, Shivalila CS, Dadon DB, Jaenisch R (2013) Multiplexed activation of endogenous genes by CRISPR-on, an RNA-guided transcriptional activator system. Cell Res 23(10):1163–1171. doi:10.1038/cr.2013.122
34. Maeder ML, Linder SJ, Cascio VM, Fu Y, Ho QH, Joung JK (2013) CRISPR RNA-guided activation of endogenous human genes. Nat Methods 10(10):977–979. doi:10.1038/nmeth.2598
35. Gilbert LA, Larson MH, Morsut L, Liu Z, Brar GA, Torres SE, Stern-Ginossar N, Brandman O, Whitehead EH, Doudna JA, Lim WA, Weissman JS, Qi LS (2013) CRISPR-mediated modular RNA-guided regulation of transcription in eukaryotes. Cell 154(2):442–451. doi:10.1016/j.cell.2013.06.044
36. Qi LS, Larson MH, Gilbert LA, Doudna JA, Weissman JS, Arkin AP, Lim WA (2013) Repurposing CRISPR as an RNA-guided platform for sequence-specific control of gene expression. Cell 152(5):1173–1183. doi:10.1016/j.cell.2013.02.022
37. Perez-Pinera P, Kocak DD, Vockley CM, Adler AF, Kabadi AM, Polstein LR, Thakore PI, Glass KA, Ousterout DG, Leong KW, Guilak F, Crawford GE, Reddy TE, Gersbach CA (2013) RNA-guided gene activation by CRISPR-Cas9-based transcription factors. Nat Methods 10(10):973–976. doi:10.1038/nmeth.2600
38. Fu Y, Sander JD, Reyon D, Cascio VM, Joung JK (2014) Improving CRISPR-Cas nuclease specificity using truncated guide RNAs. Nat Biotechnol 32(3):279–284. doi:10.1038/nbt.2808
39. Guilinger JP, Thompson DB, Liu DR (2014) Fusion of catalytically inactive Cas9 to FokI nuclease improves the specificity of genome modification. Nat Biotechnol 32:577. doi:10.1038/nbt.2909
40. Tsai SQ, Wyvekens N, Khayter C, Foden JA, Thapar V, Reyon D, Goodwin MJ, Aryee MJ, Joung JK (2014) Dimeric CRISPR RNA-guided FokI nucleases for highly specific genome editing. Nat Biotechnol 32:569. doi:10.1038/nbt.2908
41. Ran FA, Hsu PD, Wright J, Agarwala V, Scott DA, Zhang F (2013) Genome engineering using the CRISPR-Cas9 system. Nat Protoc 8(11):2281–2308. doi:10.1038/nprot.2013.143

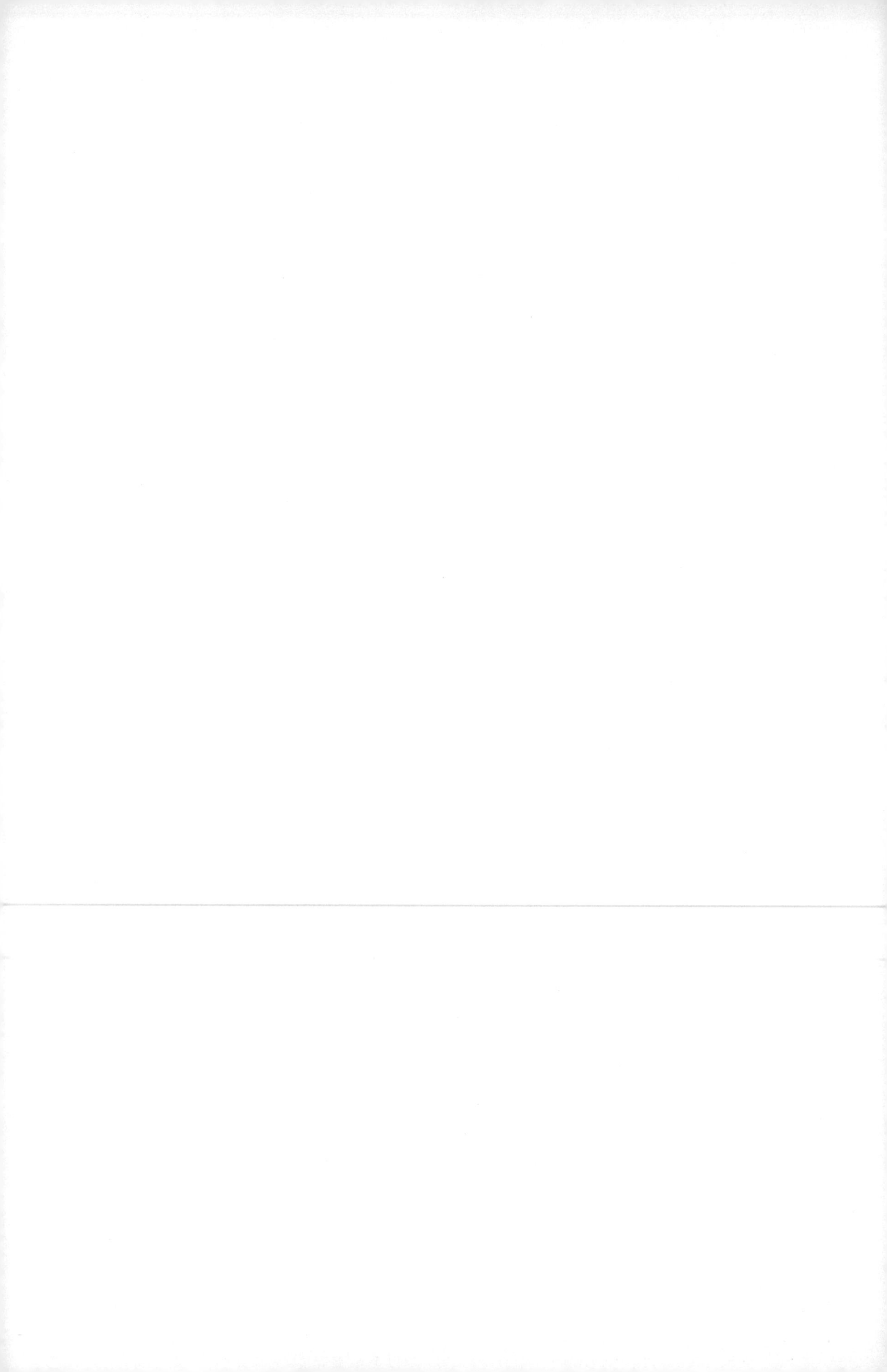

Chapter 11

Donor Plasmid Design for Codon and Single Base Genome Editing Using Zinc Finger Nucleases

Shondra M. Pruett-Miller and Gregory D. Davis

Abstract

In recent years, CompoZr zinc finger nuclease (ZFN) technology has matured to the point that a user-defined double strand break (DSB) can be placed at virtually any location in the human genome within 50 bp of a desired site. Such high resolution ZFN engineering is well within the conversion tract limitations demarcated by the mammalian DNA repair machinery, resulting in a nearly universal ability to create point mutations throughout the human genome. Additionally, new architectures for targeted nuclease engineering have been rapidly developed, namely transcription activator like effector nucleases (TALENs) and clustered regularly interspaced short palindromic repeats (CRISPR)/Cas systems, further expanding options for placement of DSBs. This new capability has created a need to explore the practical limitations of delivering plasmid-based information to the sites of chromosomal double strand breaks so that nuclease-donor methods can be widely deployed in fundamental and therapeutic research. In this chapter, we explore a ZFN-compatible donor design in the context of codon changes at an endogenous locus encoding the human RSK2 kinase.

Key words Genome editing, Zinc finger nuclease, Donor plasmid, Donor design, Cell engineering

1 Introduction

In a broad sense, genome editing using zinc finger nucleases (ZFNs) relies on targeted double stranded break (DSB) creation by the ZFN followed by DSB repair by the endogenous cellular repair machinery. A chromosomal DSB can be resolved by one of two major repair pathways: (1) nonhomologous end joining (NHEJ) and (2) homology-directed repair (HDR). In mammalian cells, the process of NHEJ is very efficient and often the preferred repair pathway [1]. However, repair by NHEJ can be error-prone and lead to insertions and/or deletions at the site of repair resulting in gene disruption. In contrast, HDR is a high-fidelity repair pathway and can be harnessed in order to create user-defined modifications within complex genomes.

Shondra M. Pruett-Miller (ed.), *Chromosomal Mutagenesis*, Methods in Molecular Biology, vol. 1239,
DOI 10.1007/978-1-4939-1862-1_11, © Springer Science+Business Media New York 2015

In the context of normal cellular function, HDR is a "copy and paste" mechanism by which a DSB is repaired using the information contained in a homologous sister chromatid. However, a user-defined mutation can be introduced by transfecting a donor plasmid that contains both the user specified mutation and sequence homologous to the regions flanking the desired site of modification. In mammalian cells, the spontaneous rate of HDR between a user defined donor plasmid and a homologous locus is very low (10^{-6}–10^{-5}) and is often outside the range for practical experimentation, even when using antibiotic selection. To overcome this limitation, several groups have shown that the creation of a DSB at or near the site of desired mutation can increase the rate of HDR by several orders of magnitude [2]. Until recently, the practical and flexible creation of a DSB at a desired locus (for example, disease SNPs) was not feasible. Due to large collective efforts in expanding engineered zinc finger repertoires, ZFNs spearheaded the first explorations into DSB-based genome editing at pre-selected loci [3, 4]. Following successful design and manufacturing of a site-specific ZFN, the next key step to create targeted mutations via HR is to design and construct a targeting donor plasmid. Here, we provide a general guide for the design of targeting donors for creating point mutations, codon changes, SNP corrections, and other small site-directed genomic modifications useful for genotyping.

2 Materials

Prepare all solutions using nuclease-free molecular grade water. Store all DNA and RNA reagents at −20 or −80 °C. Store all cell culture reagents at 4 °C.

2.1 Zinc Finger Nucleases and Donor Plasmids

1. ZFNs against the RSK2 locus were engineered by the CompoZr ZFN production group at Sigma-Aldrich (St. Louis, MO, USA) and have been described previously [5].
2. The donor plasmid targeting the human RSK2 locus was designed as described in Subheading 3 and synthesized by GeneOracle (Mountain View, CA, USA). Aside from indicated mutations (Fig. 1), the genome specific sequence within the donor covered 1,680 bp of ChrX: 20,190,035–20,191,714 based on the GRCh37/hg19 version of the human genome.

2.2 Transfection of ZFNs, Donor Plasmid, and Cell Culture

1. K562 cells.
2. Hemacytometer or other cell counting device.
3. Cell culture incubator.
4. T75 cell culture flasks.

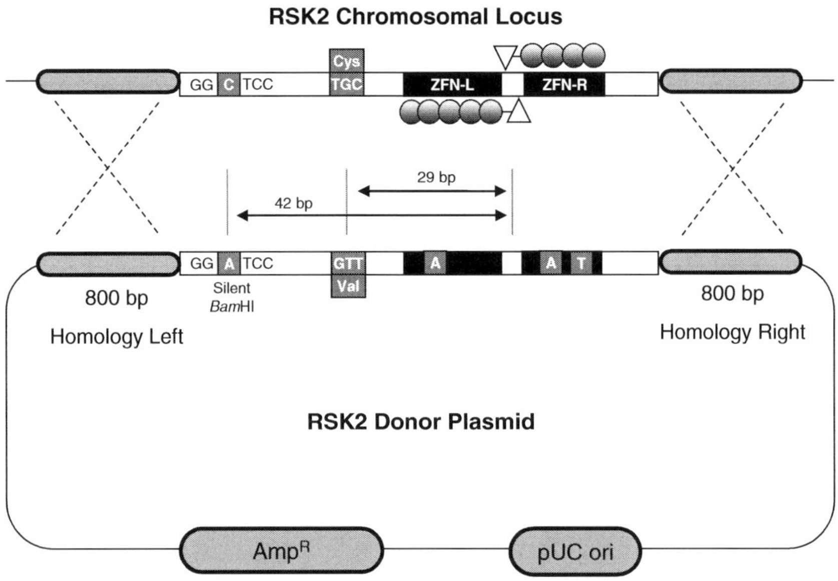

Fig. 1 Design of the RSK2 donor plasmid for creating a codon change. Three key mutations were implemented: (1) a silent mutation enabling detection and quantification via PCR and BamHI cleavage (RFLP), (2) the desired codon change (Cys-to-Val), and (3) mutations within the ZFN binding site that disrupt intracellular ZFN cleavage of the plasmid and/or cleavage of the chromosome post-integration of the donor plasmid

5. 6-well cell culture plates.
6. Cell culture medium: Iscove's Modified Dulbecco's Medium, 10 % FBS, 2 mM L-glutamine.
7. Hank's Balanced Salt Solution.
8. Nucleofector Kit V (Lonza).
9. Nucleofector II instrument (Lonza).
10. 96-well cell culture plates.
11. 5 ml round-bottom FACS tube.
12. Flow cytometer.

2.3 Genotyping of Pooled Cells and Clones for Targeted Integration Events

1. Out-out primers for RFLP analysis via BamHI:
   (a) SM165F 5′-TGCAAGCACATGAATGTATGG
   (b) SM165R 5′-ATGGAGGCAAGACACATCCT
   PCR product = 1,975 bp.
2. Junction primer sequences:
   (a) SM168R 5′-TCACAGCAAATTCCATGTTTGTT used with SM165F
   PCR product = 964 bp.
3. Restriction endonuclease BamHI.
4. Direct Load WideRange DNA Ladder (Sigma-Aldrich).

5. Mammalian Genomic DNA Miniprep Kit.
6. PCR Clean-Up Kit.
7. JumpStart Taq ReadyMix (Sigma-Aldrich).
8. Restriction endonuclease BamHI.
9. QuickExtract (Epicentre).
10. 10% TBE acrylamide gel.
11. PCR thermocycler.
12. Agarose.
13. 10× Tris-Borate-EDTA buffer: 1 M Tris Base, 1 M Boric Acid, 0.02 M EDTA (disodium salt).

3 Methods

3.1 ZFN Design and Positioning Relative to the Desired Mutation Site

If the project goal is to create a point mutation, the ZFN should be designed to cut within 100 bp of the desired mutation site if at all possible. Existing literature suggests that the penalty for moving a point mutation 100 bp away from the cut site is an approximate fourfold drop in mutation frequency [6] (*see* **Note 1**). A schematic for a donor plasmid design intended to create a codon change in the human RSK2 locus is shown in Fig. 1. In this case, a ZFN was successfully designed within 29 bases of the Cys-to-Val mutation site. For any particular gene or locus, it is best to thoroughly review published and database information about sequence variations, splice variants, etc. to ensure the ZFNs can bind and cut the target the locus of interest (*see* **Note 2**). Lastly, we recommend checking the copy number of the genomic locus of interest since many transformed cells may have in excess of two gene copies (triploid and tetraploid alleles are not uncommon, *see* **Note 3**).

3.2 Length of Homology Arms and Plasmid Backbone

When designing donor plasmids for making small mutations, it is recommended to use ~400–800 bp of homology in each arm centered close to the ZFN cut site (Fig. 1). Donor plasmids with 400–800 bp of homology in each arm have been shown to work robustly in different applications requiring point mutations [7, 8]. Increasing the homology arm lengths beyond 1 kb has generally not increased donor integration frequencies (*see* **Note 4**). Donor plasmids with shorter arm lengths (50–100 bp) have been shown to work [9], but are generally less effective across a broad range of cell types. It is reasonable to assume that drug selection methods might enable use of 50 bp homology arms in a broader range of cell types, but this has yet to be explored systematically.

For plasmid backbones, it is recommended to use small pUC-based plasmids (<2,600 bp) to limit the mass of DNA that is transfected to cells. To date, multiple variations of pUC based backbones have been used to successfully execute ZFN-based gene targeting experiments. Commercial gene synthesis companies generally prefer plasmid backbones which minimize the cost of their cloning and sequencing operations. It is acceptable to use whatever standard cloning vector vendors offer as long as the plasmid is small, lacks common restriction sites, and lacks significant sequence homology to the target region of interest.

3.3 Detecting Desired Mutations in Pooled and Clonal Cell Populations

Prior to investing the effort to derive and screen a clonal cell population, it is useful to rapidly estimate the mutation frequency of the particular ZFN-donor pair at the pooled cell level immediately following transfection (*see* **Note 5**). In the minority of cases, the mutation of interest will result in the creation or elimination of a restriction site, creating a convenient method for genotyping. In most scenarios, when restriction sites are not created or destroyed by the desired SNP, addition or ablation of a silent restriction site (i.e., a restriction fragment length polymorphism or RFLP site) somewhere in the donor will facilitate easy detection of mutant clones. Silent RFLP sites can be found by making silent point mutations within an open reading frame and scanning for new or deleted restriction enzyme sites. This process is very tedious to do by hand, and can be greatly expedited by using bioinformatics (*see* **Note 6**). The positioning of the silent RFLP site is very important: the RFLP site should be as close as possible to the desired functional mutation (or codon swap, SNP, etc.) and positioned so that the desired mutation is flanked by the RFLP site and the ZFN cut site whenever possible (Fig. 1). Mammalian homologous recombination machinery inserts donor-copied sequence directionally at decreasing frequencies when moving away from the ZFN cut site. This method of positioning the RFLP site will help ensure that when analyzing clones the desired mutation is present in all RFLP positive clones. If no RFLP site can be found with the desired positioning described above, more clones will likely need to be screened as some clones will be positive for the RFLP site but negative for the desired modification. When creating point mutations to create RFLP sites and amino acid changes, it is recommended to check the codon usage (*see* **Note 7**). In some cases, even silent mutations can have drastic effects on functional gene expression [10], so carefully consider the risks of making mutations in addition to the experimental mutation. After finding a good RFLP site, check that it is not located at another, nearby site as this will complicate downstream fragment analysis and genotyping.

3.4 Preventing Unwanted NHEJ Mutations When Creating Precise Point Mutations

If the ZFN cut site is present within an open reading frame (ORF) and the goal is to make a nearby codon change or SNP correction, it is possible that the ZFN will also cause undesirable secondary mutations at the cut site via aberrant NHEJ repair. This risk increases greatly as the distance between the ZFN cut site and desired mutation site increases. In our work with donor plasmids, some configurations resulted in 30–40 % of the correctly modified clones also having insertion-deletion (indel) mutations at the ZFN cut site, while other applications had >90 % indels. If a particular ZFN-donor combination is at high risk of contamination from indels, one solution is to incorporate mutations in the ZFN binding site. As a general practice for scenarios where the ZFN cut site and the desired mutation are separated by >50 bp, we recommend incorporating silent mutations in the donor plasmid within ZFN binding site to prevent pre-integration cleavage of the donor plasmid and/or re-cleavage of the target site post-HDR off of the plasmid. At least two nuclease-blocking mutations (NBMs) should be incorporated into the ZFN binding site—separated by at least 3 bp if they are within the same ZFN arm, or one placed in each arm (Fig. 1). The 3 bp spacing ensures each NBM lies in the binding sequence of different zinc fingers, thereby maximizing the chances for disruption of binding. Although we have not implemented NBMs in TALEN or CRISPR applications, recent data suggests that three mutations within a TALEN and CRISPR DNA-binding sequences is generally sufficient to suppress cutting [11, 12]. These mutations should be selected in a way that minimizes changes in codon usage frequency (*see* **Note 7**). This can be done by placing the NBMs at amino acids with highly degenerate codons, such as Ala, Val, Thr, Pro, Ser, Leu, Gly, and Arg. This will give more options for preserving codon usage frequency since they all have approximately four different codons to select from. NBMs have recently been used successfully in transgenic applications with oligo donors using both ZFNs [13]. Also note that the use of NBMs runs the risk of creating "locked" alleles which harbor the NBM mutations, but not the desired mutation. Such alleles cannot be targeted in a second round of transfection and single cell cloning. Such outcomes can be minimized by designing ZFNs as close as possible to the experimental mutation.

3.5 Detecting Mutations at the Pooled and Clonal Cell Level

Following transfection of the ZFN and donor construct, the next step is to assess the rate of mutation in the pooled cell population prior to single cell cloning. Two general methods may be used to detect mutations: RFLP, using restriction digest if a novel restriction site was incorporated or ablated, and/or mutation-specific junction PCR (*see* **Note 8**). Because both assays are based on PCR, it is essential to always include a "donor only" transfected sample to control for PCR artifacts, especially at the pooled-cell level when copy number of the donor plasmid will be high. For analyzing a pool of cells, the RFLP approach is the most quantitative gel-based method and best predicts success in follow-up single cell cloning efforts. For RFLP detection, two PCR primers should be designed that prime outside the region of homology present on the donor plasmid (Fig. 2). Genomic DNA is isolated 2–3 days post-transfection and can be used as a PCR template. The PCR product

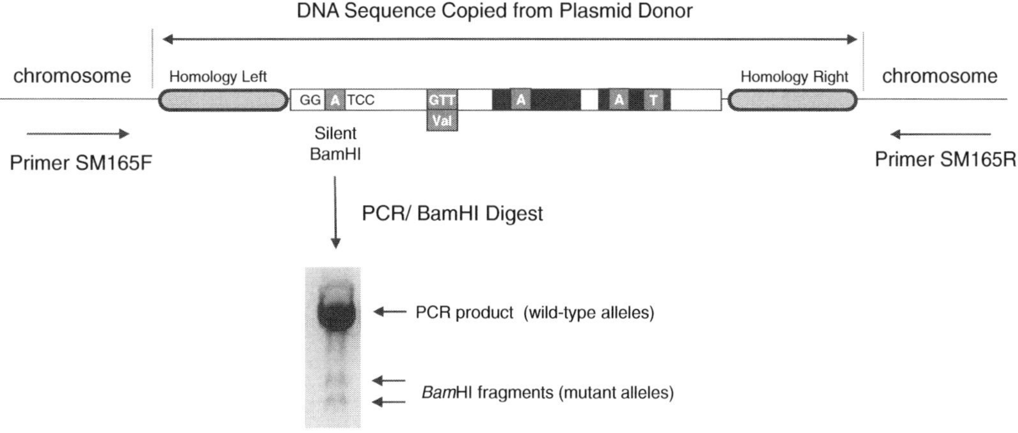

Fig. 2 Detection of modified alleles in a pool of human K562 cells by RFLP (a.k.a. out-out PCR) at 2 days post-transfection of ZFNs and donor plasmid. Note: PCR primers (SM165F and SM165R) are designed outside the region of homology contained on the donor plasmid to prevent amplification from the exogenous donor plasmid, which would result in a false positive detection of chromosomal modification

is then digested with the RFLP restriction enzyme (in this case BamHI) and resolved by electrophoresis on an agarose or polyacrylamide gel. The rate of targeted integration can be estimated by performing densitometry on the gel and comparing the intensity of the fragments released by restriction digestion against the intensity of the parental band (*see* **Note 9**).

In this particular experiment, the BamHI mutation was easily detectable by RFLP at the pooled cell level (Fig. 2), so cells were single cell cloned by flow cytometry and screened for donor recombination by junction PCR using mutation specific primers (Fig. 3). Junction PCR was used as the preliminary screen as the assay is not reagent and labor intense when screening hundreds of clones yielding a positive or negative signal with only PCR. However, junction PCR only indicates that a clone contains a modified allele but does not indicate the number of alleles that are modified within a given clone. Therefore, in this set of experiments, several positive clones were identified and queried again using the same RFLP assay applied earlier at the pooled cell level (Fig. 4). At the clonal cell level, the gel-based RFLP results become even more quantitative since there are now a limited number of alleles (usually two to four). For example, in Fig. 4, clones in which allele conversion was complete show a near complete lack of the parental PCR band (clones B7, D11, E5, and G9), and clones which are heterozygous show partial conversion (D8, E9, F10, and F5). Existing karyotype data suggests two copies of the RSK2 locus on chromosome X (*see* **Note 3**).

To further characterize the clones which are positive by RFLP, the PCR products from out-out PCR were ligated into cloning vectors and sequenced. This step separates alleles via *E. coli*-based transformation and cloning and allows for a detailed analysis of all

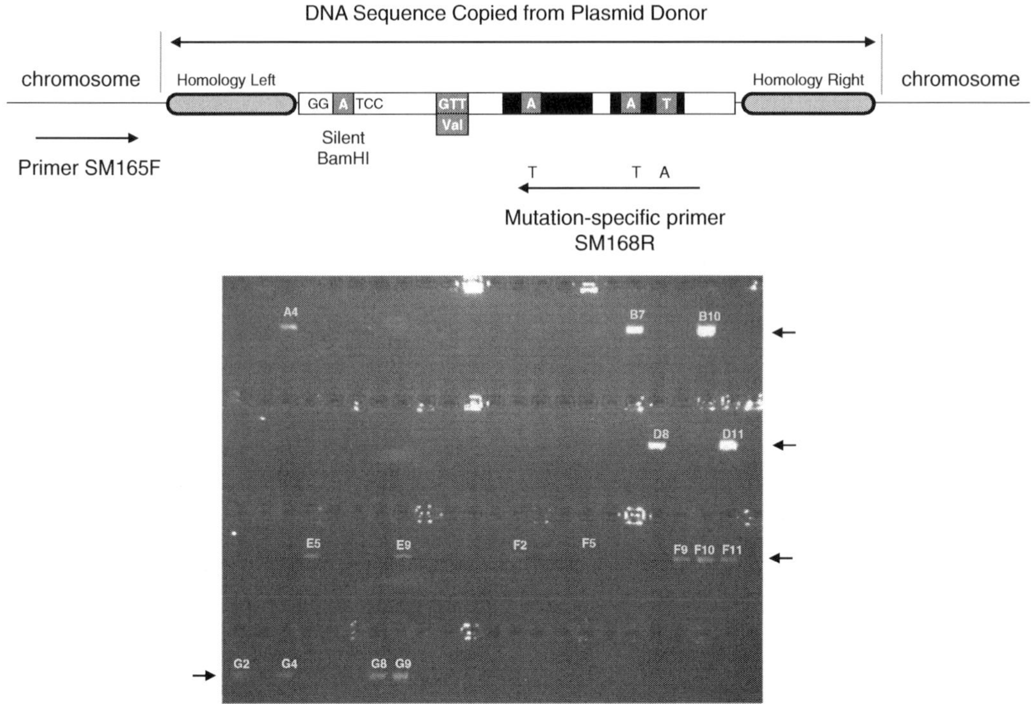

Fig. 3 Detection of modified alleles in clonal cells using junction PCR method with a mutation-specific primer (SM168R) and genome-specific primer (SM165F). Putative positive clones are labeled with *well-numbers* and *arrows*

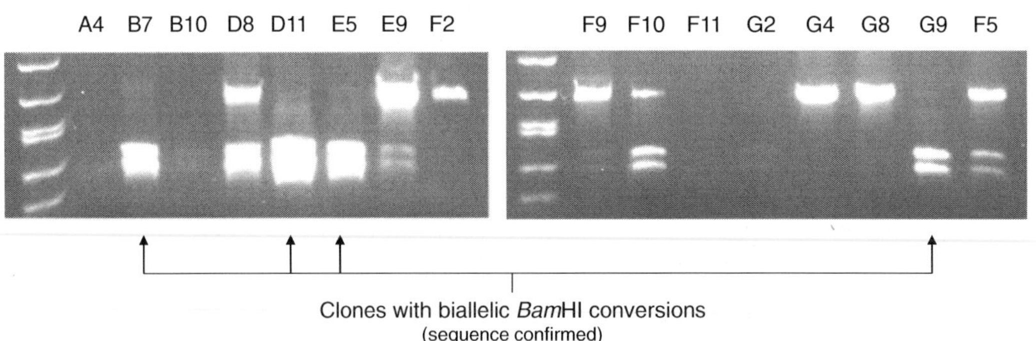

Fig. 4 Confirmation of modified alleles in clonal K562 cells by RFLP at 3-weeks post-transfection of ZFNs and donor plasmid. Clones for this assay were chosen based on positive clones from the previous junction PCR (Fig. 3)

mutations and alleles present. Upon sequencing RFLP-positive clones, we observed that donor information was incorporated in various ways ranging from complete conversion to partial conversions biased on either side of the ZFN cut site (Fig. 5). These results are consistent with previously observed results using targeted nucleases and donor plasmids [2]. Lastly, if your institution

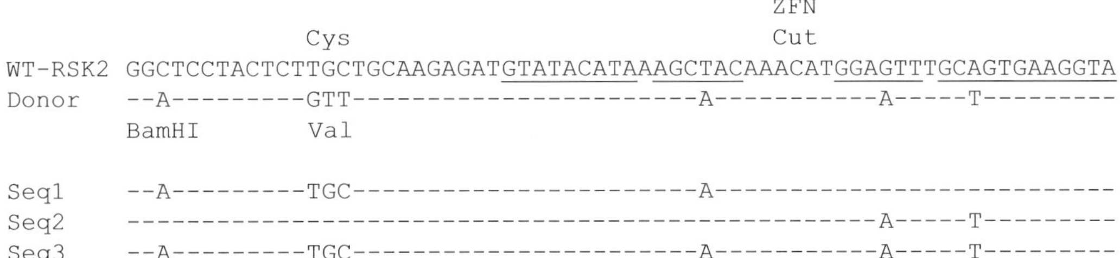

Fig. 5 Modified RSK2 sequence types (Seqs 1–3) observed in the K562 cell population post-transfection of RSK2 ZFNs and donor plasmid. Note that for sequences with partial donor incorporation, donor information is biased to the left or right of the ZFN cut site (ZFN binding regions are *underlined*), suggesting that, in repair of discrete alleles, only one side of the break primes the donor plasmid

has economical access to the required equipment and bioinformatics resources, deep sequencing technology can be used to accurately genotype cells at various stages within genome editing protocols to assess the transfer of modifications from donor plasmids [14].

4 Notes

1. Elliott et al. documented a point mutation 511 bp from a nuclease cut site, however, it occurred at a frequency that would likely require screening of >1,000 clones to isolate a mutant in the absence of selection. Using drug selection methods, it is certainly reasonable to expect that mutations can be incorporated >500 bp from the cut site, but these applications require position-restricted integration of large selectable gene cassettes that complicate ZFN design and risk affecting gene expression and regulation.
2. Before designing the nuclease or donor, it is best sequence the genomic locus in the cell type or animal of interest. Actual sequence data from biological samples may vary in unexpected ways from data published in the literature or databases. PCR amplification and sequencing prior to gene targeting work will also give valuable information about what to expect during genotyping efforts in the genome editing workflow. For example, the GC-content of some loci can make amplification of large fragments (>2 kb) very difficult.
3. Two useful databases for checking allele copy number in a wide variety of common cell types are the NCBI SKY database (http://www.ncbi.nlm.nih.gov/sky/) and the Sanger CGH viewer (http://www.sanger.ac.uk/cgi-bin/genetics/CGP/cghviewer/CghViewer.cgi).
4. Although increasing homology arm length to sizes of 5–10 kb has benefited the mouse genetics community by increasing

HDR rates in mES cells, increasing homology arms in ZFN-compatible donor plasmids beyond 1 kb has not been shown to provide significant benefits. Additionally, when designing homology arms, try to avoid repeat elements (e.g., Alu or other as cataloged by repeatmasker.org). Although this chapter is focused on small mutations, please be aware that as the amount of exogenous, nonhomologous DNA sequence between the homology arms increases, the rate of integration will likely decrease. For donors with inserts in excess of 5 kb, we generally recommend using drug selection in single cell cloning efforts. Lastly, we have observed no benefits in linearizing the donor plasmids prior to transfection and recommend using supercoiled, endotoxin-free maxi-prepped DNA.

5. If you plan to use DSB-based genome editing methods in a human cell type for which no published protocol exists, a good starting point is to test the ZFN and donor together in human K562 cells. Previous experience has shown that among human cell types, K562 cells can support extraordinarily high frequencies of HDR [15]. Thus, if your ZFN and donor pair have fundamental molecular design problems, these can be quickly resolved by preliminary experiments in pooled K562 cells prior to exploring a more challenging or unknown cell type. For mouse and rat cell culture, we recommend using Neuro2A and C6 cells, respectively, for initial tests of ZFN and donor compatibility.
6. An excellent resource for finding silent mutations which result in new restriction sites is WatCut: (http://watcut.uwaterloo.ca/watcut/watcut/template.php). Within the WatCut site, the option is termed "silent mutation analysis".
7. After a silent mutation is identified which can be used for genotyping, two methods can be used to check the quality of the new codon. First, there are numerous online codon frequency tables covering entire genomes. Second, the codon frequency within the targeted exon (or entire ORF) can be inspected and used to provide confidence in the quality of a new codon. For example, if you want to use a new codon (GCG) for Ala, check the nearest four to five Ala codons to see how frequently GCG is used.
8. When designing genotyping primers to detect single base changes, take care when the mutation is near the extreme 3′ end of the primer. In some cases, the 3′-end can be removed by contaminating exonucleases, thus destroying the SNP-level specificity of the amplification. Previous work has shown that placing a single phosphorothioate linkage at the 3′-end of primers can increase SNP detection specificity [16].
9. For quantitation, ImageJ is a freely available software package useful for performing gel-base densitometry. If the mass of the

parental PCR band is in excess of 700 ng, then the signal from the band is likely saturated and out of the linear range required for accurate quantitation. In our experience, RFLP signals that are visually observable by EtBr staining, regardless of parental band intensity, have high chances of success for clonal isolation. Fig. 2 shows a good example of overloading of the parental PCR product, with observable RFLP fragments. Despite overloading, this experiment yielded correctly targeted biallelic K562 clones at a rate of 6 of 112 clones screened (Fig. 5).

Acknowledgements

We would like to thank Morten Frödin at the University of Copenhagen for helpful advice on the functionality of the RSK2 locus.

References

1. Bollag RJ, Waldman AS, Liskay RM (1989) Homologous recombination in mammalian cells. Annu Rev Genet 23:199–225
2. Rouet P, Smith F, Jasin M (1994) Introduction of double-strand breaks into the genome of mouse cells by expression of a rare-cutting endonuclease. Mol Cell Biol 14:8096–8106
3. Bibikova M, Beumer K, Trautman JK, Carroll D (2003) Enhancing gene targeting with designed zinc finger nucleases. Science 300:764
4. Urnov FD, Miller JC, Lee YL et al (2005) Highly efficient endogenous human gene correction using designed zinc-finger nucleases. Nature 435:646–651
5. Chen F, Pruett-Miller SM, Huang Y et al (2011) High-frequency genome editing using ssDNA oligonucleotides with zinc-finger nucleases. Nat Methods 8:753–755
6. Elliott B, Richardson C, Winderbaum J et al (1998) Gene conversion tracts from double-strand break repair in mammalian cells. Mol Cell Biol 18:93–101
7. Moehle EA, Rock JM, Lee YL et al (2007) Targeted gene addition into a specified location in the human genome using designed zinc finger nucleases. Proc Natl Acad Sci U S A 104:3055–3060
8. Soldner F, Laganiere J, Cheng AW et al (2011) Generation of isogenic pluripotent stem cells differing exclusively at two early onset Parkinson point mutations. Cell 146:318–331
9. Orlando SJ, Santiago Y, DeKelver RC et al (2010) Zinc-finger nuclease-driven targeted integration into mammalian genomes using donors with limited chromosomal homology. Nucleic Acids Res 38:e152
10. Kimchi-Sarfaty C, Oh JM, Kim IW et al (2007) A "silent" polymorphism in the MDR1 gene changes substrate specificity. Science 315:525–528
11. Hsu PD, Scott DA, Weinstein JA et al (2013) DNA targeting specificity of RNA-guided Cas9 nucleases. Nat Biotechnol 31:827–832
12. Mali P, Yang L, Esvelt KM et al (2013) RNA-guided human genome engineering via Cas9. Science 339:823–826
13. Meyer M, Ortiz O, Hrabe de Angelis M et al (2012) Modeling disease mutations by gene targeting in one-cell mouse embryos. Proc Natl Acad Sci U S A 109:9354–9359
14. Yang L, Guell M, Byrne S et al (2013) Optimization of scarless human stem cell genome editing. Nucleic Acids Res 41:9049–9061
15. DeKelver RC, Choi VM, Moehle EA et al (2010) Functional genomics, proteomics, and regulatory DNA analysis in isogenic settings using zinc finger nuclease-driven transgenesis into a safe harbor locus in the human genome. Genome Res 20:1133–1142
16. Zhang J, Li K (2003) Single-base discrimination mediated by proofreading 3′ phosphorothioate-modified primers. Mol Biotechnol 25:223–228

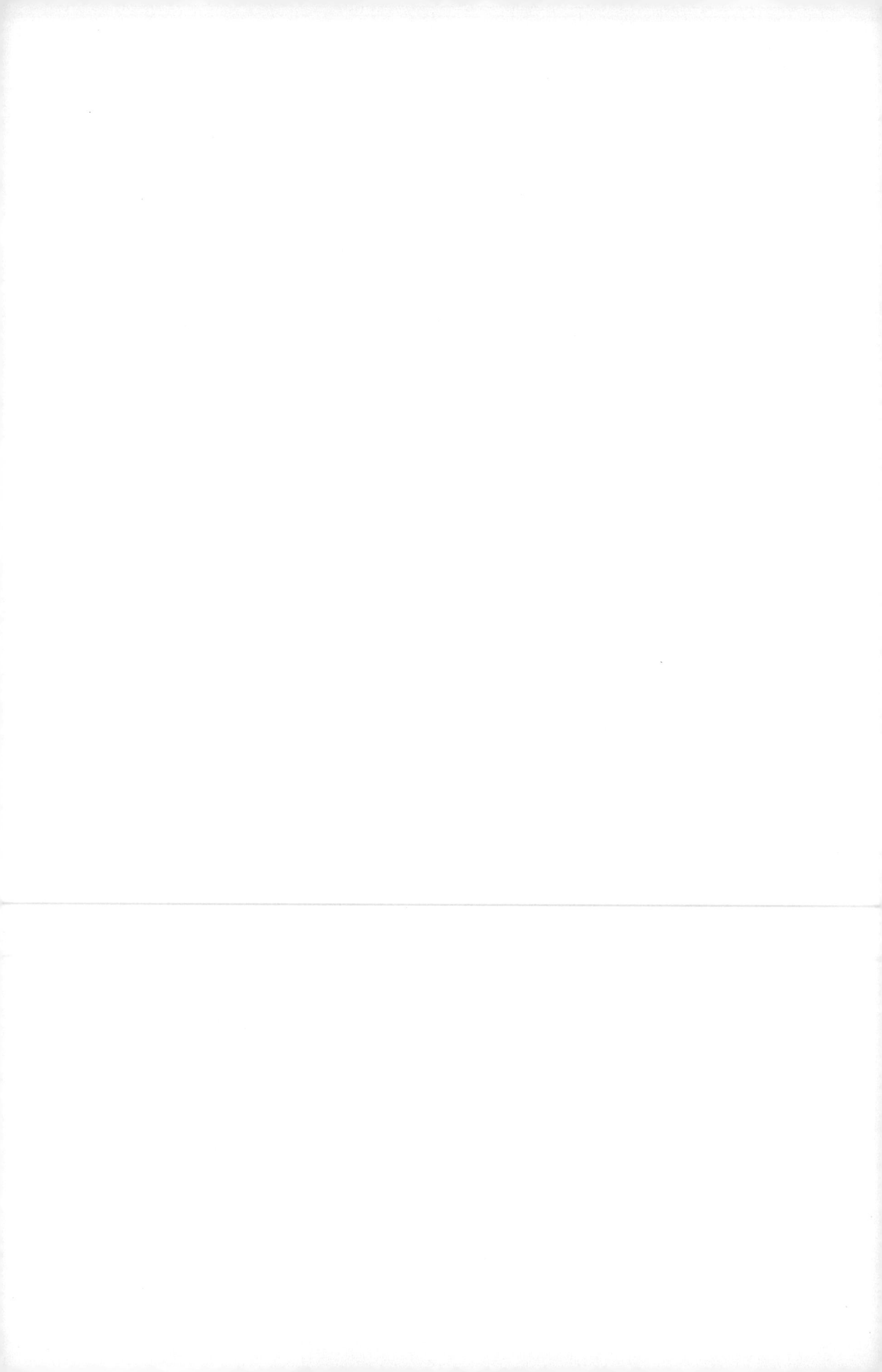

Chapter 12

Endogenous Gene Tagging with Fluorescent Proteins

John Fetter, Andrey Samsonov, Nathan Zenser, Fan Zhang, Hongyi Zhang, and Dmitry Malkov

Abstract

Human genome manipulation has become a powerful tool for understanding the mechanisms of numerous diseases including cancer. Inserting reporter sequences in the desired locations in the genome of a cell can allow monitoring of endogenous activities of disease related genes. Native gene expression and regulation is preserved in these knock-in cells in contrast to cell lines with target overexpression under an exogenous promoter as in the case of transient transfection or stable cell lines with random integration. The fusion proteins created using the modern genome editing tools are expressed at their physiological level and thus are more likely to retain the characteristic expression profile of the endogenous proteins in the cell. Unlike biochemical assays or immunostaining, using a tagged protein under endogenous regulation avoids fixation artifacts and allows detection of the target's activity in live cells. Multiple gene targets could be tagged in a single cell line allowing for the creation of effective cell-based assays for compound screening to discover novel drugs.

Key words Zinc-finger nuclease (ZFN), Genome editing, Fluorescent protein (FP), Live cells, Endogenous expression, Lamin, Actin, Tubulin, EGFR

1 Introduction

Fluorescent reporter genes (fluorescent proteins—FP) are the most popular tags used extensively to provide a visual readout of a protein of interest in cells [1]. Other reporter tag types that bind small molecule fluorescent probes can also be used [2]. Uses of tagged proteins include the study of protein abundance and localization, transcriptional and translational regulation, posttranslational modifications, protein–protein interactions, alternative splicing, RNAi-dependent effects, and others.

Conventional transfection techniques allow expressing FP-tagged recombinant proteins, but in most cases this leads to overexpression of the fusion proteins as their expression often relies on heterologous promoters. In addition, some tagged proteins are expressed from episomal or randomly integrated vectors and are

Shondra M. Pruett-Miller (ed.), *Chromosomal Mutagenesis*, Methods in Molecular Biology, vol. 1239,
DOI 10.1007/978-1-4939-1862-1_12, © Springer Science+Business Media New York 2015

therefore not controlled by the endogenous regulatory pathways leading to nonphysiological expression patterns. The overexpression might create strong artifacts in intracellular protein trafficking and protein-mediated processes [3] and might not accurately reflect the expression pattern of the endogenous locus. Moreover, standard transfection methods are not ideal for FP-tagging of multiple proteins due to extremely high heterogeneity of exogenous protein expression within a population of cells.

In contrast, the overexpression artifacts are avoided when specific reporter integration into a chromosome of a cell produces a tagged protein expressed at native level and controlled by complex endogenous regulation. One way to achieve this targeted integration (TI) into the genome is by using zinc finger nucleases (ZFNs). ZFN technology allows simultaneous tagging of multiple endogenous proteins with FPs of different colors, making this a promising approach for drug discovery.

Classical ZFNs are fusions of zinc finger proteins (ZFPs) that provide binding affinity and specificity, and the catalytic DNA-cleavage domain of FokI, a type II endonuclease. ZFNs facilitate efficient targeted editing of the genome by creating double-strand breaks (DSBs) at user-specified locations. The cell then employs the natural DNA repair mechanisms of either error-prone nonhomologous end-joining (NHEJ), single-strand annealing (SSA), or high-fidelity homologous recombination (HR) [4].

The HR repair pathway enables insertion of a reporter transgene into the targeted region. To utilize HR, a donor template is used that contains the transgene flanked by sequences homologous to the regions on either side of the cleavage site. This donor is co-delivered into the cell along with the ZFNs to fool the cell by presenting the donor in place of the sister chromatid to repair the cut. ZFNs have been shown to increase the HR rate several orders of magnitude near the cut site [5]. This approach has been used for tagging of various genes in different cell lines [3, 6–8].

2 Materials

2.1 ZFN Modification of Cell Lines

1. CompoZr® ZFNs (Designed and manufactured by Sigma-Aldrich) applying bioinformatic tools to assure that the ZFN binding site is unique (four or more mismatches for a 24 base pair recognition site) within the genome [9]. High ZFN specificity was also guaranteed by using obligate heterodimer FokI cleavage domains [10].).
2. Donor plasmid with fluorescent tag surrounded by homology arms (Genescript, Piscataway, NJ, USA). The fluorescent reporter genes BFP, GFP, and RFP used are monomeric and were obtained from Evrogen, referred to as TagBFP, TagGFP2, and TagRFP, respectively [11].

3. Human lung carcinoma A549 and osteosarcoma U2OS cell lines.
4. Medium for A549 cells: RPMI-1640 medium supplemented with 2 mM L-glutamine and 10 % fetal bovine serum.
5. Medium for U2OS: McCoy's 5A medium supplemented with 2 mM L-glutamine and 10 % fetal bovine serum.
6. Trypsin–EDTA solution.
7. MessageMAX™ T7 ARCA-Capped Message Transcription Kit (Cellscript, Inc., Madison, WI, USA).
8. Poly(A) Polymerase Tailing Kit (Epicentre Biotechnologies, Madison, WI, USA).
9. Ambion® MEGAclear™ Kit (Life Technologies, Carlsbad, CA, USA).
10. Amaxa® Nucleofector® device and appropriate Nucleofector® Kit (Lonza AG, Visp, Switzerland).
11. Hanks balanced salt solution supplemented with 2 % fetal bovine serum for microscopy imaging of cells.
12. FACS Aria III (BD Biosciences, San Jose, CA, USA).

2.2 Cell Analysis

1. Surveyor® mutation detection kit (Transgenomic Inc, Omaha, NE, USA) was used to assess ZFN cutting efficiency [12].
2. JumpStart™ REDTaq® ReadyMix™ PCR Reaction Mix (Sigma-Aldrich, St. Louis, MO, USA).

3 Methods

3.1 Initial Considerations

Before committing to a project on gene tagging, several issues must be taken into account.

1. Ensure that the signal coming from the tag can be reliably measured. Since the strength of this signal is dictated by the endogenous expression of your gene of interest (GOI), its expression level in the cell line of choice should be high enough for robust detection. The endogenous expression of the vast majority of genes including kinases and transcription factors is orders of magnitude lower than the overexpression levels to which most researchers are accustomed. An idea on the relative expression of your GOI across different cell lines can be obtained from public databases of microarray data or proteomic analysis (e.g., the NCI-60 cell line panel [13]). Since gene expression does not always correlate with protein expression, it is a good idea to evaluate the capabilities of your detection system by performing immunofluorescence experiments for the GOI in the parental cell line.

2. Fluorescent proteins are available that tend to express as dimers or as monomers. It is preferable to use a monomeric fluorescent protein to prevent interference with the target's function. High extinction coefficients and quantum yields are preferred.
3. For most studies, only one copy of a gene is tagged leaving the remaining alleles unchanged (WT). This is the case not only because heterozygous knock-ins are easier to achieve but also because tagging only one copy perturbs the system to the least extent possible. If necessary, all copies could be tagged to get a complete (homozygous) knock-in. Copies of the same gene could also be tagged with different reporters (colors) to study heterogeneity of the expression coming from different alleles.

3.2 Donor Design and Production

(*See* Chapter 11 by Davis and colleagues for additional considerations.)

Several considerations should be taken before and during donor design.

1. Take into account related pseudogenes and/or highly homologous family members as their homology to the target gene could impose constraints on ZFN/donor design.
2. It is known that genome editing can be done ~100 bp away from the cut site without drastic drops in efficiency [5, 14]. We successfully inserted reporters with up to 100 bp between the cut site and the integration site (Figs. 1a and 2a).
3. Synthesize donors using a gene synthesis company with the fluorescent protein sequence sandwiched between ~700 and 800 bp homology arms. The base vectors for the donors are derived from pUC plasmids and are AMP resistant in bacteria. The donor vectors are approximately 5 kb in size.

Fig. 1 (continued) site (*scissors*), and the tag sequence integration site (*red arrow*). Target locus names are indicated at the *top* with the chromosome number in the *parentheses*. *Right*—Schematics of corresponding loci showing the coding region (*blue*), untranslated region (*gray*), and the ZFN cut site (*scissors*). The Donors (*top*) have the homology arms of indicated length and the FP sequence (*boxed*) fused to the beginning of the target gene coding sequence (all three inserts result in N-terminal fusions). In the case of TUBA1B the first exon contains ATG only. To preserve its splice signal, the FP sequence was inserted before the ATG. Another ATG was introduced in front of FP to initiate transcription. The fusion linkers between the FP and the target protein are: GSGSGGT (lamin B1), AAGD (α-tubulin), and AGSGT (β-actin). (**b**) Imaging of the triple tagged cell line. Differential interference contrast (DIC) image (*a*) and fluorescence microscopy images of the BFP (*b*), GFP (*c*), and RFP (*d*) channels and the corresponding overlay image (*e*) of an isolated triple knock-in clone are shown. The cells were imaged live in Hanks balanced salt solution supplemented with 2 % fetal bovine serum using the corresponding filter sets and a 40×/1.3 oil objective (*see* **Note 3**). The scale bar equals 25 μm

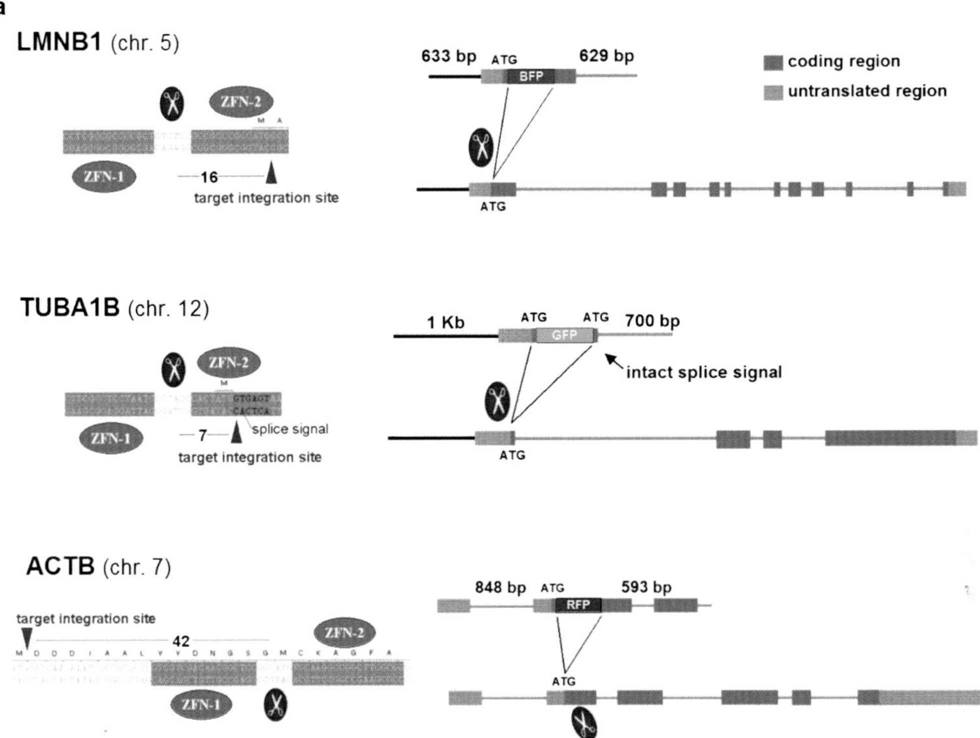

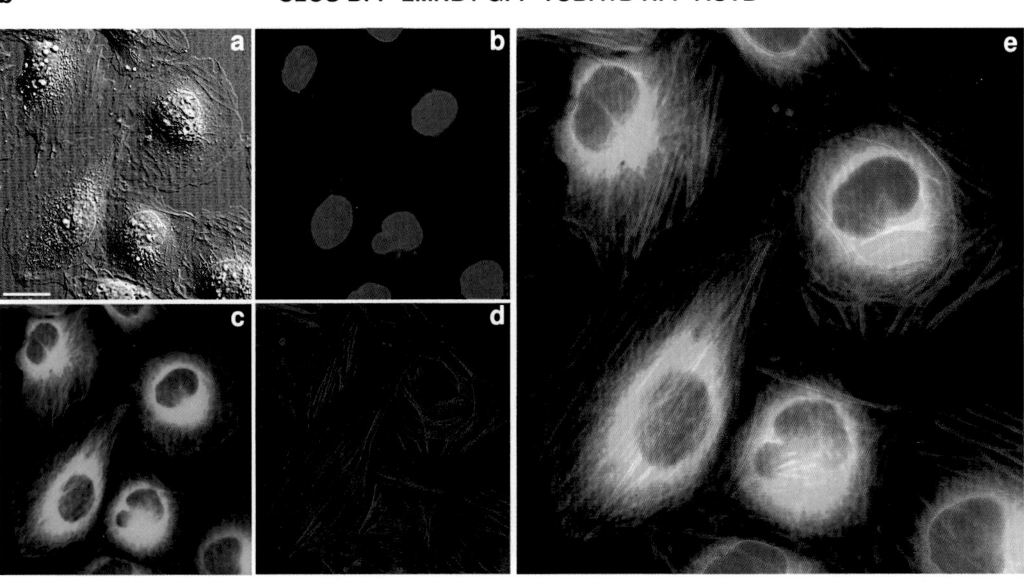

Fig. 1 Triple tagged U2OS osteosarcoma cells (BFP-LMNB1/GFP-TUBA1B/RFP-ACTB). Three genomic loci (cytoskeletal genes LMNB1, TUBA1B, and ACTB) have been endogenously tagged with fluorescent protein genes for BFP, GFP, and RFP, respectively, using ZFN technology. The integration resulted in expression of three distinct FP fusion proteins (BFP-lamin B1, GFP-α-tubulin and RFP-β-actin) from the endogenous genomic loci (*see* **Note 1**). (**a**) Schematics of the LMNB1, TUBA1B, and ACTB tagged loci. *Left*—Schematic of the genomic sequence at the target region for integration of the fluorescent tag (FP DNA) showing ZFN binding sites (*blue boxes*), the ZFN cut

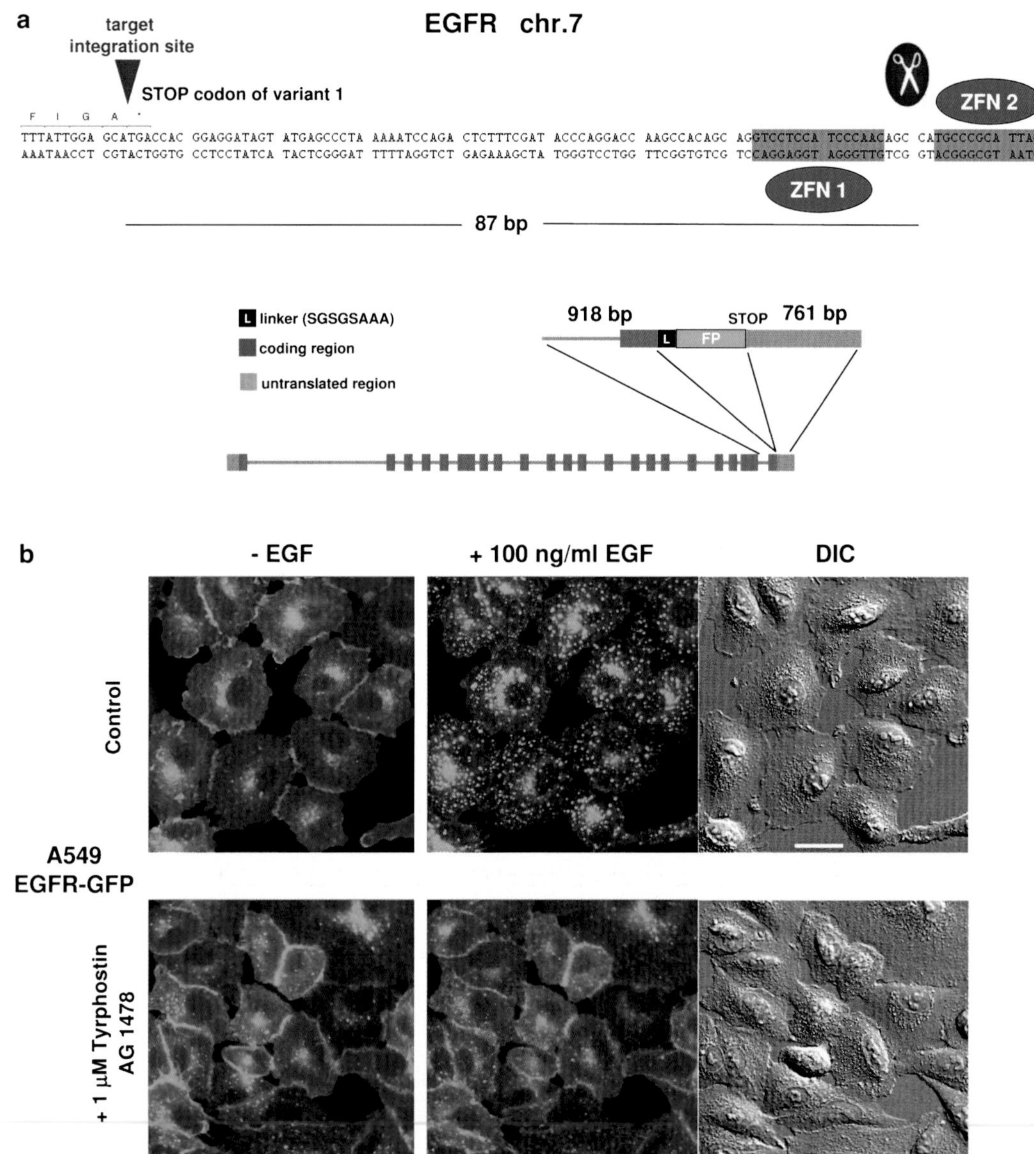

Fig. 2 EGFR Gene endogenously tagged with GFP at the C-Terminus in A549 lung carcinoma cells (*see* **Note 3**). (**a**) Schematics of the EGFR tagged locus (main splice variant). The gene sequence and relative positions for the ZFN binding site, ZFN cut site, and the targeted integration site are shown. The integration position of the donor relative to the introns and exons of the gene is indicated in the lower schematic. The donor design is shown with the homology arms of indicated length and the FP insert. (**b**) Fluorescence microscopy and differential interference contrast (DIC) images of an A549 EGFR-GFP isolated single-cell clone. The cells were imaged live before and after addition of 100 ng/mL EGF using a 40×/1.3 oil objective (*see* **Note 2**). Preincubation with 1 μM Tyrphostin AG 1478, a selective EGFR inhibitor, for 20 min prior to the addition of EGF blocks the internalization of the fusion protein. The scale bar is equal to 25 μm

4. Place a short linker consisting of small hydrophilic amino acids (a few copies of alternating glycine and serine [15]) between the FP and the target protein to improve its accessibility. The linker sequence is often shared with restriction sites that allow quickly swapping the insert of the donor. The linker can be replaced by a translational skip sequence (2A) [16, 17] if reporting of only the target's promoter activity is required.
5. Chose the location of the FP-insert (N vs. C terminus of the GOI) to ensure that the resulting FP-fusion does not block the normal localization and functionality of the protein of interest. For some target genes, the functionality can be preserved only if the reporter is inserted in the middle of the target gene rather than at the terminus (e.g., G proteins [18]). Search the literature for known cDNA fusion constructs with the target's function preserved before deciding on the insert placement.
6. The requirement for the preservation of function might place some constraints on whether all splice variants/isoforms of the target gene are tagged.

3.3 Nucleofection and Clone Isolation

1. Nucleofect donor constructs (1 μg per kb) containing each insert flanked by homology arms as described above along with ZFN mRNA (1 μg for each ZFN in the pair) designed to cut near the genomic target site (Fig. 1) into 1 million cells with the Amaxa® Nucleofector® device and appropriate Nucleofector® Kit from Lonza AG according to the product manual. To boost cutting efficiency, the cells can be placed at 30 °C or "cold-shocked" immediately after nucleofection for several days [19].
2. Successful HR-driven integration results in endogenous expression of a fluorescent fusion protein. Sort cells using fluorescence to single cells by flow cytometry and expanded into clonal populations. Multiple rounds of enrichment are required in cases of low integration rate and/or low signal before proceeding to single-cell sorting. Screen clones to select a single FP-tagged clone as a stable cell line. This line can be used as starting material for the next round of integration if tagging of multiple targets is required.
3. In addition to fluorescence, use junction PCR to detect TI with high sensitivity. Junction PCR amplifies one of the junctions at the ends of the insert with one primer sitting on the insert and the other on the genomic sequence outside of the homology arms. Sequencing at least one of the junctions can ensure the correct donor integration.
4. Perform Western Blot analysis of the generated clonal cell lines to assess the expression and function (e.g., phosphorylation) of FP fusion proteins relative to the wild type. The number of tagged vs. WT copies can be determined by digital PCR or sequencing. Southern hybridization analysis can be done to check for off-target insertions of the reporter sequences.

4 Notes

1. Tagging multiple endogenous proteins
We used ZFN technology to generate a triple reporter U2OS cell line (BFP-LMNB1/GFP-TUBA1B/RFP-ACTB) by inserting three different fluorescent reporter sequences behind the start codon of three different cytoskeletal genes (Fig. 1). All three gene loci were modified in the same cell line to allow simultaneous real-time reporting of their endogenous activity. The following loci were tagged: LMNB1 (lamin B1), TUBA1B (α-tubulin isoform 1b), ACTB (β-actin) by BFP, GFP, and RFP respectively. The integration resulted in endogenous expression of the corresponding fusion proteins: BFP-laminB1 that polymerizes to form the nuclear envelope (Fig. 1b, panel b), GFP-α-tubulin that polymerizes to form microtubules (Fig. 1b, panel c), and RFP-β-actin that polymerizes to form actin fibers (Fig. 1b, panel d). The reporter integration was done in three rounds but it should be possible to tag all three genes in one round by co-nucleofecting all ZFNs and donors together. Single cell knock-in clones were isolated to establish a stable cell line co-expressing BFP-lamin B1, GFP-α-tubulin, and RFP-β-actin from the endogenous genomic loci (Fig. 1b, panel e). At least one copy was tagged and at least one copy was not modified for all three targets (the supporting data not shown). This cell line could be used for high content screening to identify compounds that modulate cellular activity.
2. Function assessment by imaging
Fluorescent imaging of cells in the examples was done with a Nikon Eclipse TE2000-E inverted research microscope, a Photometrics CoolSNAP ES2 cooled CCD camera, and MetaMorph® software using a 40×/1.3 oil objective. Cells were imaged live in Hanks balanced salt solution supplemented with 2 % fetal bovine serum. Filtersets were BFP (ex 395–410/em 430–480), GFP (ex 450–490/em 500–550), and RFP (ex 530–560/em 590–650). The redistribution of the FP fusion proteins over time could be quantified using translocation journals in MetaMorph®. Less than 1 μM Hoechst 33342 could be used as a nuclear marker if necessary.
3. Tagging endogenous EGFR reveals its localization and activation in live cells
EGFR (Epidermal Growth Factor Receptor, also known as ErbB1) belongs to a class of receptor tyrosine kinases (RTKs). Abnormally high expression of EGFR has been correlated with many types of cancer. Overexpression of EGFR and other RTKs leads to constitutive activation of oncogenes, thus qualifying them as high-profile cancer targets. Unfortunately, a high rate of EGFR mutagenesis in cancer results in a loss of efficacy for drugs inhibiting its kinase activity [20].

Current EGFR activation and internalization assays are based on overexpression of EGFR-GFP fusion protein or detection of phosphorylated EGFR by immunohistochemistry. A major limitation of these methods is their inability to detect native EGFR in living cells. To fill this gap, we developed an A549 cell line that is modified at the endogenous EGFR locus to report activation and localization of the protein.

Using ZFNs in conjunction with a donor construct, a fluorescent transgene was inserted at the C-terminus of EGFR (Fig. 2a) and a single-clone-derived EGFR-GFP cell line was created in A549 cells (Fig. 2b). The resulting fluorescent fusion protein was expressed under endogenous regulation preserving the protein function. Activation of EGFR leads to receptor internalization in wild type cells [21]. Presence of the fluorescent tag on endogenous EGFR allows tracking of this normal internalization process after the addition of EGF to the cells (Fig. 2b). Internalization kinetics can then be quantified by image analysis. Application of a selective inhibitor of EGFR, Tyrphostin AG 1478, verifies a target-dependent effect. Having endogenous gene expression/regulation and preserved protein function for FP-tagged EGFR makes this cell line valuable for high-content screening of compound libraries to find novel modulators of EGFR activity.

References

1. Crivat G, Taraska JW (2012) Imaging proteins inside cells with fluorescent tags. Trends Biotechnol 30(1):8–16
2. Jung D, Min K, Jung J, Jang W, Kwon Y (2013) Chemical biology-based approaches on fluorescent labeling of proteins in live cells. Mol Biosyst 9(5):862–872
3. Doyon JB, Zeitler B, Cheng J, Cheng AT, Cherone JM, Santiago Y, Lee AH, Vo TD, Doyon Y, Miller JC, Paschon DE, Zhang L, Rebar EJ, Gregory PD, Urnov FD, Drubin DG (2011) Rapid and efficient clathrin-mediated endocytosis revealed in genome-edited mammalian cells. Nat Cell Biol 13(3):331–337
4. Moynahan ME, Jasin M (2010) Mitotic homologous recombination maintains genomic stability and suppresses tumorigenesis. Nat Rev Mol Cell Biol 11(3):196–207
5. Elliott B, Richardson C, Winderbaum J, Nickoloff JA, Jasin M (1998) Gene conversion tracts from double-strand break repair in mammalian cells. Mol Cell Biol 18(1):93–101
6. Goldberg AD, Banaszynski LA, Noh KM, Lewis PW, Elsaesser SJ, Stadler S, Dewell S, Law M, Guo X, Li X, Wen D, Chapgier A, DeKelver RC, Miller JC, Lee YL, Boydston EA, Holmes MC, Gregory PD, Greally JM, Rafii S, Yang C, Scambler PJ, Garrick D, Gibbons RJ, Higgs DR, Cristea IM, Urnov FD, Zheng D, Allis CD (2010) Distinct factors control histone variant H3.3 localization at specific genomic regions. Cell 140(5): 678–691
7. Sigma-Aldrich Website. Sigma-Aldrich CompoZr® cytoskeletal and pathway marker cell lines. http://www.sigmaaldrich.com/life-science/cells-and-cell-based-assays/compozr-cytoskeletal-marker-cells.html. Accessed 5 Jun 2013
8. Samsonov A, Zenser N, Zhang F, Zhang H, Fetter J, Malkov D (2013) Tagging of genomic STAT3 and STAT1 with fluorescent proteins and insertion of a luciferase reporter in the cyclin D1 gene provides a modified A549 cell line to screen for selective STAT3 inhibitors. PLoS One 8(7):e68391
9. Sigma-Aldrich Web site. CompoZr® zinc finger nuclease technology. http://www.sigmaaldrich.com/life-science/zinc-finger-nuclease-technology.html. Accessed 5 Jun 2013
10. Miller JC, Holmes MC, Wang J, Guschin DY, Lee YL, Rupniewski I, Beausejour CM, Waite

AJ, Wang NS, Kim KA, Gregory PD, Pabo CO, Rebar EJ (2007) An improved zinc-finger nuclease architecture for highly specific genome editing. Nat Biotechnol 25(7): 778–785

11. Evrogen Website. TagFPs: protein localization tags. http://evrogen.com/products/TagFPs.shtml. Accessed 5 Jun 2013
12. Kulinski J, Besack D, Oleykowski CA, Godwin AK, Yeung AT (2000) CEL I enzymatic mutation detection assay. Biotechniques 29(1): 44–46, 48
13. Moghaddas Gholami A, Hahne H, Wu Z, Auer FJ, Meng C, Wilhelm M, Kuster B (2013) Global proteome analysis of the NCI-60 cell line panel. Cell Rep 4(3):609–620
14. Porteus MH (2006) Mammalian gene targeting with designed zinc finger nucleases. Mol Ther 13(2):438–446
15. Snapp EL (2009) Fluorescent proteins: a cell biologist's user guide. Trends Cell Biol 19(11):649–655
16. Donnelly ML, Luke G, Mehrotra A, Li X, Hughes LE, Gani D, Ryan MD (2001) Analysis of the aphthovirus 2A/2B polyprotein 'cleavage' mechanism indicates not a proteolytic reaction, but a novel translational effect: a putative ribosomal 'skip'. J Gen Virol 82 (Pt 5):1013–1025
17. Funston GM, Kallioinen SE, de Felipe P, Ryan MD, Iggo RD (2008) Expression of heterologous genes in oncolytic adenoviruses using picornaviral 2A sequences that trigger ribosome skipping. J Gen Virol 89(Pt 2):389–396
18. Yu JZ, Rasenick MM (2002) Real-time visualization of a fluorescent G(alpha)(s): dissociation of the activated G protein from plasma membrane. Mol Pharmacol 61(2):352–359
19. Doyon Y, Choi VM, Xia DF, Vo TD, Gregory PD, Holmes MC (2010) Transient cold shock enhances zinc-finger nuclease-mediated gene disruption. Nat Methods 7(6):459–460
20. Quesnelle KM, Grandis JR (2011) Dual kinase inhibition of EGFR and HER2 overcomes resistance to cetuximab in a novel in vivo model of acquired cetuximab resistance. Clin Cancer Res 17(18):5935–5944
21. Sorkin A, Goh LK (2009) Endocytosis and intracellular trafficking of ErbBs. Exp Cell Res 315(4):683–696

Chapter 13

Silencing Long Noncoding RNAs with Genome-Editing Tools

Tony Gutschner

Abstract

Long noncoding RNAs (lncRNAs) are a functional and structural diverse class of cellular transcripts that comprise the largest fraction of the human transcriptome. However, detailed functional analysis lags behind their rapid discovery. This might be partially due to the lack of loss-of-function approaches that efficiently reduce the expression of these transcripts. Here, I describe a method that allows a specific and efficient targeting of the highly abundant lncRNA *MALAT1* in human (lung) cancer cells. The method relies on the site-specific integration of RNA-destabilizing elements mediated by Zinc Finger Nucleases (ZFNs).

Key words Cancer, CRISPR, Genome engineering, Homologous recombination, MALAT1, LncRNA, Single cell analysis, TALEN, Zinc finger nuclease

1 Introduction

LncRNAs represent a novel and exciting class of transcripts usually defined by their size (>200 nucleotides) and the lack of an open reading frame of significant length (<100 amino acids). Several studies link the expression of these transcripts to human diseases, e.g., cancer [1]. Functional analysis using RNA interference-mediated knockdown approaches are a common strategy to infer a gene's cellular role. However, these widely used approaches have multiple limitations [2] and might have limited efficiency for lncRNA research due to the intracellular localization (nuclear) and secondary structure of a large fraction of lncRNA molecules.

To overcome these limitations, a novel gene targeting method was developed to reduce the expression of the lncRNA *MALAT1* in human A549 lung cancer cells [3]. *MALAT1* is a ~8 kb long, highly abundant, nuclear transcript which was originally discovered in a screen for lung cancer metastasis associated genes [4, 5]. The targeting method relies on the site-specific integration of a selection marker (here: *GFP*) and RNA-destabilizing elements or

Shondra M. Pruett-Miller (ed.), *Chromosomal Mutagenesis*, Methods in Molecular Biology, vol. 1239, DOI 10.1007/978-1-4939-1862-1_13, © Springer Science+Business Media New York 2015

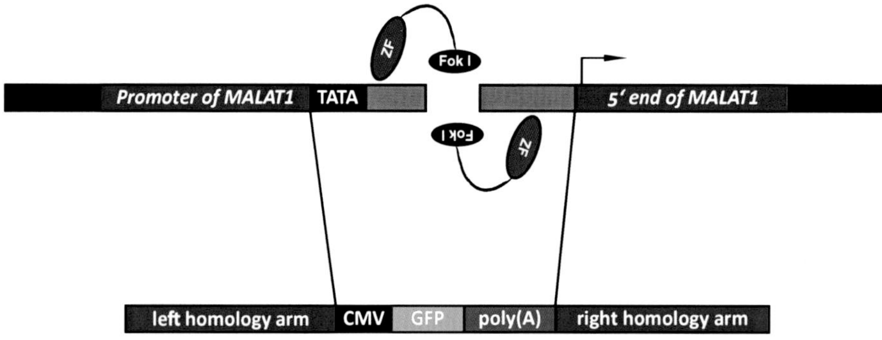

Fig. 1 Targeting non-protein-coding genes with gene-editing tools like ZFNs requires some special considerations as ncRNA genes have no open reading frame. Thus, a strategy that relies on the homologous recombination-mediated repair was developed. In general, the gene of interested (here: *MALAT1*) is targeted at its 5′-end to avoid the generation of truncated RNA transcripts that still might possess a function. A CMV-promoter driven fluorescent selection marker (GFP) followed by a poly(A) signal is cloned into a donor plasmid. However, other selection strategies and resistance markers can be used (e.g., Puromycin or Neomycin selection). Homology arms (about 800 nt each) flank the "CMV-GFP-poly(A)"-cassette and allow site-specific integration after ZFN-induced DNA double-strand breakage. Use of poly(A) signals for silencing of downstream sequences is recommended for RNA Polymerase II transcripts only. If the ncRNA of interested is transcribed by RNA Polymerase I or III, an alternative RNA destabilizing element as described previously [3] might be used

transcriptional stop signals, e.g., poly(A) signals, into the promoter region of the *MALAT1* gene. The integration is mediated by ZFNs that specifically introduce a DNA double-strand break (DSB) [6]. The induced DNA damage activates the cellular repair pathways, namely, Nonhomologous end joining (NHEJ) or Homologous Recombination (HR). By providing an appropriate template (donor plasmid) the HR pathway can be used to repair the DSB and to integrate exogenous DNA sequences (Fig. 1). Application of this method to human lung cancer cells yielded a stable, specific and more than 1,000-fold reduction of *MALAT1* expression and functional analysis established *MALAT1* as an active regulator of lung cancer metastasis [7]. Importantly, the methods' concept is of broad applicability and allows targeting of protein-coding genes as well as other lncRNAs using any kind of recently developed genome targeting tools, e.g., ZFNs, TALENs, or the CRISPR/Cas9 system.

2 Materials

Store all components according to manufacturer's recommendations. Use ultrapure water for nucleic acid analysis. ZFNs are commercially available from Sigma-Aldrich. Alternative methods were described that allow homemade generation of ZFNs [8, 9] or fast assembly of TALENs [10]. CRISPR/Cas9 plasmids are available from Addgene.

2.1 Cloning

1. Plasmid containing a selection marker of choice, e.g., *Green fluorescent protein* (*GFP*) followed by a poly(A) signal, e.g., bovine growth hormone (bGH) poly(A) signal.
2. Genomic DNA from cell line(s) subjected to modifications.
3. Genomic DNA isolation kit.
4. Proofreading DNA Polymerase.
5. Cloning primer for homology arms with appropriate restriction sites.
6. Agarose and agarose gel chamber.
7. Gel purification kit.
8. Restriction enzymes needed for cloning of homology arms.
9. PCR purification kit.
10. T4 DNA Ligase.
11. Competent bacteria.
12. LB-Medium: 5 g/L yeast extract, 10 g/L Tryptone, 10 g/L NaCl.
13. LB-Agar plates: LB-Medium with 15 g/L Agar.
14. Antibiotics, e.g., Ampicillin, Kanamycin.
15. Plasmid DNA preparation kits.

2.2 Cell Culture and Transfection

1. Cell line of choice.
2. Appropriate complete cell culture medium for cell line of interest containing supplements, serum, and antibiotics where appropriate.
3. Transfection reagent of choice.
4. Cell culture plates (96-well, 24-well, 6-well, 10 and 15 cm).
5. 0.05 or 0.25 % Trypsin-EDTA.
6. Phosphate-buffered saline (PBS).
7. 12 × 75 mm tube with cell strainer cap.
8. Conical centrifuge tubes.

2.3 Single Cell Analysis

1. Cell sorter.
2. Power SYBR Green Cells-to-CT Kit (Life Technologies, Carlsbad, CA, USA).
3. qPCR primer for reference and target gene.
4. DirectPCR lysis reagent (Peqlab, Wilmington, DE, USA) or mammalian genomic DNA MiniPrep Kit.
5. Integration-PCR primer spanning the nuclease cleavage site.
6. DNA-Polymerase of choice suitable for genotyping PCR.
7. PCR strip tubes or 96-well PCR plates and adhesive films.
8. Thermocycler.

3 Methods

The targeting approach requires cloning of a donor plasmid (Subheading 3.1), its transfection into cells together with ZFNs (or any other gene editing tool) (Subheading 3.2). After cell expansion, cells need to be enriched using Fluorescence Activated Cell Sorting (FACS) (Subheading 3.3). FACS is also used to distribute single cells into 96-wells for clonal growth. Finally, cell clones are analyzed for site-specific integration events and target gene expression levels (Subheading 3.4). *See* Fig. 2 for a protocol workflow. Design and cloning of gene-specific ZFNs or other gene-editing tools is highly user-specific and will not be covered here.

3.1 Cloning of a Donor Plasmid

1. Use proofreading DNA polymerases and genomic DNA to PCR amplify about 800 nt long left and right homology arms (*see* **Note 1**).
2. Run PCR program for 30 cycles and with an elongation time of 1 min per 1 kb.
3. Load PCR products on an agarose gel (1 % w/v) and let run at 5–8 V/cm.
4. Purify PCR products using a Gel Extraction kit according to manufacturer's recommendations. Elute in 30 μL pre-warmed water (50–60°C). Measure concentration of PCR products.
5. Use about 400 ng of PCR product and incubate for 1 h at 37 °C with appropriate restriction enzymes.
6. Purify PCR products using a PCR purification kit according to manufacturer's recommendations. Elute in 20 μL pre-warmed water (50–60 °C) and determine concentrations.
7. In parallel, prepare the donor plasmid accordingly by digesting and purifying the plasmid with the same reagents and protocols.
8. Clone the first homology arm into the donor plasmid by ligating the PCR product and the prepared plasmid using T4 DNA ligase. Use a 3:1 M ratio (PCR–Plasmid) for optimal ligation efficiency.
9. Transform competent *E.coli*, e.g., by heat shock (42 °C for 30–45 s, on ice for 2 min)
10. Streak *E. coli* on LB plates containing appropriate antibiotics.
11. Incubate plates for 12–16 h at 37 °C.
12. Pick single colonies and inoculate 2.5–5 mL LB-Medium containing antibiotics.
13. Grow colonies for 8–12 h and isolate plasmid DNA using a Mini-Prep kit.
14. Sequence-verify your clone harboring the first homology arm.

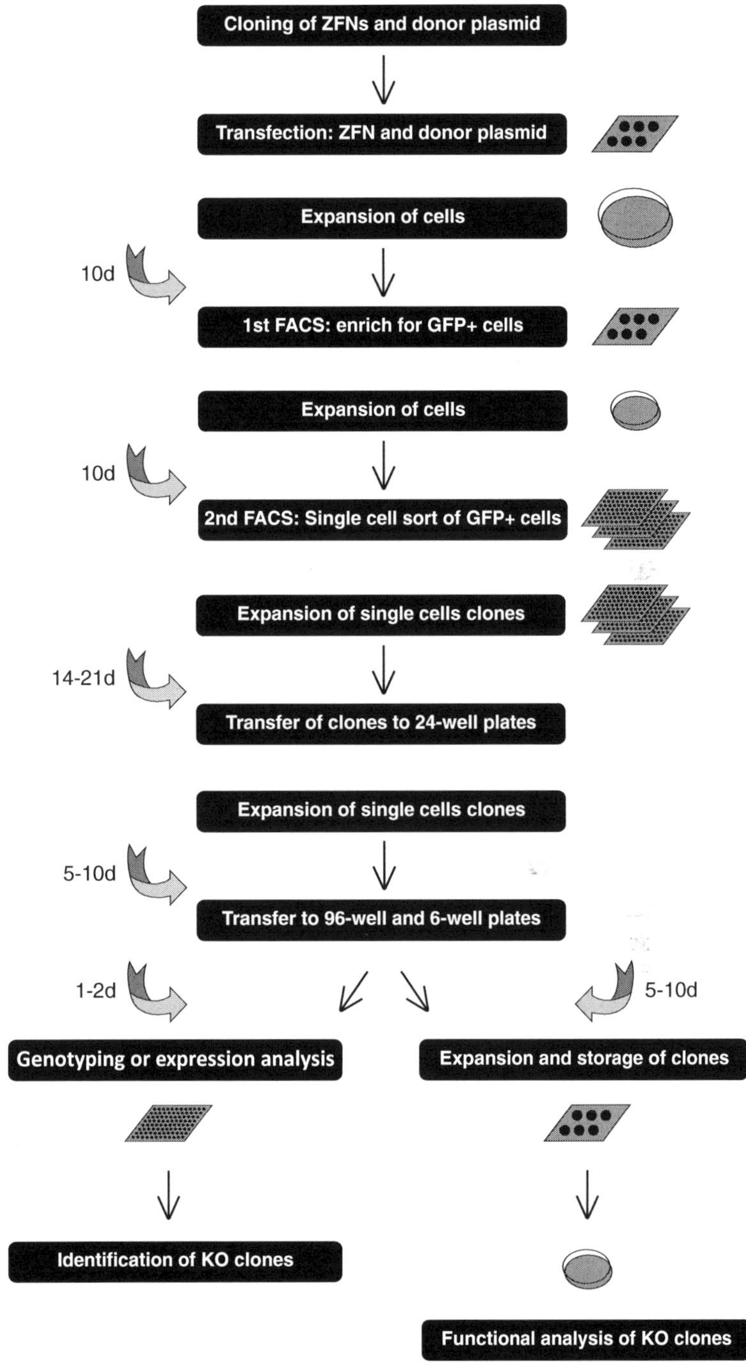

Fig. 2 Workflow for lncRNA knockout. Single, homozygous clones can be obtained within 6–8 weeks after ZFN and donor plasmid transfection

15. Continue cloning the second homology arm into the plasmid obtained above.

    Repeat **steps 7–14** accordingly.

16. Use 20–40 μL of starting culture used for Mini-Prep and inoculate 25–35 mL LB-Medium containing antibiotics.
17. Perform Plasmid DNA isolation using a Midi-Prep kit.

3.2 Transfection of ZFNs and Donor Plasmid

The optimal transfection protocol highly depends on the cell line that is subjected to manipulations. Transfection conditions should thus be optimized in advance. The protocol introduced here was successfully applied to human A549 lung cancer cells.

1. Seed cells ($2–3 \times 10^5$ per 6-well) in 2 mL cell culture medium (+10 % FBS, no antibiotics) (*see* **Note 2**).
2. The next day, prepare plasmid mix by combining 3 μg donor plasmid and 0.5 μg of ZFN plasmid each (1 μg ZFN plasmids in total) (*see* **Note 3**).
3. Combine plasmid mix (4 μg) with 8 μL Turbofect transfection reagent (Thermo Scientific) in serum-/antibiotics-free cell culture medium (final volume = 200 μL). Mix briefly.
4. Incubate for 15 min at room temperature.
5. Add transfection mix dropwise to cells and shake plate back and forth for equal distribution.
6. Incubate cells for 4–6 h with transfection mix.
7. Remove medium and add fresh, complete growth medium to cells.
8. Cells might be evaluated for GFP expression prior to further processing.

3.3 Cell Sorting

1. Expand cells for 10 days after donor and ZFN plasmid transfection.
2. Remove medium, wash cells once with PBS and add Trypsin–EDTA.
3. Incubate cells at 37 °C and allow for detach (5–15 min).
4. Resuspend cells in complete cell culture medium and transfer into conical centrifuge tube.
5. Spin down cells at $500 \times g$ for 5 min.
6. Completely remove cell culture medium and resuspend cell pellet in 2–4 mL PBS/FBS (1 % v/v) by pipetting up and down (*see* **Note 4**).
7. Pipet cells into BD Falcon 12 × 75 mm Tubes using the cell strainer cap to filter the cell suspension.
8. Perform **steps 2**–**7** with GFP-negative wild-type cells.
9. Put cells on ice and continue with cell sorting.
10. Use GFP-negative cells to adjust instrument settings and set threshold for GFP-selection.

11. Perform cell sorting to enrich for GFP-positive cells. Sort cells into 1.5 mL reaction tubes containing 50–100 μL complete cell culture medium (*see* **Note 5**).
12. Spin down cells in a tabletop centrifuge (800 × *g*, 5 min) and remove supernatant.
13. Resuspend cells in complete growth medium and seed into appropriate cell culture plates (*see* **Note 6**).
14. Expand cells for about 10 days to obtain at least one confluent 10 cm plate for further processing.
15. Add 200 μL complete growth medium per well into 96-well plate. Prepare 5–10 plates per cell line/construct/ZFN (*see* **Note 7**).
16. Prepare cells and adjust instrument settings as described in **steps 2–10**.
17. Sort GFP-positive cells into 96-well plates. GFP-negative wild-type cells might be sorted as well to obtain appropriate negative control clones for subsequent biological experiments.
18. Incubate cells at 37 °C. Add 100 μL complete medium after 5–7 days (*see* **Note 8**).

3.4 Cell Clone Analysis

1. About 7–10 days after sorting inspect 96-well plates and mark wells that contain cells.
2. Replace cell culture medium in respective wells by carefully removing the old medium using a 200 μL—pipet and sterile tips.
3. Continuously inspect 96-wells and mark wells that contain cells.
4. About 14–21 days after cell sorting first single cell clones might be ready for transfer into 24-well plates: Remove medium, wash once with PBS and add about 40 μL Trypsin–EDTA per 96-well. After incubation at 37 °C inspect cells for complete detachment. Resuspend cell clones in about 150 μL complete medium and transfer into 24-wells containing additional 500 μL complete growth medium.
5. After another 5–10 days, cells in 24-well plates might be confluent and are assigned an identification number. Then, cell clones are simultaneously transferred to 96-well and 6-well plates: Remove medium, wash once with PBS and add about 100 μL Trypsin–EDTA per 24-well. After incubation at 37 °C inspect cells for complete detachment. Resuspend cell clones in about 400 μL complete medium and transfer 100 μL into 96-well and 400 μL into a 6-well containing additional 2 mL complete growth medium.

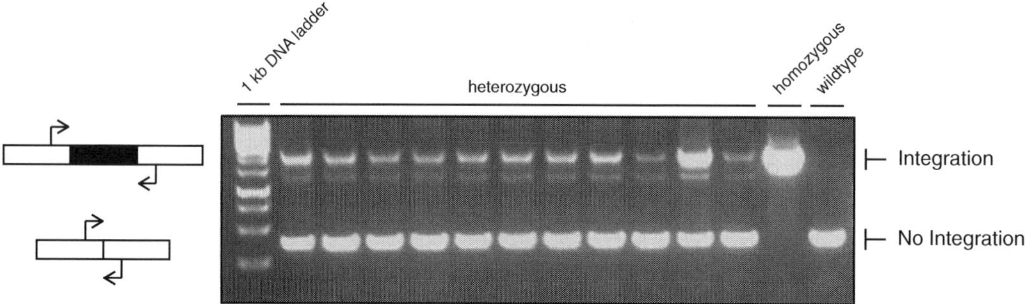

Fig. 3 Genotyping of cell clones by Integration-PCR. Primers cover the ZFN cleavage site. Monoallelic and biallelic integration events can be detected due to the different product sizes. In this example, 1 out of 12 clones harbored a biallelic integration of the selection marker after the selection process and thus showed a strong reduction in lncRNA expression (not shown)

6. The next day, cells in 96-wells are subjected to gene expression or genotyping analysis using the Power SYBR Green Cells-to-Ct kit (Life Technologies) or the DirectPCR lysis reagent (Peqlab) or GenElute mammalian genomic DNA MiniPrep Kit (Sigma-Aldrich) according to manufacturer's recommendations respectively.
7. For genotyping analysis an Integration-PCR is performed using primer pairs that span the ZFN cleavage site (*see* **Note 9**). A site-directed integration will lead to a longer PCR product (Fig. 3) (*see* **Note 10**).
8. Corresponding positive, homozygous clones in the 6-well plates are further expanded and transferred to 10 cm plates (*see* **Note 11**).
9. Single cell clones might be frozen and stored in liquid nitrogen.

4 Notes

1. Homology arms should be cloned from the same cell line that will be used for genome editing due to potential single nucleotide polymorphisms (SNPs). Homologous recombination strongly depends on perfect homology and can be impaired by SNPs.
2. The cell line(s) used for ZFN-mediated integration of exogenous DNA must possess a certain homologous recombination rate. Several cell lines might be tested, if no integration events are detected.
3. Although not absolutely required, linearization of the donor plasmid might increase integration rates. Please note that linearized plasmids are less stable and thus a modified transfection

protocol might be used. In this case, ZFN plasmids might be transfected prior to the donor plasmid to allow ZFN protein expression.

4. Careful pipetting should be performed to prevent disruption of cells while obtaining a single cell suspension, which is critical for subsequent single cell sorting. Addition of EDTA (1 mM final conc.) to the PBS/1 % FBS solution might be beneficial to prevent cell aggregation.
5. A total of 1–3 % of GFP-positive cells can be anticipated, but this rate might vary and depends on multiple parameters. Depending on the instrument and exact settings up to 4×10^5 cells can be sorted into one 1.5 mL reaction tube.
6. Antibiotics should be added to the cell culture medium after cell sorting to avoid contaminations.
7. The cell lines' capability to grow as a single cell colony should be tested beforehand. If a cell sorter (e.g., BD Bioscience FACS Aria II) is used, optimal sorting conditions should be determined in advance. Roughly, 10–40 single cell colonies can be expected per 96-well plate.
8. Some cell lines might show an improved single cell growth, if conditioned medium or medium with higher serum concentration is used (max. 20 % v/v). If conditioned medium is used, sterile filter before applying to single cells to avoid contaminations.
9. Alternatively, a Junction-PCR can be performed for genotyping. Here, one primer anneals to a sequence region outside the homology arms and the second primer specifically binds to the newly integrated (exogenous) sequence, e.g., the selection marker (here: *GFP*).
10. Different amounts of donor plasmid should be tested, if high rates of random, nonspecific donor plasmid integrations are observed, i.e., GFP-positive cells that lack a site-specific integration of the donor plasmid. Also, an efficient counter selection strategy could be applied, e.g., cloning the herpes simplex virus thymidine kinase gene outside the homology arms. Nonspecific integration and expression of this suicide gene confers sensitivity towards ganciclovir [11].
11. In theory, targeted integration on both chromosomes is necessary to obtain an efficient gene knockdown. However, cancer cells might show diverse degrees of gene amplifications and deletions. Also, epigenetically silenced or imprinted genes as well as genes localized on the X or Y-chromosomes represent exceptions of the rule. Thus, a single, site-specific integration might already lead to an efficient silencing. On the other hand, multiple integration events must occur simultaneously

in human polyploid cells (e.g., hepatocytes, heart muscle cells, megakaryocytes) or in amplified chromosome regions to significantly impair target gene expression.

Acknowledgement

The author wishes to acknowledge the support of his colleagues at the German Cancer Research Center (DKFZ) Heidelberg who helped to establish this method and to set up the protocol. A special thanks goes to Matthias Groß and Dr. Monika Hämmerle for critical reading of the manuscript. T.G. is supported by an Odyssey Postdoctoral Fellowship sponsored by the Odyssey Program and the CFP Foundation at The University of Texas MD Anderson Cancer Center.

References

1. Gutschner T, Diederichs S (2012) The hallmarks of cancer: a long non-coding RNA point of view. RNA Biol 9(6):703–719. doi:10.4161/rna.20481
2. Jackson AL, Linsley PS (2010) Recognizing and avoiding siRNA off-target effects for target identification and therapeutic application. Nat Rev Drug Discov 9(1):57–67. doi:10.1038/nrd3010
3. Gutschner T, Baas M, Diederichs S (2011) Noncoding RNA gene silencing through genomic integration of RNA destabilizing elements using zinc finger nucleases. Genome Res 21(11):1944–1954. doi:10.1101/gr.122358.111
4. Gutschner T, Hammerle M, Diederichs S (2013) MALAT1—a paradigm for long noncoding RNA function in cancer. J Mol Med 91(7):791–801. doi:10.1007/s00109-013-1028-y
5. Ji P, Diederichs S, Wang W, Boing S, Metzger R, Schneider PM, Tidow N, Brandt B, Buerger H, Bulk E, Thomas M, Berdel WE, Serve H, Muller-Tidow C (2003) MALAT-1, a novel noncoding RNA, and thymosin beta4 predict metastasis and survival in early-stage non-small cell lung cancer. Oncogene 22(39):8031–8041. doi:10.1038/sj.onc.1206928
6. Miller JC, Holmes MC, Wang J, Guschin DY, Lee YL, Rupniewski I, Beausejour CM, Waite AJ, Wang NS, Kim KA, Gregory PD, Pabo CO, Rebar EJ (2007) An improved zinc-finger nuclease architecture for highly specific genome editing. Nat Biotechnol 25(7):778–785. doi:10.1038/nbt1319
7. Gutschner T, Hammerle M, Eissmann M, Hsu J, Kim Y, Hung G, Revenko A, Arun G, Stentrup M, Gross M, Zornig M, MacLeod AR, Spector DL, Diederichs S (2013) The noncoding RNA MALAT1 is a critical regulator of the metastasis phenotype of lung cancer cells. Cancer Res 73(3):1180–1189. doi:10.1158/0008-5472.CAN-12-2850
8. Fu F, Voytas DF (2013) Zinc Finger Database (ZiFDB) v2.0: a comprehensive database of C(2)H(2) zinc fingers and engineered zinc finger arrays. Nucleic Acids Res 41(Database issue):D452–D455. doi:10.1093/nar/gks1167
9. Sander JD, Dahlborg EJ, Goodwin MJ, Cade L, Zhang F, Cifuentes D, Curtin SJ, Blackburn JS, Thibodeau-Beganny S, Qi Y, Pierick CJ, Hoffman E, Maeder ML, Khayter C, Reyon D, Dobbs D, Langenau DM, Stupar RM, Giraldez AJ, Voytas DF, Peterson RT, Yeh JR, Joung JK (2011) Selection-free zinc-finger-nuclease engineering by context-dependent assembly (CoDA). Nat Methods 8(1):67–69. doi:10.1038/nmeth.1542
10. Cermak T, Doyle EL, Christian M, Wang L, Zhang Y, Schmidt C, Baller JA, Somia NV, Bogdanove AJ, Voytas DF (2011) Efficient design and assembly of custom TALEN and other TAL effector-based constructs for DNA targeting. Nucleic Acids Res 39(12):e82. doi:10.1093/nar/gkr218
11. Moolten FL, Wells JM (1990) Curability of tumors bearing herpes thymidine kinase genes transferred by retroviral vectors. J Natl Cancer Inst 82(4):297–300

Chapter 14

Gene Editing Using ssODNs with Engineered Endonucleases

Fuqiang Chen, Shondra M. Pruett-Miller, and Gregory D. Davis

Abstract

Gene editing using engineered endonucleases, such as zinc finger nucleases (ZFNs), transcription activator-like effector nucleases (TALENs), and clustered regularly interspaced short palindromic repeats (CRISPR)/Cas9 nucleases, requires the creation of a targeted, chromosomal DNA double-stranded break (DSB). In mammalian cells, these DSBs are typically repaired by one of the two major DNA repair pathways: nonhomologous end joining (NHEJ) or homology-directed repair (HDR). NHEJ is an error-prone repair process that can result in a wide range of end-joining events that leads to somewhat random mutations at the site of DSB. HDR is a precise repair pathway that can utilize either an endogenous or exogenous piece of homologous DNA as a template or "donor" for repair. Traditional gene editing via HDR has relied on the co-delivery of a targeted, engineered endonuclease and a circular plasmid donor construct. More recently, it has been shown that single-stranded oligodeoxynucleotides (ssODNs) can also serve as DNA donors and thus obviate the more laborious and time-consuming plasmid vector construction process. Here we describe the use of ssODNs for making defined genome modifications in combination with engineered endonucleases.

Key words Single-stranded oligodeoxynucleotides, ssODNs, Engineered endonucleases, Zinc finger nuclease, ZFN, CRISPR, cas9, Double-stranded break, Homology-directed repair

1 Introduction

Engineered endonucleases, such as zinc finger nucleases (ZFNs), transcription activator-like effector nucleases (TALENs), and clustered regularly interspaced short palindromic repeats (CRISPR)/Cas9 nucleases, can be tailor-made to target and cut virtually any sequence of interest within complex genomes with high efficiencies [1]. Unlike the DSBs caused by cellular reactive oxygen species, ionizing radiation, ultraviolet light, or other chemical agents, the DSBs generated by endonucleases are better defined [2]. For instance, cleavage by FokI-tethered engineered endonucleases (ZFNs and TALENs) produces nascent 5′ overhangs with 5′ phosphate group and 3′ hydroxyl group [3], while cleavage by Cas9 endonucleases creates nascent blunt ends [4].

Shondra M. Pruett-Miller (ed.), *Chromosomal Mutagenesis*, Methods in Molecular Biology, vol. 1239,
DOI 10.1007/978-1-4939-1862-1_14, © Springer Science+Business Media New York 2015

In host cells these DSBs may undergo further end resection before they are repaired by nonhomologous end joining (NHEJ) or by homology-directed repair (HDR) pathways [5, 6]. NHEJ is error prone and can generate a wide range of random mutations at the cut site, typically dominated by small insertions and/or deletions (indels). Therefore, gene disruption and ultimately gene knockout may be identified from the resulting mutation repertoire. In contrast, HDR is a template-dependent process [6] and thus can be harnessed for user-defined gene editing when a well-designed DNA donor is co-delivered together with engineered endonucleases.

Traditional gene targeting has been achieved using plasmid or viral vectors as DNA donors even before the advent of engineered endonucleases. Previous work has demonstrated that the co-delivery of ZFNs in combination with plasmid-based targeting constructs enables efficient HDR-mediated gene modification in several eukaryotic systems [7–9]. These plasmid donors typically contain homology arms of 200–800 bp, each flanking the desired DNA modification or insertion event [10]. While these plasmid donors are considerably smaller and simpler to build than more traditional gene targeting vectors, which may have up to 10 kb in each homology arm; their construction and preparation can still be expensive and time consuming. In contrast, ssODNs can be designed and synthesized within just a few days. This simplicity and flexibility enables the researcher to expedite gene editing practice with reduced cost.

In the absence of a targeted DSB, gene editing via ssODNs is possible but extremely inefficient [11] and various attempts on chemical modification to enhance the stability of ssODNs or to stimulate the integration have not fundamentally altered the trend [12, 13]. In contrast, genome sequence conversion by ssODNs is greatly enhanced when there is an induced DSB nearby and various models have been proposed regarding the mechanisms [6, 14–16]. Using ZFNs to create targeted DSBs, we demonstrated that high gene editing efficiencies in mammalian cells can be achieved with ssODNs and identified factors that can significantly influence the efficiency [17]. In brief, the frequency of the DSB (e.g., the cutting efficiency of the nuclease) and the distance between the DSB and the editing site are the primary determinants. These findings correspond with what has been previously reported with plasmid donors [18]. Therefore, it is highly recommended to use an endonuclease that can efficiently create DSBs at or as close as possible to the intended gene editing site. Furthermore, in designing ssODNs, attention must be placed on the local sequence characteristics of DSBs in order to maximize editing efficiency. As mentioned above, ZFNs, TALENs, and CRISPR/Cas9 nucleases result in different types of DSBs: 5′

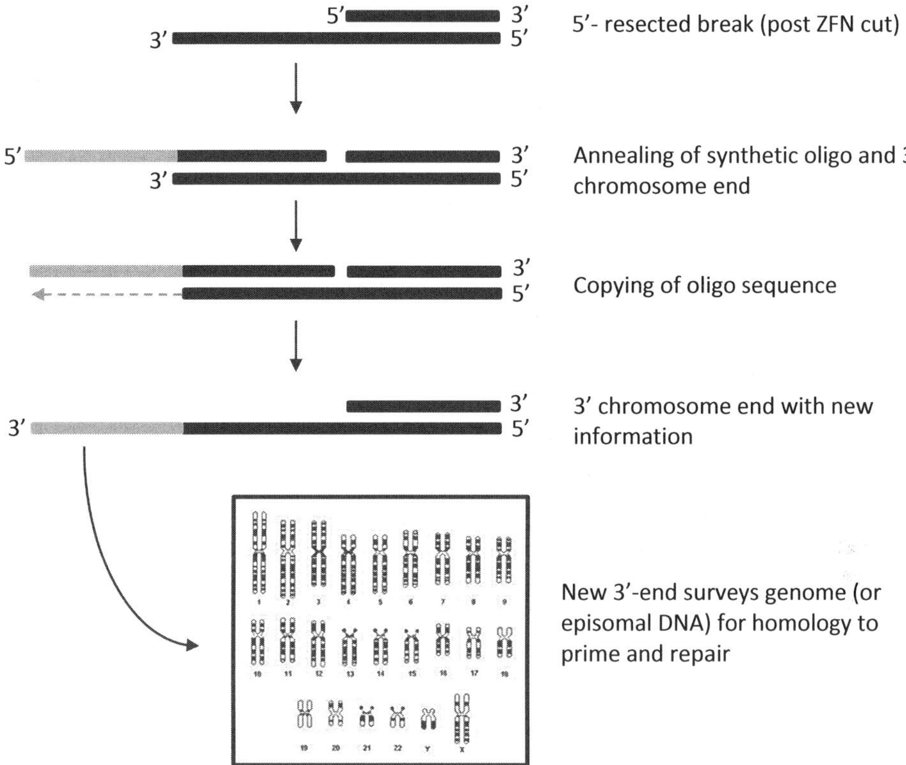

Fig. 1 Proposed mechanism for copying of synthetic oligo information by resected double-strand break ends. This scheme explains resulting transfer of oligo information observed in previous attempts to create point mutations, insertions, and deletions [17, 22]

overhangs for FokI-based ZFNs and TALENs and blunt ends for CRISPR/Cas9 nucleases. However, regardless of the type of overhang that results, repair by the HDR pathway involves 5′ to 3′ end resection of such overhangs, which ultimately results in nascent 3′ single stranded ends being exposed [19]. It is hypothesized that these nascent 3′ ends act as "primers" and are then extended by the endogenous cellular machinery using the supplied ssODN donor as a "template" (Fig. 1). Thus, mismatches between the nascent 3′ end or "primer" and the ssODN donor or "template", especially at the last few bases of the nascent 3′ end, may result in reduced editing efficiencies. In addition, the length of homology arms can also affect editing efficiency to some degree. To illustrate the design and use of ssODNs in gene editing, here we describe the use of ssODNs for introducing insertion, creating precise genomic DNA deletion, and making codon conversion, using ZFNs or CRISPR/Cas9 nucleases as DSB-inducing agents.

2 Materials

Prepare all solutions using nuclease-free molecular grade water. Store all DNA and RNA reagents at −20 °C or −80 °C. Store all cell culture reagents at 4 °C. Strictly follow all waste disposal guidelines including biohazard disposal regulations.

2.1 Cell Transfection Components

1. RSK2 BamHI site insertion ssODN: 5′TATTGGAGTTGG CTCCTACTCTGTTTGCAAGAGATGTATACATAAA GCTACAGGATCCAACATGGAGTTTGCAGTGAAGGTA AATTTTTTTTATTTAAAATGCAATTCAT3′.
2. AAVS1 1 kb deletion ssODN: 5′CCTGGACTTTGTCTCC TTCCCTGCCCTGCCCTCTCCTGAACCTGAGCCAGCT CCCATAGCCTAGGGACAGGATTGGTGAC AGAAAAGCCCCATCCTTAGG3′.
3. RSK2 Cys436Val codon conversion ssODN: 5′GGATA TGAAGTAAAAGAAGATATTGGAGTTGGaTCCTAC TCTGTTgttAAGAGATGTATACATAAAGCaACAAAC ATGGAaTTTGCAGTGAAGGTAAATTTTTTTT ATTTAAAATGCAATTCATA3′.
4. RSK2 BamHI site insertion forward primer: 5′GCATG CTGAGTAACATACTTCCCTA3′.
5. RSK2 BamHI site insertion reverse primer: 5′GAGGTGT AAACTGCTACTGCTCTG3′.
6. AAVS1 1 kb deletion forward primer: 5′GACCCATGCA GTCCTCCTTAC3′.
7. AAVS1 1 kb deletion reverse primer: 5′GGCTCCATC GTAAGCAAACC3′.
8. AAVS1 wild-type forward primer: 5′TTCGGGTCA CCTCTCACTCC3′.
9. AAVS1 wild-type reverse primer: 5′GGCTCCATCGTA AGCAAACC3′.
10. RSk2 Cys436Val codon conversion forward primer: 5′GCAT GCTGAGTAACATACTTCCCTA3′.
11. RSK2 Cys436Val codon conversion reverse primer: 5′GAGG TGTAAACTGCTACTGCTCTG3′.
12. RSK2 Val specific qPCR forward primer: 5′GGAG TTGGATCCTACTCTGTTGTT3′.
13. RSK2 Cys specific qPCR forward primer: 5′GGAGT TGGCTCCTACTCTGTTTG3′.
14. RSK2 qPCR common reverse primer: 5′CTGGTA CTGACATGCATGAACAAG-3′.
15. Cas9 plasmid (Sigma-Aldrich, St. Louis, MO, USA).

16. RSK2 guide RNA plasmid (Sigma-Aldrich).
17. AAVS1 ZFN plasmid (Sigma-Aldrich).
18. AAVS1 ZFN mRNA (Sigma-Aldrich) (*see* **Note 9**).
19. K562 cells.
20. Cell culture incubator.
21. T75 cell culture flasks.
22. 6-Well cell culture plates.
23. Cell culture medium: Iscove's Modified Dulbecco's Medium, 10 % FBS, 2 mM L-glutamine.
24. Hanks' Balanced Salt Solution.
25. Nucleofector Kit V (Lonza, Visp, Switzerland).
26. Nucleofector instrument (Lonza).

2.2 Gene Editing Assay and Single Cell Cloning Components

1. Mammalian Genomic DNA Miniprep Kit.
2. PCR Clean-Up Kit.
3. JumpStart Taq ReadyMix (Sigma-Aldrich).
4. JumpStart SYBR Green ReadyMix (Sigma-Aldrich).
5. Restriction endonuclease BamHI.
6. QuickExtract (Epicentre, Madison, WI, USA).
7. 10 % acrylamide gel.
8. 3 % agarose gel.
9. PCR thermocycler.
10. Real-time PCR system.
11. 96-Well cell culture plates.
12. Flow cytometer.

3 Methods

The procedures for carrying out ssODN-mediated gene editing will be described through three different examples: (1) targeted insertion at the human RSK2 locus using Cas9 nuclease; (2) precise genomic DNA deletion at the human AAVS1 locus using ZFNs; and (3) codon conversion at the human RSK2 locus using ZFNs. All procedures are carried out at room temperature unless stated otherwise.

3.1 ssODN Design and Preparation

1. Design a 110-nt ssODN for integration of a BamHI restriction site at the human RSK2 locus. The design details are presented in Fig. 2. Because the integration site is 1 bp upstream of the cleavage site, the antisense strand is designated as the genomic DNA "primer" after Cas9 cleavage. If the sense strand is des-

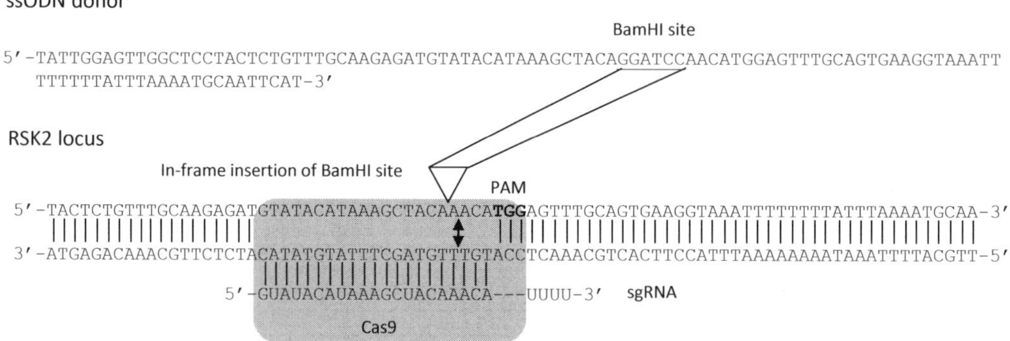

Fig. 2 ssODN design for an integration of a BamHI restriction site at the human RSK2 locus using CRISPR/Cas9. Cas9 is represented by a *grey rectangle*, and single-guide RNA (sgRNA) is represented by the target-specific sequence (also referred to as a protospacer) and the polyU track. CRISPR/Cas9 cleavage sites are indicated by a *double arrow*, and the integration site is indicated by a *triangle*

ignated as the genomic DNA "primer" after Cas9 cleavage and the antisense strand is used as ssODN homology arms, an undesirable mismatch at the 3′ end will occur (*see* **Note 1**).

2. Design a 100-nt ssODN for creating a precise 1 kb genomic DNA deletion at the human AAVS1 locus. The design details are presented in Fig. 3. The antisense strand is designated as the genomic DNA "primer" after ZFN cleavage. The deletion ssODN consists of a distal segment that defines the distal deletion border and a proximal segment that defines the deletion border at the cleavage site [17] and serves as an annealing partner for the genomic DNA 3′ end after cleavage (*see* **Note 2**).
3. Design a 125-nt ssODN for making a Cys436Val codon change in the human RSK2 kinase. The design details are presented in Fig. 4. Again, the antisense strand is designated as the genomic DNA "primer" after ZFN cleavage. The ssODN also carries a silent mutation for creating a diagnostic BamHI site and a silent ZFN-blocking mutation (ZBM) on the ZFN left and right binding sites, respectively, for preventing the ZFNs from re-cutting modified allele (*see* **Note 3**). The codon mutation will render the RSK2 kinase insensitive to the pharmacological kinase inhibitor fmk and thus produce a biochemical phenotype [17, 20, 21].
4. Synthesize ssODNs at 0.2–1.0 μmol scale (*see* **Note 4**) and purify by PAGE. Synthesize all other PCR primers each at 0.025 μmol scale and purify by desalt.
5. Centrifuge ssODN containing tubes for 30 s and reconstitute each to a final concentration of 100 μM in 10 mM Tris, pH 7.6 (*see* **Note 5**).

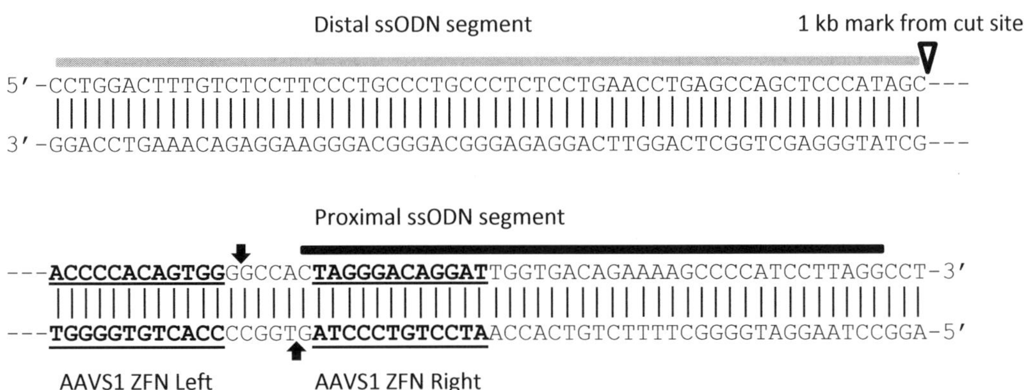

ssODN donor

5′-CCTGGACTTTGTCTCCTTCCCTGCCCTGCCCTCTCCTGAACCTGAGCCAGCTCCCATAGC
CTAGGGACAGGATTGGTGACAGAAAAGCCCCATCCTTAGG-3′

Fig. 3 ssODN design for a precise deletion of a 1 kb genomic DNA fragment at the human AAVS1 locus using ZFNs. ZFN-binding sites are *boldfaced* and *underlined*, and the expected ZFN cleavage sites are indicated by *arrows*. The exact 1 kb deletion borders from the cleavage site on the antisense strand to the *triangle mark*. Distal and proximal ssODN segments are indicated by a *grey bar* and a *black bar*, respectively. The two segments are joined together to form the 1 kb deletion ssODN

ssODN donor

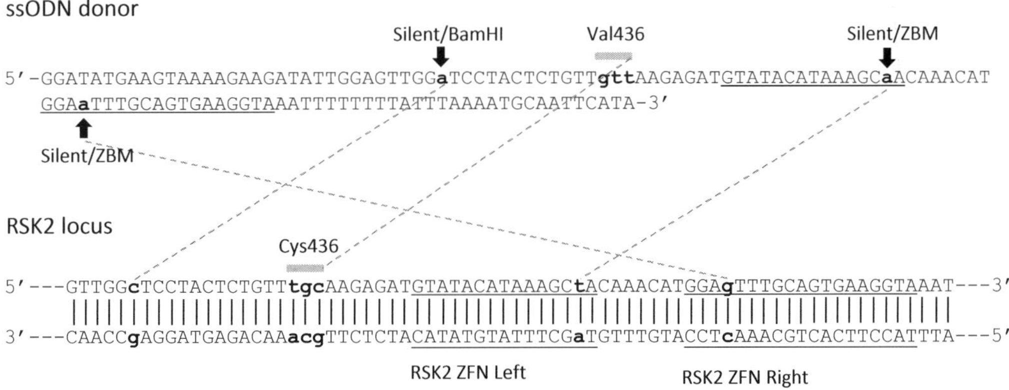

Fig. 4 ssODN design for a Cys436Val codon conversion at the human RSK2 locus using ZFNs. ZFN-binding sites are *underlined*. Silent mutations on the ssODN are introduced to create a diagnostic BamHI site and two ZBMs (ZFN-blocking mutations), one on each ZFN-binding site

3.2 Transfection

1. Culture K562 cells in Iscove's Modified Dulbecco's Medium, supplemented with 10 % FBS and 2 mM L-glutamine. Maintain the culture at 37 °C and 5 % CO_2. Split the culture every 2–3 days or when the culture reaches 1–2 million cells per mL (*see* **Note 6**).

2. Split the culture to 0.25 million cells per mL 1 day before transfection. The culture will reach about 0.5 million cells per mL for transfection the next day (*see* **Note 7**).
3. Prepare a transfection solution for BamHI site insertion at the RSK2 locus in a 1.5-mL microcentrifuge tube by adding 5 μg of Cas9 plasmid DNA, 2 μg of RSK2 guide RNA plasmid DNA, and 3 μL of 100 μM of BamHI site insertion ssODN (*see* **Note 8**). Also prepare an oligo-only transfection control (3 μL of 100 μM of BamHI site insertion ssODN). Keep the tubes on ice.
4. Prepare a transfection solution for 1 kb deletion at the AAVS1 locus in a 1.5-mL microcentrifuge tube by adding 2–4 μg of each AAVS1 ZFN mRNA (or 2.5 μg of each ZFN plasmid DNA) and 3 μL of 100 μM of AAVS1 I kb deletion ssODN (*see* **Note 9**). Also prepare an oligo-only transfection control (3 μL of 100 μM of AAVS1 1 kb deletion ssODN). Keep the tubes on ice.
5. Prepare a transfection solution for Cys436Val codon conversion at the RSK2 locus in a 1.5-mL microcentrifuge tube by adding 2–4 μg of each RSK2 ZFN mRNA (or 2.5 μg of each ZFN plasmid DNA) and 3 μL of 100 μM of RSK2 Cys436Val codon conversion ssODN (*see* **Note 9**). Also prepare an oligo-only transfection control (3 μL of 100 μM of RSK2 Cys436Val codon conversion ssODN). Keep the tubes on ice.
6. Add 2 mL of fresh culture medium to each well of a 6-well plate and equilibrate the medium at 37 °C and 5 % CO_2 for at least 20 min before transfection.
7. Set the Nucleofector program to T-016 (*see* **Note 10**). Unseal the cuvettes and disposable transfer pipettes and place them inside a cell culture hood.
8. Count the cells with a hemocytometer and transfer the required amount of cells to a sterile 50-mL conical centrifuge tube. Each transfection reaction will require 0.5–1.0 million cells (*see* **Note 11**). Centrifuge the cells at 200 × *g* at room temperature for 5 min. Carefully remove the culture medium by aspiration.
9. Resuspend the cell pellet in 20 mL of Hank's Balanced Salt Solution. Centrifuge at 200 × *g* at room temperature for 5 min and carefully remove the wash fluid by aspiration. Repeat the wash step once.
10. Carefully remove all wash fluid after the second wash step. Gently resuspend the cell pellet in Nucleofection Solution V to 0.5–1.0 million cells per 100 μL (*see* **Note 12**).
11. Pipette 100 μL of the cells into a 1.5 mL tube containing a transfection solution and mix thoroughly but gently by pipetting

up and down three times. Transfer the whole content to a nucleofection cuvette and perform the transfection immediately using the T-016 program (*see* **Note 13**).

12. Add pre-equilibrated culture medium to the transfected cells inside the cuvette with a disposable transfer pipette and immediately transfer the whole content into a well containing 2 mL of pre-equilibrated culture medium. Culture the cells at 37 °C and 5 % CO_2 immediately after nucleofection.

3.3 Gene Editing Analysis

1. Extract genomic DNA 2–3 days after transfection using a Mammalian Genomic DNA Miniprep Kit (*see* **Note 14**).
2. RSK2 BamHI site insertion analysis. PCR amplify the genomic DNA with the RSK2 BamHI site insertion forward primer (5′GCATGCTGAGTAACATACTTCCCTA3′) and reverse primer (5′GAGGTGTAAACTGCTACTGCTCTG3′) using JumpStart Taq ReadyMix with the following cycling condition: 98 °C for 2 min for initial denaturation; 34 cycles of 98 °C for 15 s, 61 °C for 30 s, and 72 °C for 30 s; and a final extension at 72 °C for 5 min. Purify PCR products with a PCR Clean-Up Kit. Digest purified PCR products (about 600 ng per digestion) with BamHI (20 units) at 37 °C for 2 h. Resolve digestion products on a 10 % acrylamide gel. Two BamHI digestion fragments with the expected sizes (194 and 240 bp) will be observed on the positive transfection reaction (Fig. 5).

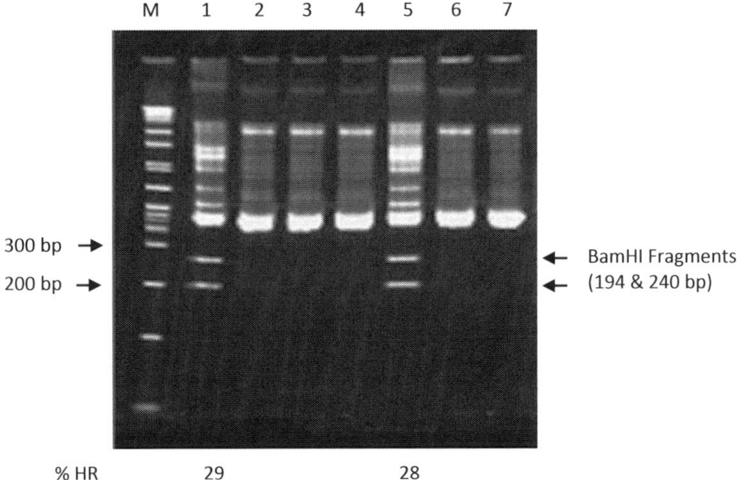

Fig. 5 Restriction digestion analysis of an ssODN-mediated BamHI site integration at the human RSK2 locus. K562 cells were transfected with an ssODN together with Cas9 plasmid and single guide RNA (sgRNA) plasmid. Genomic DNA was harvested 2 or 8 days after transfection and PCR amplified. PCR products were digested with BamHI and resolved on a 10 % acrylamide gel. *Lane 1*: ssODN + Cas9 + sgRNA at day 2. *Lane 2*: ssODN + nuclease deficient Cas9 + sgRNA at day 2. *Lane 3*: ssODN only at day 2. *Lane 4*: Blank control at day 2. *Lane 5*: ssODN + Cas9 + sgRNA at day 8. *Lane 6*: ssODN + nuclease deficient Cas9 + sgRNA at day 8. *Lane 7*: ssODN only at day 8. *M* Wide range DNA markers

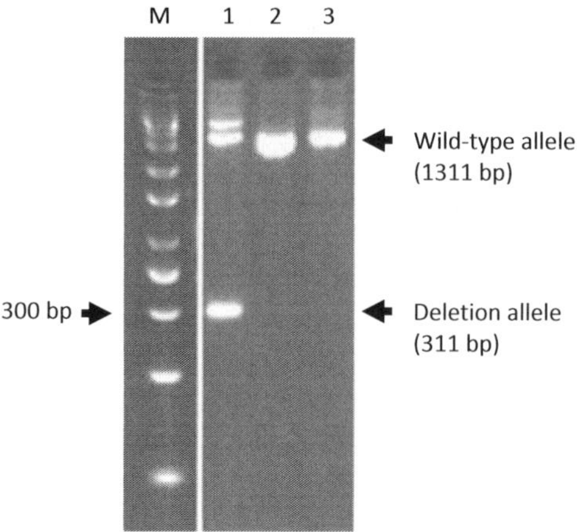

Fig. 6 PCR analysis of an ssODN-mediated 1 kb genomic DNA deletion at the human AAVS1 locus. K562 cells were transfected with an ssODN together with ZFN mRNA. Genomic DNA was harvested 2 days after transfection and PCR amplified. PCR products were resolved on a 3 % agarose gel. *Lane 1*: ssODN + ZFN mRNA. *Lane 2*: ssODN only. *Lane 3*: ZFN mRNA only. *M* Wide range DNA marker

3. AAVS1 1 kb deletion analysis. PCR amplify the genomic DNA with the AAVS1 1 kb deletion forward primer (5′GACCCATGCAGTCCTCCTTAC3′) and reverse primer (5′GGCTCCATCGTAAGCAAACC3′) using JumpStart Taq ReadyMix with the following cycling condition: 98 °C for 2 min for initial denaturation; 34 cycles of 98 °C for 15 s, 62 °C for 30 s, and 72 °C for 30 s; and a final extension at 72 °C for 5 min. Resolve PCR products on a 3 % agarose gel (*see* **Note 15**). A deletion allele fragment (311 bp) and a wild-type allele fragment (1,311 bp) and will be observed on the positive transfection reaction (Fig. 6).
4. RSK2 Cys436Val codon conversion analysis. PCR amplify the genomic DNA with the RSK2 codon conversion forward primer (5′GCATGCTGAGTAACATACTTCCCTA3′) and reverse primer (5′GAGGTGTAAACTGCTACTGCTCTG3′) using JumpStart Taq ReadyMix with the following cycling condition: 98 °C for 2 min for initial denaturation; 34 cycles of 98 °C for 15 s, 61 °C for 30 s, and 72 °C for 30 s; and a final extension at 72 °C for 5 min. Purify PCR products with a PCR Clean-Up Kit. Digest purified PCR products (600 ng each) with BamHI (20 units) at 37 °C for 2 h. Resolve digestion products on a 10 % acrylamide gel. Two BamHI digestion fragments with the expected sizes (146 and 282 bp) will be observed on the positive transfection reaction [17]. The BamHI digestion serves as a convenient indirect assay for the codon conversion (*see* **Note 16**).

3.4 Single-Cell Cloning and Genotyping

1. Sort the cells into 96-well culture plates containing 100 μL per well of pre-warmed culture medium using a flow cytometer (*see* **Note 17**). Add additional 100 μL of pre-warmed culture medium to each well after sorting.
2. Maintain the plates in an incubator at 37 °C and 5 % CO_2 and grow the cells to approximately 1.0 million cells per mL (*see* **Note 18**).
3. Consolidate the clones in fresh 96-well plates and make a duplicate copy for each clone. Save one copy for clone isolation and expansion later (*see* **Note 19**).
4. Centrifuge the plate at 200 × *g* for 10 min to pellet the cells and decant the medium to a waste container immediately after removing the plate from the centrifuge. Press the plate upside down on a stack of clean paper towel for 30 s to remove residual medium (*see* **Note 20**).
5. Lyse the cells by adding 50–100 μL of QuickExtract to each well and incubating the plate on a thermocycler with the following temperature profile: 65 °C for 15 min, 98 °C for 20 min, and hold at 20 °C or 4 °C (for overnight). Keep the cell lysate at −20 °C for long term storage.
6. AAVS1 1 kb deletion genotyping. Screen the clones for deletion allele using the AAVS1 1 kb deletion forward primer (5′GACC CATGCAGTCCTCCTTAC3′) and reverse primer (5′GGCTCC ATCGTAAGCAAACC3′). Screen the clones for wild-type allele using the AAVS1 wild-type forward primer (5′TTCGG GTCACCTCTCACTCC3′) and reverse primer (5′GGCTC CATCGTAAGCAAACC3′). Perform the PCR amplification each with 2 μL of cell lysate in a 25-μL reaction using JumpStart Taq ReadyMix with the following cycling condition: 98 °C for 2 min for initial denaturation; 34 cycles of 98 °C for 15 s, 62 °C for 30 s, and 72 °C for 30 s; and a final extension at 72 °C for 5 min. Resolve PCR products on a 3 % agarose gel to identify clones with biallelic 1 kb deletion (Fig. 7).
7. RSK2 Cys436Val codon conversion genotyping by qPCR. Screen the clones for Cys436Val conversion allele using the RSK2 Val specific qPCR forward primer (5′GGAGT TGGATCCTACTCTGTTGTT3′) and the RSK2 qPCR common reverse primer (5′CTGGTACTGACATGCATGAA CAAG-3′). Screen the clones for wild-type allele using the Cys specific qPCR forward primer (5′ GGAGTTGGCTCCT ACTCTGTTTG-3′) and the RSK2 qPCR common reverse primer. Perform qPCR amplification each with 2 μL of cell lysate in a 25 μL reaction using JumpStart SYBR Green ReadyMix on a Mx3000P Real-Time PCR System with the following cycling condition: 98 °C for 2 min for initial denaturation; 40 cycles of 98 °C for 15 s, 62 °C for 30 s, and 72 °C

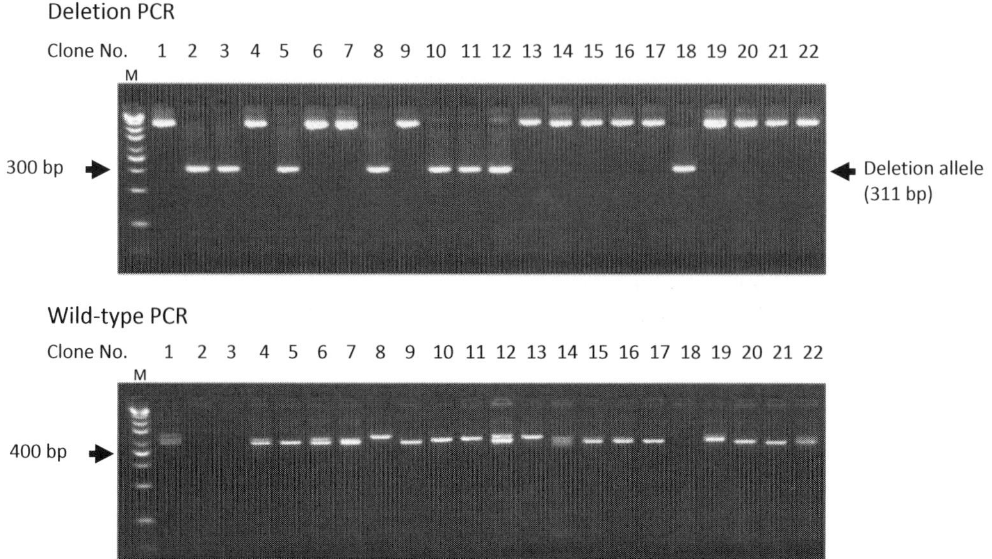

Fig. 7 Single-cell cloning of an ssODN-mediated 1 kb genomic DNA deletion at the human AAVS1 locus. K562 cells were transfected with an ssODN and ZFN mRNA and sorted by flow cytometry. Single-cell clones were analyzed by deletion PCR for detecting the presence of 1 kb deletion allele and by wild-type PCR for detecting the presence of wild-type and/or NHEJ-modified alleles. Biallelic 1 kb deletion was identified on clones 2, 3, and 18. M: Wide range DNA molecular markers

for 30 s; followed by a dissociation curve segment (95 °C, 1 min; 55 °C, 30 s; 95 °C, 30 s). Identify candidate biallelic Cys436Val clones and further verify the clones by DNA sequencing.

4 Notes

1. The insertion site is one bp from the Cas9 cleavage site, which is an optimal distance between a DSB and an editing site. ssODN-mediated gene editing efficiency greatly depends on the distance between the DSB and the editing site. CRISPR/Cas9 nucleases require only the presence of a PAM sequence (NGG) for targeting and cutting and theoretically its frequency is once every 8 bp. The high targeting density is advantageous for creating DSBs at the proximity of gene editing sites. In general, editing efficiency decreases significantly when an editing site is more than 20 bp away from the cleavage site. The length of the homology arms of the ssODN is 52 nt each, which is optimal for efficient integration. In general, the homology arm may vary from 30 to 60 nt without greatly affecting the efficiency. In using ZFNs for DSBs, we found that a drastic drop in efficiency occurred when the homology arm was less than 20 nt in length.

2. The proximal deletion border should be as close to the cleavage site as possible. However, it can be moved forward from the cleavage site (toward the distal deletion border), but a recessive deletion border from the cleavage site (further away from the distal deletion border) will result in drastic reduction in deletion efficiency even just by a few bases. Deletion efficiency will also proportionally decrease as the deletion fragment size increases. We found that co-delivering of human RAD52 mRNA (2 μg) and the addition of a NHEJ inhibitor (20 μM NU7026) in the plating medium significantly increased the efficiency of large deletions (2–20 kb).
3. The Cys436Val codon conversion site is 27 bp from the ZFN cut site. This is a more challenging editing distance. However, the RSK ZFN pair are highly active and thus it compensates for the distance. The silent mutation to create the diagnostic BamHI restriction site is helpful for assay but not essential. The silent mutations (nuclease blocking modifications or NBMs) on the ZFN-binding sites are also not essential, but they can reduce the re-cutting frequency of codon converted allele.
4. For ssODNs ≤110 nt in length, we recommend a 0.2 μmol synthesis scale, which is usually enough for more than ten transfection reactions. For ssODNs ≥ 120 nt, we recommend a 1.0 μmol synthesis scale. Oligo synthesis yields may drop drastically if the length exceeds 110 nt.
5. ssODNs can also be dissolved in nuclease-free water. The concentration also can be increased to 200 μM if needed to accommodate the volume of transfection solution.
6. When working with adherent cells, split the culture every 2–3 days or when the culture researches approximately 80 % confluency.
7. For adherent cells, split the culture 2 days before transfection. Seed the culture at about 20 % confluency so that the culture will be at about 60–80 % confluency at the time of cell preparation for transfection.
8. Prepare Cas9 plasmid and RSK2 guide RNA plasmid at a concentration ≥1 μg/μL to limit the total transfection solution volume to ≤10 μL as per manufacturer's recommendations.
9. We found that ZFN mRNA gave better editing efficiency than ZFN plasmid DNA when using ssODNs as donors. We synthesized ZFN mRNA by in vitro transcription and poly(A) tailing using MessageMAX T7 ARCA-Capped Message Transcription Kit and A-Plus Poly(A) Polymerase Tailing Kit (Cellscript, Madison, WI, USA). ZFN plasmid DNA (Sigma-Aldrich, St. Louis, MO, USA) was digested with XbaI or XhoI and purified by phenol/chloroform extraction and ethanol precipitation. Two

micrograms of the digested DNA was used for each 40-μL in vitro transcription reaction. It is important to perform the reaction and subsequent purification in an RNase-free environment.

10. Different cell lines require different nucleofection programs. Follow the instructions from the manufacturer when working with other cell lines.
11. Up to two million K562 cells can be used per nucleofection, but it may reduce the transfection efficiency slightly. We found that 0.6–1.0 million cells per nucleofection were suitable for A549, HCT116, HEK293, HepG2, and U2OS, and that 1.2–2.0 million cells per nucleofection were required for MCF7.
12. Do not leave cells in a nucleofection solution for a prolonged time (i.e., >30 min).
13. When using ZFN mRNA for delivery, the mixing of transfection solution with cells and the subsequent nucleofection must be carried out rapidly to prevent ZFN mRNA from degradation.
14. When using a small amount of transfected cells for gene editing assay, prepare the genomic DNA using QuickExtract. Draw 100–200 μL of cells to a PCR tube and centrifuge briefly to pellet the cells. Remove the culture medium and lyse the cells by adding 50–100 μL of QuickExtract to each tube and incubating the tube on a thermocycler with the following temperature profile: 65 °C for 15 min, 98 °C for 20 min, and hold at 20 °C or 4 °C (for overnight).
15. For large deletions, the wild-type allele may not be amplified under the PCR condition. The short extension time is intended to favor the amplification of deletion allele.
16. The efficiency of the BamHI site integration is indicative of the codon conversion efficiency in the cell population, but the two integration events are not always linked in individual cells.
17. If a flow cytometer is not available, perform the single cell cloning by limiting dilution. Dilute the culture to 0.3–0.5 cells per 200 μL using culture medium, and transfer 200 μL of the diluted culture to each well of a 96-well plate using a multichannel pipette. Mix the diluted culture frequently during the process. Identify single cell clones under microscope after the cell colonies have formed (2–3 weeks after limiting dilution).
18. When working with adherent cells, grow the culture to about 60 % confluency before clone consolidation. The time it takes for cells to grow out from a single cell clone to 60 % confluency varies widely between cell lines. For example, HEK293 cells take about 3–4 weeks, while MCF10A cells take about 10 weeks.
19. Freeze the replicate plate at −80 °C for later clone recovery if necessary.

20. For adherent cells, detach cells with trypsin/EDTA at 37 °C for 5–10 min (depending on cell types). Neutralize the trypsin with 100 μL of medium per well. Make a replicate plate for each clone for clone isolation and expansion later. Prepare cell lysate for PCR or qPCR assay using QuickExtract as described in **steps 4** and **5** of the Subheading 3.4.

References

1. Gaj T, Gersbach CA, Barbas CF III (2013) ZFN, TALEN, and CRISPR/Cas-based methods for genome engineering. Trends Biotechnol 31:397–405
2. Pfeiffer P, Goedecke W, Obe G (2000) Mechanisms of DNA double-strand break repair and their potential to induce chromosomal aberrations. Mutagenesis 15:289–302
3. Orlando SJ, Santiago Y, DeKelver RC et al (2010) Zinc-finger nuclease-driven targeted integration into mammalian genomes using donors with limited chromosomal homology. Nucleic Acids Res. doi:10.1093/nar/gkq512v
4. Jinek M, Chylinski K, Fonfara I et al (2012) A programmable dual-RNA-guided DNA endonuclease in adaptive bacterial immunity. Science 337:816–821
5. Lieber MR (2010) The mechanism of double-strand DNA break repair by the nonhomologous DNA end-joining pathway. Annu Rev Biochem 79:181–211
6. Jensen NM, Dalsgaard T, Jakobsen M et al (2011) An update on targeted gene repair in mammalian cells: methods and mechanisms. J Biomed Sci. doi:10.1186/1423-0127-18-10
7. Bibikova M, Beumer K, Trautman JK et al (2003) Enhancing gene targeting with designed zinc finger nucleases. Science 300:764
8. Urnov FD, Miller JC, Lee YL et al (2005) Highly efficient endogenous human gene correction using designed zinc-finger nucleases. Nature 435:646–651
9. Geurts AM, Cost GJ, Freyvert Y et al (2009) Knockout rats via embryo microinjection of zinc-finger nucleases. Science 325:433
10. Moehle EA, Rock JM, Lee YL et al (2007) Targeted gene addition into a specified location in the human genome using designed zinc finger nucleases. Proc Natl Acad Sci U S A 104:3055–3060
11. Campbell CR, Keown W, Lowe L et al (1989) Homologous recombination involving small single-stranded oligonucleotides in human cells. New Biol 1:223–227
12. Igoucheva O, Alexeev V, Yoon K (2001) Targeted gene correction by small single-stranded oligonucleotides in mammalian cells. Gene Ther 8:391–399
13. Rios X, Briggs AW, Christodoulou D et al (2012) Stable gene targeting in human cells using single-strand oligonucleotides with modified bases. PLoS ONE. doi:10.1371/journal.pone.0036697
14. Storici F, Snipe JR, Chan GK et al (2006) Conservative repair of a chromosomal double-strand break by single-strand DNA through two steps of annealing. Mol Cell Biol 26:7645–7657
15. Liu J, Majumdar A, Liu J et al (2010) Sequence conversion by single strand oligonucleotide donors via non homologous end joining in mammalian cells. J Biol Chem 285:23198–23207
16. Radecke S, Radecke F, Cathomen T et al (2010) Zinc-finger nuclease-induced gene repair with oligodeoxynucleotides: wanted and unwanted target locus modifications. Mol Ther 18:743–753
17. Chen F, Pruett-Miller SM, Huang Y et al (2011) High-frequency genome editing using ssDNAoligonucleotides with zinc-finger nucleases. Nat Methods 8:753–755
18. Rouet P, Smih F, Jasin M (1994) Introduction of double-strand breaks into the genome of mouse cells by expression of a rare-cutting endonuclease. Mol Cell Biol 14:8096–8106
19. Huertas P (2010) DNA resection in eukaryotes: deciding how to fix the break. Nat Struct Mol Biol 17(1):11–6
20. Cohen MS, Zhang C, Shokat KM et al (2005) Structural bioinformatics-based design of selective, irreversible kinase inhibitors. Science 308:1318–1321
21. Doehn U, Hauge C, Frank SR et al (2009) RSK is a principal effector of the RAS-ERK pathway for eliciting a coordinate promotile/invasive gene program and phenotype in epithelial cells. Mol Cell 35:511–522
22. Meyer M, Ortiz O, Hrabe de Angelis M et al (2012) Modeling disease mutations by gene targeting in one-cell mouse embryos. Proc Natl Acad Sci U S A 109:9354–9359

Chapter 15

Genome Editing in Human Pluripotent Stem Cells Using Site-Specific Nucleases

Kunitoshi Chiba and Dirk Hockemeyer

Abstract

Human embryonic stem cells (hESCs) and induced pluripotent stem cells (iPSCs) (Thomson, Science 282:1145–1147, 1998; Takahashi et al. Cell 131:861–872, 2007), collectively referred to as pluripotent stem cells (hPSCs), are currently used in disease modeling to address questions specific to humans and to complement our insight gained from model organisms (Soldner et al. Cell 146:318–331, 2011; Soldner and Jaenisch, Science 338:1155–1156, 2012). Recently, genetic engineering using site-specific nucleases has been established in hPSCs (Hockemeyer et al. Nat Biotechnol 27:851–857, 2009; Hockemeyer et al., Nat Biotechnol 29:731–734, 2011; Zou et al., Cell Stem Cell 5:97–110, 2011; Yusa et al., Nature 478:391–394, 2011; DeKelver et al., Genome Res 20:1133–1142, 2010), allowing a level of genetic control previously limited to model systems. Thus, we can now perform targeted gene knockouts, generate tissue-specific cell lineage reporters, overexpress genes from a defined locus, and introduce and repair single point mutations in hPSCs. This ability to genetically engineer pluripotent stem cells will significantly facilitate the study of human disease in a defined genetic context. Here we outline protocols for efficient gene targeting in hPSCs.

Key words Genome editing, Site-specific nucleases, ZFN, TALEN, CRISPR/Cas9, Human pluripotent stem cells, hESCs, hiPSC

1 Introduction

Genetic manipulations of hPSCs [1, 2] by conventional homologous recombination have proven to be too inefficient for routine applications. This inefficiency is attributable to a lower frequency of homologous recombination (HR) events in human cells, which poses a challenge in gene targeting of hPSCs [4]. Pioneering experiments using endonucleases, such as SCE-I [14, 15], have shown homology-directed repair (HDR) events to be stimulated by a DNA double-strand break (DSB) made in close proximity to the recombination site [16]. Since SCE-1 has a fixed recognition site, it has limited use in gene targeting across the genome. These early experiments, however, led to the development of engineered site-specific DNA nucleases that introduce a DSB at

Shondra M. Pruett-Miller (ed.), *Chromosomal Mutagenesis*, Methods in Molecular Biology, vol. 1239,
DOI 10.1007/978-1-4939-1862-1_15, © Springer Science+Business Media New York 2015

a defined genomic position, thus facilitating gene editing at that site. The first site-specific DNA nuclease to be engineered was the zinc-finger nuclease (ZFN), which is a chimeric protein composed of zinc-finger DNA-binding domains and a FOK1 cleavage domain originally found in *Flavobacterium okeanokoites* [5, 17]. The zinc finger DNA binding motif establishes specificity to about 3 bases and can be arrayed to recognize longer targeting sequences. ZFNs are designed as pairs with appropriately spaced DNA binding sites that permit the dimerization of the FOK1-nuclease domains. As this dimerization is required for nuclease activity, ZFNs can be engineered to form obligatory heterodimers, which restricts catalytic activity and prevents nonspecific DNA cleavage. The most advanced ZFN libraries currently use an array of up to 6 zinc fingers in a ZFN, allowing for an 18 bp recognition site. An alternative gene-editing tool, known as transcription activator-like effector nucleases (TALENs), has been used to genetically modify hPSCs [6]. Like ZFNs, TALENs consist of a DNA-binding domain and a FOK1 nuclease domain [18, 19]. However, unlike ZFNs, the DNA-binding domain of TALENs is adapted from transcription activator-like effectors (TALEs), which were originally discovered in the plant bacterial pathogen *Xanthomonas* and found to alter transcription in plant hosts. TALE DNA-binding is accomplished via repeat domains so that one repeat domain specifies the binding to one base in the TALE DNA recognition sequence. The base preference of each repeat domain is determined by two adjacent amino acids called a repeat variable di-residue (RVD) [20]. By combining different RVDs, TALENs can be designed to target nearly any sequence. Publically available RVDs (Addgene) can be systematically re-assembled using the Golden-Gate assembly method, making TALEN generation easy and accessible [21].

In addition to the aforementioned gene-editing tools, the RNA-guided nuclease adapted from Cas9 is an emerging genome engineering method with wide ranging applications. Cas9 is derived from clustered regularly interspaced short palindromic repeats (CRISPR) and is involved in the "adaptive immune system" of microbes [22]. The CRISPR system is composed of the Cas nuclease and noncoding RNA elements called guide RNAs. The guide RNAs possess distinct repeat elements derived from exogenous DNA targets known as protospacers and protospacer adjacent motifs (PAMs) [23]. The most commonly used CRISPR/Cas9 system is derived from *Streptococcus pyogenes* and uses the PAM sequence 5'-NGG-3′. The 20 bp of sequence immediately preceding the PAM sequence is fused with the other trans activating RNA components, and the resulting chimeric RNA is used as a single-guide RNA (sgRNA) [24–26]. While ZFNs, TALENs, and the

CRISPR/Cas9 systems have been shown to facilitate genome editing in hPSCs, each approach has its pros and cons, which may dictate the choice for a given experiment [27–29].

Several genes have now been successfully targeted in human cells using site-specific nucleases. The easiest approach involves disruption of gene function by nonhomologous end-joining (NHEJ) [30, 31], while more complex modification of the genome are generally achieved by creating a site-specific DSB with a nuclease and providing a donor DNA template to promote resolution by HDR [5, 6]. Several proof-of-concept experiments illustrate the utility of this approach. Previously, we established the AAVS1 locus as a "safe harbor" locus to enable the ectopic expression of genes in hPSCs [32]. This locus is dispensable in hPSCs, does not show strong variegation and silencing effects, and can be used for the robust expression of genes in hPSCs and differentiated cell types. In addition, we demonstrated that endogenous genes, such as Oct4, could be modified using site-specific nucleases with a repair donor plasmid that integrates a GFP reporter gene [5, 6]. Further, to introduce a so-called "scarless" point mutation with single base pair resolution, we used homology-directed repair with single strand oligonucleotide coupled with FAC-sorting [3] (details are also described in **Note 4**). Additionally, we demonstrated that site-specific nucleases could be used for a targeted chromosomal deletion in hPSCs [33–35]. As shown in prior studies, there are innumerable applications for gene targeting using site-specific nucleases [7–13]. Here we describe some of the fundamental protocols to achieve efficient gene targeting in hPSCs.

2 Materials

2.1 Cel-1 Assay

1. Lysis buffer: 100 mM Tris–HCl pH 8.0, 5 mM EDTA pH 8.0, 0.2 % SDS, 200 mM NaCl, 0.1 mg/ml Proteinase K (this should be added fresh).
2. 2-Propanol.
3. 70 % ethanol.
4. Tris–EDTA buffer: 10 mM Tris–HCl pH 8.0, 0.1 mM EDTA pH 8.0.
5. PCR primers.
6. PrimeSTAR HS DNA Polymerase (Takara Bio Inc., Otsu, Japan).
7. 2 % agarose gel: agarose in 1× Tris–acetate–EDTA (TAE) buffer by diluting 50× TAE stock.

8. Gel and PCR Cleanup kit.
9. 10× Standard Taq Reaction buffer (New England Biolabs, Ipswich, MA, USA).
10. 150 mM $MgCl_2$.
11. Surveyor® nuclease (Transgenomic, Omaha, NE, USA or *see* **Note 4** for homemade Cel-1 nuclease).

2.2 Electroporation

1. WIBR3 human ES cells [36].
2. Mitomycin C inactivated mouse embryonic fibroblast (MEF) feeder cells.
3. hESC media (500 ml): 380 ml DMEM/F12, 75 ml Fetal Bovine Serum (FBS) performance plus (Gibco/Life Technologies, Carlsbad, CA, USA), 25 ml KnockOut Serum Replacement (Gibco/Life Technologies), 5 ml GlutaMax (Gibco/Life Technologies), 5 ml 100× MEM Nonessential Amino Acids Solution, 5 ml 10,000 U/ml Penicillin-Streptomycin, 50 μl 55 mM 2-Mercaptoethanol, 80 μl 25 μg/ml basic Fibroblast growth factor (bFGF/FGF2).
4. 6-well Clear TC-Treated multiple well plates.
5. Rho Kinase (ROCK)-inhibitor (Y-27632) (10 μM final).
6. 1× Phosphate buffered saline (PBS) pH 7.4.
7. 0.25 % Trypsin–EDTA.
8. Gene Pulser Cuvette 0.4 cm electrode gap (Bio-Rad, Hercules, CA, USA).
9. Gene Pulser Xcell Eukaryotic System (Bio-Rad).
10. Mitomycin C inactivated DR-4 MEFs [37].
11. ZFNs, TALENs, or CRISPR/Cas9s expression plasmid.
12. Repair-donor DNA (plasmid or single strand oligoDNA).

2.3 Identifying and Isolating Targeted Cells

1. Puromycin, Neomycin, Hygromycin.
2. pEGFP-N2 (Addgene, Cambridge, MA, USA).
3. Fluorescence activated cell sorting (FACS).

2.4 Colony Picking

1. Dissection microscope equipped in tissue culture cabinet.
2. Mitomycin C inactivated mouse embryonic fibroblast (MEF) feeder cells.
3. 12-well Clear TC-Treated multiple well plates.
4. Pasteur pipette.
5. Matrigel (BD Biosciences, San Jose, CA, USA).

3 Methods

3.1 Evaluating Site-Specific Nuclease Activity by Measuring Nonhomologous End-Joining (NHEJ) Frequencies Using the Cel-1 Assay

To test the activity of site-specific nucleases they should first be transfected into HEK 293 T cells using standard protocols and evaluated using the Cel-1 assay. Active nucleases will induce a double-strand break (DSB) that is repaired by NHEJ when a homology donor DNA template is not available for HDR. As DNA repair by NHEJ is imperfect, it can result in insertions or deletions (indels) at the DSB site. The Cel-1 assay is designed to measure frequency of indels generated by NHEJ [38]. To this end, DNA from cells that were exposed to the nuclease is extracted, and the genomic region around the DSB-site is amplified by PCR (amplicon size 350 bp-1 kb). The PCR product is denatured before slowly being re-annealed. Due to indels created by NHEJ, the PCR amplicons are heterogeneous and when re-annealed form heteroduplexed double-stranded DNA containing mismatches. Cel-1 nuclease cleaves specifically at sites of these base pair mismatches. Since heteroduplex formation results from base pair mismatches, Cel1 nuclease can be used to digest DNA molecules into smaller fragments resolvable on an agarose gel. Thus the frequency of indels caused by NHEJ, and therefore efficiency of site-specific nucleases, can be inferred by the relative increase of Cel-1-digested fragments.

1. Prepare ZFN, TALEN or CRISPR expression plasmids (*see* **Note 1**).
2. Transfect 500 ng of each left and right ZFN/TALEN plasmids, or 500 ng of Cas9 expression plasmid and 500 ng of sgRNA expression plasmid (total 1 μg) into 20 % confluent HEK 293 T cells on 12-well (*see* **Note 2**). Transfection of a GFP expression plasmid is required for the control of Cel-1 assay to estimate nonspecific digestion by the Cel-1 nuclease.
3. Isolate genomic DNA 72 h after transfection. Add Proteinase K (100 μg/ml) to lysis buffer (400 μl/well of a 12-well plate) and incubate at 37 °C for at least 3 h.
4. Precipitate genomic DNA with 400 μl of isopropanol and wash DNA pellet with 70 % ethanol. Dissolve the DNA pellet in 150 μl Tris–EDTA buffer at pH 8.0. Ensure that DNA is fully dissolved (if needed incubate DNA at 37 °C overnight or until dissolution).
5. Amplify targeted locus by PCR using high fidelity DNA polymerase (amplicon sizes ranging between ~350 bp and ~1 kb).
6. Verify that only a single PCR product is present using agarose gel electrophoresis, should there be several bands, optimize PCR conditions. Purify PCR product by gel extraction or PCR purification methods. After purification elute DNA in 35 μl of water.

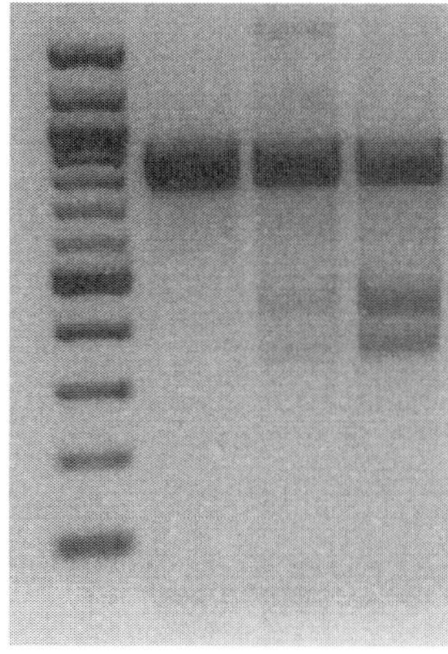

Fig. 1 An example gel image from a Cel-1 assay. This image shows an approximation of digestion efficiency for site-specific nucleases. Based on this information, number of colonies for picking is determined. The lanes shown above correspond to GFP, TALEN, and CRISPR/Cas9 expression plasmids that were transiently transfected in HEK293T cells. GFP was used as a control for transfection and Cel-1 nuclease digestion. The expected size of untargeted product is 797 bp, while correctly targeted amplicons have two additional DNA bands at 351 and 446 bp, indicating cleavage by Cel-1. While the TALENs and CRISPR/Cas9 used in this assay were designed to target the same site, the difference in digested band intensity is indicative of higher digestion efficiency in DNA targeted with CRISPRs

7. Use between 500 ng and 2 μg of DNA in14.4 μl and add 1.8 μl 10× Standard Taq Reaction buffer in a PCR tube.
8. Generate heteroduplexes by denaturing and slowly annealing the PCR product. Use the following heteroduplex program (*see* **Note 3**):95 °C for 5 min, 95–85 °C at −2 °C/s, 85–25 °C at −0.1C°/s, hold at 4 °C.
9. Add 1.8 μl of 150 mM $MgCl_2$ (final conc.15 mM) and 0.5 μl of the Cel-1 nuclease (*see* **Note 4**).
10. Incubate at 42 °C for 30 min.
11. Separate DNA fragments by agarose gel electrophoresis (Fig. 1).

3.2 Repair-Donor DNA Preparation

There are several ways to insert, convert or delete sequences of genomic DNA via site-specific nucleases and subsequent homology directed repair. Two of the most commonly used methods are described here:

3.2.1 Using Plasmid DNA as a Repair Template (Donor)

The general structure of a repair plasmid comprises an insertion sequence, such as antibiotic-resistance cassettes, flanked by each right and left homology arm (*see* **Note 5**). Homology arms can vary in length ranging from 50 bp [33] to 750 bp [5, 6] depending on the size of the insertion. In our experience, longer homology (~750 bp) arms increase, to some extent, the editing efficiencies of larger insertions, e.g., simultaneous insertion of a GFP-reporter and an antibiotic-resistance gene [33].

3.2.2 Single Strand Oligodeoxynucleotides (ssODN)

ssODN are commonly used for scarless genome engineering approaches such as the introduction of point mutations or the repair of disease causing mutations in patient specific induced pluripotent stem cells (iPSCs). The length of ssODN should be around 100–120 bp. Mutations should be close to the recognition site of the site-specific nuclease [39, 40].

3.3 Electroporation

There are several ways to deliver targeting plasmids into hPSCs. Transient transfection and electroporation are the most commonly used. In addition to being robust and cost-effective, a further advantage of electroporation is that a high concentration of cells and plasmids can be handled in a single step. A flowchart of the targeting procedures using electroporation is shown in Fig. 2.

1. Culture hPSCs on mitomycin C-inactivated mouse embryonic fibroblast (MEF) feeder layers (2.4×10^6 cells/9.5 cm^2) in hESC medium until cells are semi-confluent. We generally use

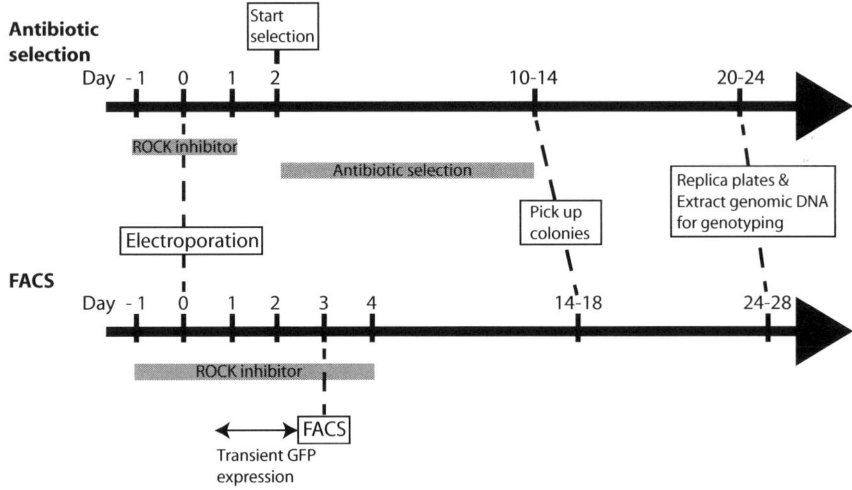

Fig. 2 Timeline for gene targeting using electroporation. Only significant procedures are indicated in the figure. The day before and after electroporation (Days-1 & 0), hPSCs are cultured in ROCK inhibitor-containing hESC media. For FACS selection, cells are continuously cultured with ROCK inhibitor an additional day after sorting (Day 4). Antibiotic selection starts 48 h after electroporation (Day 2). During antibiotic selection, DR-4 feeders or Matrigel plates are used for hPSC culture. 8–12 days after antibiotic selection or 10–14 days after FACS, colonies are ready for picking. The exact day for picking colonies depends on the colony size, confluence, and differentiation status of hPSCs. 7–10 days after picking, colonies are again ready for picking. Replica plates are made for continuous culture, while the original plates are used for genomic DNA extraction for genotyping

1-2 × 10^7 hPSCs (this equals about one to two 6-well plates that are about 50 % confluent). For hESC/iPSC culture, media should be changed every day.

Day-1

2. Culture hPSCs in hESC medium containing 10 μM ROCK-inhibitor 24 h before electroporation.
3. Prepare two 6-well plates of DR-4 feeder cells plated at 2.4 × 10^6 cells/9.5 cm^2 (*see* **Note 6**).

Day 0

4. Wash the cells with 1x PBS and collect cells by trypsinization with 0.25 % Trypsin–EDTA. Neutralize trypsin with hESC media, triturate cells into a single cell suspension and collect cells in a 50 ml conical tube. Note: feeder cells might come off as a sheet of cells and are removed by letting them sediment to the bottom of the tube.
5. Transfer the supernatant containing single hPSCs to a new tube and pellet by centrifugation at 188 × *g* and 4 °C.
6. Resuspend 1–2 × 10^7 cells in 500 μl of 1× PBS.
7. Prepare 10 μg of ZFNs, TALENs (5 μg each left or right nuclease) or 10 μg of CRISPR plasmid (5 μg each Cas9 and sgRNA) and 40 μg of a repair template plasmid in 300 μl of 1× PBS.
8. Combine hPSC suspension with plasmid and transfer into a Gene Pulser Cuvette. Put the cuvette on ice for 3–5 min.
9. Electroporate using the Gene Pulser Xcell System; 250 V, 500 F, 0.4 cm cuvettes. Subsequently plate the entire content of the cuvette onto two 6-well feeder plates (DR-4) in the hESC media containing 10 μM ROCK inhibitor.

Day 1

10. Change media with hESC media containing 10 μM ROCK inhibitor.

Day 2

11. Change media with hESC media without ROCK inhibitor.

Day 3

12. Start selection, described in the next section

3.4 Identifying Targeted Cells

There are multiple ways to select targeted cells. The easiest method is to use antibiotics in conjunction with the selectable marker. For each hPSC line, a killing curve should be established prior to the targeting experiment. When antibiotic selection cannot be used, e.g., in the scarless insertion of a point mutation with a ssODN, we

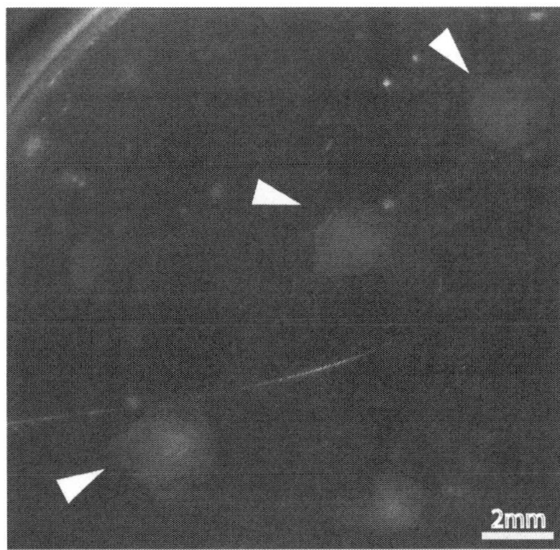

Fig. 3 An example of colony size for picking. Colonies ready for picking are indicated with *white arrows*. The suitable size for picking is 2-4 mm

isolate the targeted cells by co-electroporation of a plasmid expressing a fluorescent protein and subsequent fluorescence-activated cell sorting (FACS).

3.4.1 Using Antibiotic Selection to Identify Targeted hPSC Clones

1. Electroporate the cells with site-specific nucleases and a repair plasmid containing antibiotic resistance genes.
2. Three days after electroporation, exchange with hESC medium containing an antibiotic (puromycin: 0.5 μg/ml, neomycin 70 μg/ml, or hygromycin 35 μg/ml).
3. Keep the cells in the selection media for 10–14 days. Colonies become apparent about 6–8 days after the start of selection and can be picked as single cell-derived colonies 10–14 days after start of selection. Colonies suitable for picking are shown in Fig. 3.

3.4.2 Isolating Targeted Clones Using FACS

1. Co-electroporate 10 μg of the site-specific nucleases with 35 μg of the repair template together with 5 μg of an eGFP-expression plasmid (e.g., pEGFP-N2) (*see* **Note 7**).
2. Three days after electroporation, isolate eGFP-expressing cells by FACS.
3. Plate the single-cell suspension at low density (we usually plate cells at several densities ranging from 250 cells/well to 3,000 cells/well of a 6-well plate) on MEF feeder plates in hESC medium supplemented with 10 μM ROCK inhibitor for the first 24 h.
4. Culture cells for 10–14 days, pick clones when colonies are at least 2 mm.

3.5 Colony Picking

1. Prepare MEF feeders on 12-well plates in advance. We base the number of colonies that we pick on the efficiency of the site-specific nuclease in the Cel-1 assay (*see* above) and the targeting

strategy. Targeting of the OCT4 and PPP12R2 genes using gene trap strategies yielded high targeting efficiencies and low background of random integration. In such cases, we usually isolate 12–24 clones. When targeting genes that are not expressed in hPSCs, we use autonomous selection cassettes. Targeting efficiencies using this strategy can range widely, but correlate strongly with the efficiencies found in the Cel-1 assay. For site-specific nucleases that have a robust Cel-1 efficiency (5–10 % of the PCR-product being sensitive to Cel-1 digestion) picking 60–72 clones is usually sufficient. Depending on the strategy, targeting with ssODNs can require the isolation of up to 200 single cell-derived clones.

2. Use a dissection microscope mounted in a biosafety cabinet, identify and pick undifferentiated hPSC colonies using a p20 pipette tip or a Pasteur pipette. Break up individual colonies into 15–25 pieces and transfer these to one well of the 12-well feeder plates.
3. Culture cells until they become 75 % confluent; this usually takes 9–10 days.
4. Before isolating genomic DNA from these hPSC clones, generate a replica plate. Use this replica plate to maintain the cells while isolating genomic DNA from the parental plate for genotyping. To generate replica plates of the picked hPSCs, prepare the same number of 12-well MEF feeder plates as described above. Break up a colony from individual wells into about 15–25 clumps and transfer it to the new feeder plates. Make sure to pick up undifferentiated colonies. Normally, differentiated colonies look denser and possess an ambiguous border delineating the hPSC colony from the feeder layer.
5. Keep the replica plates in culture until completion of genotyping.

3.6 Genotyping

Genotyping is required not only for confirmation of successful gene targeting but also for identifying a potential random integration of plasmid DNA that might also confer antibiotic resistance or transgene expression. We generally use Southern blotting for this purpose, probing with two different probes, hybridizing to sequences located either external or internal to the homology sequences of the donor repair template. We also use Southern blotting to identify whether clones are heterozygously or homozygously targeted. For higher resolution genotyping, such as detection of point mutations or small deletions, PCR and subsequent sequencing are encouraged.

1. Lyse cells from the confluent 12-well plates once the replica plates have been made and extract genomic DNA, as described above for the Cel-1 assay. The extracted genomic DNA is a

mixture of hPSCs and MEF feeder cells. However, the amount of MEF DNA compared to hPSC DNA is small and its contribution can be ignored in most genotyping strategies.

2. Use 15 μl (10 %) of genomic DNA for restriction enzyme digestion. For efficient digestion, use of highly concentrated restriction enzymes is recommended (60–100 U/reaction). The protocol of Southern blotting used here is described in Molecular Cloning—A Laboratory manual (Cold Spring Harbor Laboratory Press). As mentioned above, both external and internal probes should be used to validate successful targeting.
3. If necessary, PCR and subsequent sequencing are useful for detection of small modifications such as introduction of point mutations or deletion of fragments. At least one primer should be designed outside of the homology arms. PCR analysis can be used to pre-screen hPSC clones prior to Southern blot analysis to reduce the number of candidates that have to be analyzed.

4 Notes

1. Design of ZFNs, TALENs and CRISPRs.
   ZFNs are commercially available from Sigma-Aldrich or can be generated using available open source libraries, e.g., "OPEN"-source (Oligomerized Pool Engineering) constructed by the Zinc Finger Consortium [41]. TALENs can be designed and assembled as described in [19, 20]. CRISPRs are available from Addgene, and targeting constructs can be made as described previously [25]. Each of the described methods has advantages and disadvantages. For example, CRISPRs are easier to assemble than ZFNs and TALENs, although more off-target effects have been reported with CRISPR targeting [26, 27]. TALENs have lower binding specificity for guanine [19], thus for targeting of G-rich regions, using CRISPRs may result in more efficient cutting. Which methods are suitable for the experiment will depend on experimental requirements and constraints.
2. Functional test of designed site-specific nucleases.
   Before using a novel site-specific nuclease in an hPSC gene targeting experiment, we generally evaluate its efficiency in a common tumor cell line such as HEK293T. These experiments can guide the choice of donor plasmid and editing strategy as well as the choice of how many single-cell derived colonies that need to be isolated and characterized to achieve a successful targeting event.
3. Heteroduplex formation.
   For successful heteroduplex formation, the initial PCR product should be a single band. This avoids high background and

nonspecific degradation for the PCR product. If the targeting region contains repeated sequences or is biased for specific bases, such as a GC-rich promoter region, it is sometimes difficult to anneal heteroduplexes. In these special cases, indels can be detected by sequencing of subcloned PCR amplicons.

4. Purification of Cel-1 nuclease
   Cel-1 nuclease can be purified from celery as reported previously [42].
5. Selection cassette.
   Puromycin, Neomycin, and Hygromycin are commonly used for selection in hPSCs. When these selection markers are used as autonomous selection cassettes, they are integrated into the genome together with an ectopic promoter and polyadenylation (polyA) signal to drive their expression (e.g., CMV or phosphoglycerate kinase (PGK) promoter and bovine growth hormone polyadenylation sequence (polyA)). Selection cassettes are typically flanked by LoxP sites, which can be removed by Cre recombinase-meditated excision. An alternative approach is to target the antibiotic-resistance gene such that its expression will be controlled from the regulatory elements of the endogenous genes. These so called "gene trap" approaches often use either a polycistronic 2A peptide sequence [43] or internal ribosomal entry site (IRES) that links the reading frame of the gene of interest with the expression of the selectable marker.
6. DR-4 MEF feeders
   DR-4 MEF feeder cells are derived from MEFs prepared from the DR-4 mouse strain (Jackson lab stock #003208). DR-4 mice harbor three independently segregating loci that collectively express drug resistance genes: neomycin/G418, puromycin, and hygromycin resistance. These three resistance genes are controlled by the promoter and polyadenylation signal of the murine 3-phosphoglycerate kinase gene (*Pgk1*), which ensures high levels of expression in both embryonic stem (ES) cells and mouse embryonic fibroblasts (MEFs). DR-4 cells also carry a mutation in the *hprt* gene located on the X-chromosome, which confers resistance to 6-thioguanine and abolishes endogenous hprt function. MEFs prepared from the DR-4 mice display resistance to concentrations of the drugs G418, puromycin, 6-thioguanine, and hygromycin well above those used normally for the selection of drug-resistant ES cells. If DR-4 feeder cells are not available, hPSCs can be kept in feeder-free systems using Matrigel as a substrate and defined media as previously described [44].
7. Sorting of targeted hPSCs by FACS.
   Instead of using an independent GFP-expressing plasmid, fluorescent protein-fused Cas9 can be used for sorting of targeted hPSCs by FACS [31].

References

1. Thomson JA (1998) Embryonic stem cell lines derived from human blastocysts. Science 282(5391):1145–1147
2. Takahashi K, Tanabe K, Ohnuki M, Narita M, Ichisaka T, Tomoda K, Yamanaka S (2007) Induction of pluripotent stem cells from adult human fibroblasts by defined factors. Cell 131(5):861–872
3. Soldner F, Laganiere J, Cheng AW, Hockemeyer D, Gao Q, Alagappan R, Khurana V, Golbe LI, Myers RH, Lindquist S et al (2011) Generation of isogenic pluripotent stem cells differing exclusively at two early onset Parkinson point mutations. Cell 146(2):318–331
4. Soldner F, Jaenisch R (2012) Medicine. iPSC disease modeling. Science 338(6111):1155–1156
5. Hockemeyer D, Soldner F, Beard C, Gao Q, Mitalipova M, DeKelver RC, Katibah GE, Amora R, Boydston EA, Zeitler B et al (2009) Efficient targeting of expressed and silent genes in human ESCs and iPSCs using zinc-finger nucleases. Nat Biotechnol 27(9):851–857
6. Hockemeyer D, Wang H, Kiani S, Lai CS, Gao Q, Cassady JP, Cost GJ, Zhang L, Santiago Y, Miller JC et al (2011) Genetic engineering of human pluripotent cells using TALE nucleases. Nat Biotechnol 29(8):731–734
7. Zou J, Maeder M, Mali P, Pruett-Miller S, Thibodeau-Beganny S, Chou B, Chen G, Ye Z, Park I, Daley G et al (2009) Gene targeting of a disease-related gene in human induced pluripotent stem and embryonic stem cells. Cell Stem Cell 5(1):97–110
8. Yusa K, Rashid ST, Strick-Marchand H, Varela I, Liu P-Q, Paschon DE, Miranda E, Ordóñez A, Hannan NRF, Rouhani FJ et al (2011) Targeted gene correction of α1-antitrypsin deficiency in induced pluripotent stem cells. Nature 478(7369):391–394
9. DeKelver RC, Choi VM, Moehle EA, Paschon DE, Hockemeyer D, Meijsing SH, Sancak Y, Cui X, Steine EJ, Miller JC et al (2010) Functional genomics, proteomics, and regulatory DNA analysis in isogenic settings using zinc finger nuclease-driven transgenesis into a safe harbor locus in the human genome. Genome Res 20(8):1133–1142
10. Doyon JB, Zeitler B, Cheng J, Cheng AT, Cherone JM, Santiago Y, Lee AH, Vo TD, Doyon Y, Miller JC et al: Rapid and efficient clathrin-mediated endocytosis revealed in genome-edited mammalian cells. Nature Cell Biology 2011, 13(3):331–337
11. Forster R, Chiba K, Schaeffer L, Regalado SG, Lai CS, Gao Q, Kiani S, Farin HF, Clevers H, Cost GJ et al: Human Intestinal Tissue with Adult Stem Cell Properties Derived from Pluripotent Stem Cells. Stem Cell Reports 2014, 2(6):838–852
12. Grassart A, Cheng AT, Hong SH, Zhang F, Zenzer N, Feng Y, Briner DM, Davis GD, Malkov D, Drubin DG: Actin and dynamin2 dynamics and interplay during clathrin-mediated endocytosis. The Journal of Cell Biology 2014, 205(5):721–735
13. Sexton AN, Regalado SG, Lai CS, Cost GJ, O'Neil CM, Urnov FD, Gregory PD, Jaenisch R, Collins K, Hockemeyer D: Genetic and molecular identification of three human TPP1 functions in telomerase action: recruitment, activation, and homeostasis set point regulation. Genes & Development 2014, 28(17):1885–1899
14. Bibikova M, Beumer K, Trautman JK, Carroll D (2003) Enhancing gene targeting with designed zinc finger nucleases. Science 300(5620):764
15. Porteus MH, Baltimore D (2003) Chimeric nucleases stimulate gene targeting in human cells. Science 300(5620):763
16. Jasin M (1996) Genetic manipulation of genomes with rare-cutting endonucleases. Trends Genet 12(6):224–228
17. Urnov FD, Miller JC, Lee Y-L, Beausejour CM, Rock JM, Augustus S, Jamieson AC, Porteus MH, Gregory PD, Holmes MC (2005) Highly efficient endogenous human gene correction using designed zinc-finger nucleases. Nature 435(7042):646–651
18. Boch J, Scholze H, Schornack S, Landgraf A, Hahn S, Kay S, Lahaye T, Nickstadt A, Bonas U (2009) Breaking the code of DNA binding specificity of TAL-type III effectors. Science 326(5959):1509–1512
19. Moscou MJ, Bogdanove AJ (2009) A simple cipher governs DNA recognition by TAL effectors. Science 326(5959):1501
20. Miller JC, Tan S, Qiao G, Barlow KA, Wang J, Xia DF, Meng X, Paschon DE, Leung E, Hinkley SJ et al (2010) A TALE nuclease architecture for efficient genome editing. Nat Biotechnol 29(2):143–148
21. Cermak T, Doyle EL, Christian M, Wang L, Zhang Y, Schmidt C, Baller JA, Somia NV, Bogdanove AJ, Voytas DF (2011) Efficient design and assembly of custom TALEN and other TAL effector-based constructs for DNA targeting. Nucleic Acids Res 39(12):e82
22. Deveau H, Garneau JE, Moineau S (2010) CRISPR/Cas system and its role in phage-bacteria interactions. Annu Rev Microbiol 64(1):475–493

23. Jinek M, Chylinski K, Fonfara I, Hauer M, Doudna JA, Charpentier E (2012) A programmable dual-RNA-guided DNA endonuclease in adaptive bacterial immunity. Science 337(6096):816–821
24. Jinek M, East A, Cheng A, Lin S, Ma E, Doudna J (2013) RNA-programmed genome editing in human cells. Elife 2:e00471
25. Mali P, Yang L, Esvelt KM, Aach J, Guell M, DiCarlo JE, Norville JE, Church GM (2013) RNA-guided human genome engineering via Cas9. Science 339(6121):823–826
26. Cong L, Ran FA, Cox D, Lin S, Barretto R, Habib N, Hsu PD, Wu X, Jiang W, Marraffini LA et al (2013) Multiplex genome engineering using CRISPR/Cas systems. Science 339(6121):819–823
27. Fu Y, Foden JA, Khayter C, Maeder ML, Reyon D, Joung JK, Sander JD (2013) High-frequency off-target mutagenesis induced by CRISPR-Cas nucleases in human cells. Nat Biotechnol 31(9):822–826
28. Hsu PD, Scott DA, Weinstein JA, Ran FA, Konermann S, Agarwala V, Li Y, Fine EJ, Wu X, Shalem O et al (2013) DNA targeting specificity of RNA-guided Cas9 nucleases. Nat Biotechnol 31(9):827–832
29. Ran FA, Hsu PD, Lin C-Y, Gootenberg JS, Konermann S, Trevino AE, Scott DA, Inoue A, Matoba S, Zhang Y et al (2013) Double nicking by RNA-guided CRISPR Cas9 for enhanced genome editing specificity. Cell 154(6):1380–1389
30. Perez EE, Wang J, Miller JC, Jouvenot Y, Kim KA, Liu O, Wang N, Lee G, Bartsevich VV, Lee Y-L et al (2008) Establishment of HIV-1 resistance in CD4+ T cells by genome editing using zinc-finger nucleases. Nat Biotechnol 26(7):808–816
31. Reyon D, Tsai SQ, Khayter C, Foden JA, Sander JD, Joung JK (2012) FLASH assembly of TALENs for high-throughput genome editing. Nat Biotechnol 30(5):460–465
32. Smith JR, Maguire S, Davis LA, Alexander M, Yang F, Chandran S, ffrench-Constant C, Pedersen RA (2008) Robust, persistent transgene expression in human embryonic stem cells is achieved with AAVS1-targeted integration. Stem Cells 26(2):496–504
33. Lee HJ, Kim E, Kim JS (2010) Targeted chromosomal deletions in human cells using zinc finger nucleases. Genome Res 20(1):81–89
34. Orlando SJ, Santiago Y, Dekelver RC, Freyvert Y, Boydston EA, Moehle EA, Choi VM, Gopalan SM, Lou JF, Li J et al (2010) Zinc-finger nuclease-driven targeted integration into mammalian genomes using donors with limited chromosomal homology. Nucleic Acids Res 38(15):e152
35. Yang H, Wang H, Shivalila CS, Cheng AW, Shi L, Jaenisch R (2013) One-step generation of mice carrying reporter and conditional alleles by CRISPR/Cas-mediated genome engineering. Cell 154(6):1370–1379
36. Lengner CJ, Gimelbrant AA, Erwin JA, Cheng AW, Guenther MG, Welstead GG, Alagappan R, Frampton GM, Xu P, Muffat J et al (2010) Derivation of Pre-X inactivation human embryonic stem cells under physiological oxygen concentrations. Cell 141(5):872–883
37. Tucker KL, Wang Y, Dausman J, Jaenisch R (1997) A transgenic mouse strain expressing four drug-selectable marker genes. Nucleic Acids Res 25(18):3745–3746
38. Guschin DY, Waite AJ, Katibah GE, Miller JC, Holmes MC, Rebar EJ (2010) A rapid and general assay for monitoring endogenous gene modification. Methods Mol Biology 649: 247–256
39. Chen F, Pruett-Miller SM, Huang Y, Gjoka M, Duda K, Taunton J, Collingwood TN, Frodin M, Davis GD (2011) High-frequency genome editing using ssDNA oligonucleotides with zinc-finger nucleases. Nat Methods 8(9):753–755
40. Radecke S, Radecke F, Cathomen T, Schwarz K (2009) Zinc-finger nuclease-induced gene repair with oligodeoxynucleotides: wanted and unwanted target locus modifications. Mol Ther 18(4):743–753
41. Maeder ML, Thibodeau-Beganny S, Osiak A, Wright DA, Anthony RM, Eichtinger M, Jiang T, Foley JE, Winfrey RJ, Townsend JA et al (2008) Rapid “Open-Source” engineering of customized zinc-finger nucleases for highly efficient gene modification. Mol Cell 31(2):294–301
42. Yang B, Wen X, Kodali NS, Oleykowski CA, Miller CG, Kulinski J, Besack D, Yeung JA, Kowalski D, Yeung AT (2000) Purification, cloning, and characterization of the CEL I nuclease. Biochemistry 39(13):3533–3541
43. Donnelly ML, Hughes LE, Luke G, Mendoza H, ten Dam E, Gani D, Ryan MD (2001) The ‘cleavage’ activities of foot-and-mouth disease virus 2A site-directed mutants and naturally occurring ‘2A-like’ sequences. J Gen Virol 82(Pt 5):1027–1041
44. Ludwig TE, Levenstein ME, Jones JM, Berggren WT, Mitchen ER, Frane JL, Crandall LJ, Daigh CA, Conard KR, Piekarczyk MS et al (2006) Derivation of human embryonic stem cells in defined conditions. Nat Biotechnol 24(2):185–187

Chapter 16

Strategies to Increase Genome Editing Frequencies and to Facilitate the Identification of Edited Cells

Matthew Porteus

Abstract

The power of genome editing is increasingly recognized as it has become more accessible to a wide range of scientists and a wider range of uses has been reported. Nonetheless, an important practical aspect of the strategy is develop methods to increase the frequency of genome editing or methods that enrich for genome-edited cells such that they can be more easily identified. This chapter discusses several different approaches including the use of cold-shock, exonucleases, surrogate markers, specialized donor vectors, and oligonucleotides to enhance the frequency of genome editing or to facilitate the identification of genome-edited cells.

Key words Genome editing, Engineered nuclease, Homologous recombination, Nonhomologous end joining, Homing endonuclease, Zinc finger nuclease, TAL effector nuclease, CRISPR/Cas9

1 Genome Editing via Nonhomologous End Joining or by Homologous Recombination

There are several different approaches to genome editing, but in this chapter, I focus on double-strand break (DSB)-mediated genome editing. In DSB-mediated genome editing, a nuclease is designed to recognize a specific site in the genome where it will create a DNA DSB. Every cell has mechanisms to both recognize the existence of a DSB and redundant pathways to repair the DSB. The two major repair pathways of DSB repair are non-homologous end-joining (NHEJ) and homologous recombination. In NHEJ, which is fundamentally a stitching process, the two free ends are held in close proximity, processed to create ends that can be ligated together, and then ligated together. In homologous recombination, which is fundamentally a "copy and paste" process, the ends are processed to create free 3′ ends, which then invade into a homologous undamaged DNA template and using this undamaged template new DNA is synthesized, creating a new block of fresh DNA that is then recombined into the region

Shondra M. Pruett-Miller (ed.), *Chromosomal Mutagenesis*, Methods in Molecular Biology, vol. 1239,
DOI 10.1007/978-1-4939-1862-1_16,

surrounding the DSB. Both NHEJ and HR are usually quite accurate in the repair of a DSB but NHEJ is a lower fidelity mechanism of repair because if base damage or loss occurs at the site of the DSB, NHEJ can not recapture that lost sequence information. In addition, NHEJ is a lower fidelity form of repair because at some low frequency the processing of the ends to ligate them together leads to small insertions and deletions at the site of the DSB. When insertions/deletions occur at the site of the nuclease-induced DSB this is genome editing by "mutagenic NHEJ." HR is considered a high-fidelity form of repair because it uses highly accurate DNA polymerases and proofreading to recreate the region around the DSB from an undamaged template, thus leading to complete accuracy in the repair of the DSB. The usual template for repair by HR is the undamaged sister-chromatid but HR can be inaccurate, however, when a donor DNA template is provided to the cell. For reasons that are not well understood, at some frequency the cell will utilize a provided donor sequence as the template for the repair of the DSB by homologous recombination if there is sufficient identity between the damaged genomic region and the homology arms of the donor DNA molecule. But when the donor DNA is designed to have sequence differences between the homology arms then the donor DNA sequence differences are incorporated into the genomic DNA. In this way, the HR machinery is highly faithful to the donor DNA template but is mutagenic with respect to the endogenous genomic target. Using this strategy both single nucleotide changes and large blocks of DNA (>10 KB) can be incorporated at the site of the DSB. Genome editing by HR gives both spatial and nucleotide precision in contrast to editing by mutagenic NHEJ which gives precise spatial but not nucleotide resolution.

Because the mechanism of genome editing by mutagenic NHEJ vs. HR there are different mechanisms to enrich for each of these outcomes.

2 Creating Highly Active Nucleases

The most important method to increase the frequency of genome editing is to design highly active site-specific nucleases. There are currently four fundamental platforms for engineering site-specific nucleases: (1) homing endonucleases (HEs) [1, 2]; (2) zinc finger nucleases (ZFNs) [3, 4]; (3) TAL-effector nucleases (TALENs) [5, 6]; and (4) RNA-guided endonucleases (RGENs from the CRISPR/Cas9 system) [7–9]. In addition, a fifth hybrid platform in which a TAL effector DNA binding domain is fused to a homing endonuclease to create a "Mega-Tal" has been described although experience with this platform is much more limited than with the other platforms [10]. While homing endonucleases and ZFNs were essential in the early development of genome editing, the challenges in redesigning HEs and ZFNs to be highly active at new target sites have been significant and now only researchers with significant expertise in these platforms continue to use them. Despite the challenges in re-engineering highly active and specific ZFNs, ZFNs remain the only platform that has entered human clinical trials. In contrast, TALENs can be made using a simple "Golden Gate" system [11] by any lab that is sophisticated in molecular biology and high school students can make RGENs. In addition to their ease of design, TALENs and RGENs also have the property of having a relatively high "hit" rate in that screening just a few will usually result in the identification of an engineered nuclease that gives high rates of genome modification (>30 % mutagenic NHEJ in K562 cells for example).

While RGENs are simpler to make and thus it is easier to screen for ones that have the highest on-target activity, TALENs do seem to be more specific. Nonetheless, several investigators have performed whole genome sequencing on cell lines created using RGENs and not found evidence that the nucleases induced off-target mutations [12, 13]. Furthermore, given the relatively high spontaneous mutation rate of cells grown in culture, the transient mutational burden created by an engineered nuclease is likely to be minimal.

As a practical matter, therefore, the best approach to creating high frequencies of genome editing is to engineer several different RGEN and/or TALEN variants and screen for the ones with the highest on-target activity. The methods described below simply augment the genome editing frequency created by the engineered nuclease. These methods can to some extent "rescue" a poor initial nuclease but the best approach is to invest in engineering a highly active nuclease initially.

3 Increasing Mutagenic NHEJ by Increasing the Expression of the Engineered Nuclease

Once active nucleases are generated, the next most important factor in generating high rates of genome editing is to achieve sufficient levels of nuclease expression. For genome editing by mutagenic NHEJ, higher levels of nuclease expression generates higher frequencies of on-target, as well as off-target activity. For genome editing by HR, there is a Goldilocks effect whereby just the right amount of nuclease needs to be expressed in order to maximize genome editing by HR [14]. To achieve optimal levels of nuclease expression, therefore, requires a careful examination of delivery method (whether viral or non-viral) and of which promoter drives the expression of the nuclease. While the CMV promoter is broadly used and broadly useful, other promoters are more active in different cell types and thus if low genome editing efficiencies are achieved, switching promoter elements to increase nuclease expression levels is one possible solution.

One simple method to increase genome editing by mutagenic NHEJ is to use "cold shock" of cells by culturing cells expressing ZFNs or TALENs at 30 °C [15, 16]. This treatment can dramatically increase the frequency of mutagenic NHEJ. The mechanism by which cold shock increases mutagenic NHEJ is partially understood in that it increases the expression level of the ZFN/TALEN probably by slowing their degradation through the proteasome. But cold shock may also have a direct effect on decreasing the fidelity of the NHEJ process itself. The effect of cold shock on the genome editing frequency using RGENs has not been published but it would be predicted that cold shock would also increase the frequency of mutagenic NHEJ induced by RGENs because of the effect on both Cas9 expression and the effect on the fidelity of NHEJ.

There are two caveats to the use of cold shock as a method of increasing the frequency of genome editing. The first is that cold shock decreases the frequency of genome editing by homologous recombination, providing support that the reduced temperature affects genome editing by directly affecting the repair process itself. The second is that many cell types, particularly primary cell types, do not tolerate prolonged exposure to 30 °C. Thus, using cold shock to increase the frequency of genome editing by NHEJ requires a careful calibration in the cell type being used.

4 Using Exonucleases to Enhance Mutagenic NHEJ

A clever approach to increasing genome editing by mutagenic NHEJ but not HR is to co-express an exonuclease that processes the ends of the DSB into forms that can only be repaired by the NHEJ machinery in a mutagenic fashion [17]. For homing endonucleases, this approach can dramatically convert nucleases that have undetectable levels of on-target activity to ones that generate high levels of activity. This best studied exonuclease is TREX2, a 3′ to 5′ single-strand exonuclease that trims 3′ overhangs to blunt overhangs. The effect of TREX2 is most pronounced for genome editing using homing endonucleases but can also be observed for ZFN- and TALEN-induced editing. The effect of TREX2 on RGEN mediated genome editing has not been reported but given that RGENs are thought to create blunt ends, one would expect minimal enhancement by TREX2. TREX2, however, may dramatically increase the mutagenic NHEJ frequency when paired "nickase" RGENs are used which are designed to create nicks on opposite DNA strands that would result in 3′ single-strand overhangs.

5 Using Surrogate Reporters to Identify Populations that Are Enriched in Genome Editing

As discussed above, a critical factor to achieving high levels of genome editing is to assure sufficient levels of nuclease expression and optimizing the delivery method and promoter/enhancer elements are basic steps to achieving this goal. Once these steps have been optimized, however, there are further strategies that can be utilized to identify target cell populations in which higher frequencies of genome editing occur. All of these methods are fundamentally based on identifying cell populations in which the components of genome editing have been introduced as sufficient levels.

The simplest approach is to co-transfect a selectable reporter gene, such as a fluorescent reporter gene or a cell surface marker, along with the nucleases and then use fluorescent activated cell sorting (FACS) or magnetic bead based sorting to purify cells that have been transfected [13, 18, 19]. If one uses FACS, for example, one can even purify cells with the highest levels of fluorescent protein expression as those mark the cells with the highest levels of nuclease expression and thereby the highest levels of genome editing by mutagenic NHEJ. This strategy can be refined by placing the reporter gene cassette on the same plasmid as the nuclease expression cassette thus assuring that every cell that expresses the reporter is also expressing a nuclease. A further refinement of the strategy is to directly link the expression of the nuclease to the reporter gene through a 2A ribosomal skip peptide such that the reporter and nuclease will be expressed as separate peptides or a fluorescent protein-nuclease fusion protein in which case the two will be expressed as a single polypeptide.

The above strategies are useful methods to identify cells that express the nuclease of choice and can be easily used to enrich for genome-edited cells. A related but more sophisticated approach is to use a surrogate reporter to identify cells that have shown nuclease activity [19, 20]. In this approach, a reporter is generated in which nuclease target sites are placed on a surrogate plasmid and this plasmid is co-transfected into the cell type of interest. The surrogate reporter plasmid is designed such that if the nucleases cut the plasmid, a reporter gene (either a fluorescent protein or an antibiotic resistance gene) is activated. The activation of the reporter gene can then be used to enrich for cells that have undergone genome editing at the endogenous genomic target site. That is, activity of the nucleases on an extra-chromosomal reporter plasmid does identify cells that have a higher probability of being modified at the endogenous genomic target. Using this strategy dramatic enrichments in genome edited cells, particularly those modified by mutagenic NHEJ, can be achieved.

6 Designing Donor Constructs to Enrich for Genome Edited Cells

By designing the donor construct appropriately, one can enrich for cells that have been edited by homologous recombination. The general structure of a donor vector consists of homology arms (both 5′ and 3′ that are identical to the genomic sequence that flank the nuclease induced DSB) surrounding the defined genetic change one wants to induce at the site of the DSB. The defined genetic change can be a simple nucleotide change or consist of the insertion of large cassette of genes. The simplest method to enrich for cells that have been modified by homologous recombination is to use donor vectors that are similar in structure to targeting vectors that are used to modify mouse embryonic stem cells. In these donor vectors, a selectable marker (usually an antibiotic resistance gene but it could also be a fluorescent protein or unique cell surface marker) is placed between the homology arms. Initially all transfected cells will express the selectable marker but after ~10 days of division the only cells that will continue to express the selectable marker are ones in which the donor vector has inserted into the genome, either through homologous recombination into the desired target site or by random integration to sites elsewhere in the genome. Individual colonies can then be screened by PCR or Southern blotting to distinguish targeted integrants from random integrants. After nuclease mediated targeted gene insertion, the number of colonies with targeted integrations is usually at least 10 % and often >90 % of the colonies that express the selectable marker. If the selectable marker is flanked by Lox sites it can then be excised by the Cre recombinase leaving a single Lox site "scar" behind or if the selectable marker is flanked by a TA dinucleotide it can be excised using transposon mediated excision creating a "scarless" excision [21].

Another strategy to enrich for targeted integrants compared to random integrants is to place the selectable marker in the donor vector such that only after integration by homologous recombination will it acquire a promoter to drive expression. Because the selectable marker does not have a promoter, if it integrates randomly it will not express the selectable marker unless it happens by rare chance to integrate next to a promoter. A key aspect of this strategy is that the promoter being used to drive expression of the selectable marker after targeted genome editing must be strong enough in that particular cell type for selection or identification to occur. Many endogenous promoters, for example, are not strong enough to drive the expression of fluorescent proteins to sufficient levels to distinguish from untargeted cells by flow cytometry because the endogenous gene is not strongly expressed.

Finally, an ideal strategy to enrich for genome-edited cells is when the modified cells have a selective advantage over unmodified cells. This circumstance only occurs rarely but with a thorough understanding of the biology of the system can sometimes be designed. Examples of when modified cells have a natural selective advantage without having to use selectable markers (an engineered method to give cells a selective advantage) include screening for cells with mutations in the HPRT gene by growing them in 6-thioguanine or for mutations in the PIG-A gene by challenging cells with the drug aerolysin [22, 23].

7 Oligonucleotide-Mediated Genome Editing

Genome editing can also be induced by combining an engineered nuclease with an oligonucleotide [13, 24, 25]. The mechanism of oligonucleotide mediated genome editing does not proceed through the classic NHEJ or HR pathways, is not well understood. The proposed mechanism is though a combination of NHEJ and HR in which the oligonucleotide anneals to the 3′ single-strand tails on one side of the break, which then forms a template for the DNA polymerase to make short stretch of DNA that is homologous to the other side of the DSB. This end is then joined to the other side of the break through end joining or by recombination. While the mechanism of oligonucleotide-mediated end joining after an engineered nuclease induced DSB is not well understood, it can be an effective way to create small changes at the site of the DSB, within tens of nucleotides of the DSB, or create defined deletions. Enriching or oligonucleotide mediated genome editing events can be done as described above by sorting for cells that have been highly transfected or by using extra-chromosomal nuclease reporter plasmid thereby enriching for cells that have undergone genome editing.

8 Summary

In the last decade genome editing has moved from being something that could only be used in specialized cell types such as yeast and murine embryonic stem cells to a technology that can be used in every cell biological system available. Nonetheless, strategies that enhance the frequency of genome editing continue to make this approach to genome engineering for both research and therapeutic purposes even more accessible. The most important strategy is to engineer highly active nucleases that create DNA double-strand breaks at specific genomic sites. But even after creating highly active nucleases, other strategies can further enhance either the frequency of genome editing or give method to enrich for edited cells. In this chapter I have reviewed the use of cold-shock, exonucleases, surrogate markers, specialized donor vectors, and oligonucleotides; all of which are useful methods to scientists wishing to use genome editing in their work.

References

1. Paques F, Duchateau P (2007) Meganucleases and DNA double-strand break-induced recombination: perspectives for gene therapy. Curr Gene Ther 7(1):49–66
2. Chan SH, Stoddard BL, Xu SY (2011) Natural and engineered nicking endonucleases–from cleavage mechanism to engineering of strand-specificity. Nucleic Acids Res 39(1):1–18. doi:10.1093/nar/gkq742
3. Porteus MH, Carroll D (2005) Gene targeting using zinc finger nucleases. Nat Biotechnol 23(8):967–973. doi:10.1038/nbt1125
4. Urnov FD, Rebar EJ, Holmes MC, Zhang HS, Gregory PD (2010) Genome editing with engineered zinc finger nucleases. Nat Rev Genet 11(9):636–646. doi:10.1038/nrg2842
5. Christian M, Cermak T, Doyle EL, Schmidt C, Zhang F, Hummel A, Bogdanove AJ, Voytas DF (2010) Targeting DNA double-strand breaks with TAL effector nucleases. Genetics 186(2):757–761. doi:10.1534/genetics.110.120717
6. Doyle EL, Stoddard BL, Voytas DF, Bogdanove AJ (2013) TAL effectors: highly adaptable phytobacterial virulence factors and readily engineered DNA-targeting proteins. Trends Cell Biol 23(8):390–398. doi:10.1016/j.tcb.2013.04.003
7. Damian M, Porteus MH (2013) A crisper look at genome editing: RNA-guided genome modification. Mol Ther 21(4):720–722. doi:10.1038/mt.2013.46
8. Cong L, Ran FA, Cox D, Lin S, Barretto R, Habib N, Hsu PD, Wu X, Jiang W, Marraffini LA, Zhang F (2013) Multiplex genome engineering using CRISPR/Cas systems. Science 339(6121):819–823. doi:10.1126/science.1231143
9. Mali P, Esvelt KM, Church GM (2013) Cas9 as a versatile tool for engineering biology. Nat Methods 10(10):957–963. doi:10.1038/nmeth.2649
10. Boissel S, Jarjour J, Astrakhan A, Adey A, Gouble A, Duchateau P, Shendure J, Stoddard BL, Certo MT, Baker D, Scharenberg AM (2013) megaTALs: a rare-cleaving nuclease architecture for therapeutic genome engineering. Nucleic Acids Res. doi:10.1093/nar/gkt1224
11. Cermak T, Doyle EL, Christian M, Wang L, Zhang Y, Schmidt C, Baller JA, Somia NV, Bogdanove AJ, Voytas DF (2011) Efficient design and assembly of custom TALEN and other TAL effector-based constructs for DNA targeting. Nucleic Acids Res 39(12):e82. doi:10.1093/nar/gkr218
12. Wang H, Yang H, Shivalila CS, Dawlaty MM, Cheng AW, Zhang F, Jaenisch R (2013) One-step generation of mice carrying mutations in multiple genes by CRISPR/Cas-mediated genome engineering. Cell 153(4):910–918. doi:10.1016/j.cell.2013.04.025
13. Ding Q, Regan SN, Xia Y, Oostrom LA, Cowan CA, Musunuru K (2013) Enhanced efficiency of human pluripotent stem cell genome editing through replacing TALENs with CRISPRs. Cell Stem Cell 12(4):393–394. doi:10.1016/j.stem.2013.03.006
14. Pruett-Miller SM, Connelly JP, Maeder ML, Joung JK, Porteus MH (2008) Comparison of zinc finger nucleases for use in gene targeting in mammalian cells. Mol Ther 16(4):707–717. doi:10.1038/mt.2008.20
15. Doyon Y, Choi VM, Xia DF, Vo TD, Gregory PD, Holmes MC (2010) Transient cold shock enhances zinc-finger nuclease-mediated gene disruption. Nat Methods 7(6):459–460. doi:10.1038/nmeth.1456
16. Carlson DF, Fahrenkrug SC, Hackett PB (2012) Targeting DNA with fingers and TALENs. Mol Therapy Nucleic Acids 1:e3. doi:10.1038/mtna.2011.5
17. Certo MT, Gwiazda KS, Kuhar R, Sather B, Curinga G, Mandt T, Brault M, Lambert AR, Baxter SK, Jacoby K, Ryu BY, Kiem HP, Gouble A, Paques F, Rawlings DJ, Scharenberg AM (2012) Coupling endonucleases with DNA end-processing enzymes to drive gene disruption. Nat Methods 9(10):973–975. doi:10.1038/nmeth.2177
18. Certo MT, Ryu BY, Annis JE, Garibov M, Jarjour J, Rawlings DJ, Scharenberg AM (2011) Tracking genome engineering outcome at individual DNA breakpoints. Nat Methods 8(8):671–676. doi:10.1038/nmeth.1648
19. Kim H, Kim MS, Wee G, Lee CI, Kim H, Kim JS (2013) Magnetic separation and antibiotics selection enable enrichment of cells with ZFN/TALEN-induced mutations. PLoS One 8(2):e56476. doi:10.1371/journal.pone.0056476
20. Kim H, Um E, Cho SR, Jung C, Kim H, Kim JS (2011) Surrogate reporters for enrichment of cells with nuclease-induced mutations. Nat Methods 8(11):941–943. doi:10.1038/nmeth.1733
21. Yusa K (2013) Seamless genome editing in human pluripotent stem cells using custom endonuclease-based gene targeting and the piggyBac transposon. Nat Protoc 8(10):2061–2078. doi:10.1038/nprot.2013.126

22. Zou J, Maeder ML, Mali P, Pruett-Miller SM, Thibodeau-Beganny S, Chou BK, Chen G, Ye Z, Park IH, Daley GQ, Porteus MH, Joung JK, Cheng L (2009) Gene targeting of a disease-related gene in human induced pluripotent stem and embryonic stem cells. Cell Stem Cell 5(1):97–110. doi:10.1016/j.stem.2009.05.023
23. Russell DW, Hirata RK (1998) Human gene targeting by viral vectors. Nat Genet 18(4):325–330. doi:10.1038/ng0498-325
24. Chen F, Pruett-Miller SM, Huang Y, Gjoka M, Duda K, Taunton J, Collingwood TN, Frodin M, Davis GD (2011) High-frequency genome editing using ssDNA oligonucleotides with zinc-finger nucleases. Nat Methods 8(9):753–755. doi:10.1038/nmeth.1653
25. Yang L, Guell M, Byrne S, Yang JL, De Los AA, Mali P, Aach J, Kim-Kiselak C, Briggs AW, Rios X, Huang PY, Daley G, Church G (2013) Optimization of scarless human stem cell genome editing. Nucleic Acids Res 41(19):9049–9061. doi:10.1093/nar/gkt555

Chapter 17

Using Engineered Endonucleases to Create Knockout and Knockin Zebrafish Models

Victoria M. Bedell and Stephen C. Ekker

Abstract

Over the last few years, the technology to create targeted knockout and knockin zebrafish animals has exploded. We have gained the ability to create targeted knockouts through the use of zinc finger nucleases (ZFNs), transcription activator-like effector nucleases (TALENs) and clustered regularly interspaced short palindromic repeats/CRISPR associated system (CRISPR/Cas). Furthermore, using the high-efficiency TALEN system, we were able to create knockin zebrafish using a single-stranded DNA (ssDNA) protocol described here. Through the use of these technologies, the zebrafish has become a valuable vertebrate model and an excellent bridge between the invertebrate and mammalian model systems for the study of human disease.

Key words TALEN, Genome engineering, Zebrafish, Homology directed repair, HDR

1 Introduction

Over the last 40 years, the zebrafish (*Danio rerio*) has gained notable momentum as a valuable nonmammalian vertebrate model system, particularly in understanding developmental mechanisms. Its exogenously fertilized embryos allow real-time, in vivo observation of development from the single-cell stage. Furthermore, as a vertebrate the zebrafish have many similarities with humans, including the nervous system, skin, cartilage and bone, blood and vasculature, kidney, liver, pancreas, gut, and innate and adaptive immune systems. This combination of features makes the zebrafish an excellent model for studying development, human disease and for high-throughput drug studies. The zebrafish life cycle and housing requirements tend to make experiments conducted in this system faster and cheaper than contemporary mammalian animals, making the zebrafish an excellent bridge model system between invertebrates and mammalian systems.

Advancements in genetic manipulation technologies have helped the zebrafish to approach its full potential as a vertebrate

Shondra M. Pruett-Miller (ed.), *Chromosomal Mutagenesis*, Methods in Molecular Biology, vol. 1239,
DOI 10.1007/978-1-4939-1862-1_17, © Springer Science+Business Media New York 2015

development and disease model system. The first steps were transient over-and under-expression experiments. Simple overexpression experiments involved adding exogenous DNA and messenger RNA through microinjection [1]. Next, the application of morpholino antisense technology enabled targeted, transient decrease in gene expression, often called a "knockdown," of most genes during embryonic development [2]. Random chemical and retroviral mutagenesis followed by DNA analyses were the first well-used methods to screen for mutations in desired loci (TILLING [3]; retrovirus [4]). However, true targeted knockout and knockin experiments, allowing for long-term, heritable genomic changes remained out of reach in this animal system for many years.

Three major molecular technologies altered the zebrafish genome engineering landscape: zinc finger nucleases (ZFNs; Fig. 1a; [5, 6]), transcription activator-like effector nucleases (TALENs; Fig. 1b; [7, 8]; for review *see* [9]) and clustered regularly interspaced short palindromic repeats/CRISPR associated system (CRISPR/Cas; Fig. 1c; [10, 11]. All three systems are active in zebrafish and can be used to create targeted lesions in the genome. ZFNs and TALENs employ custom-designed, locus-specific proteins fused to a sequence-independent nuclease domain to generate targeted mutations, while CRISPRs (Fig. 1c) use a different, RNA-guided mechanism.

The protein-based ZFN (Fig. 1a) and TALEN (Fig. 1b) systems use sequence-specific binding motifs to create targeted double-stranded breaks (DSBs; for review *see* [12]). The core DNA binding motif is attached to a nuclease, usually Fok I, which must homodimerize to catalyze a DSB [13, 14]. Therefore, for nuclease homodimerization to occur, ZFNs and TALENs are designed as a pair of nuclease-guiding proteins that bind to both sides of the intended DSB site. The localized dimerization requirement increases nuclease specificity in this approach.

Targeted DSBs facilitate the engineering of site-specific knockouts because changes may occur through cellular repair processes. The most common mechanism that introduces genomic alterations is the nonhomologous end joining (NHEJ) pathway that typically deletes or adds random sequences to the DNA around the DSB. If the custom restriction enzymes target an open reading frame in a gene, this often creates out-of-frame proteins [15, 16].

ZFNs were the first site-specific knockout system whose in vivo use was pioneered in zebrafish (Fig. 1a; [17, 18]). Each lab-made ZFN typically consists of three fingers [6]. Each ZF motif is ~30 amino acids long and recognizes three bases (Fig. 1a; [19]). A single ZFN consists of three motifs recognizing nine bases. Constructing a highly functional ZFN is a technically complex endeavor because each finger can influence the binding of its neighboring finger. These interactions make it difficult to predict how well an entire ZFN will bind. Therefore, both bacterial [18]

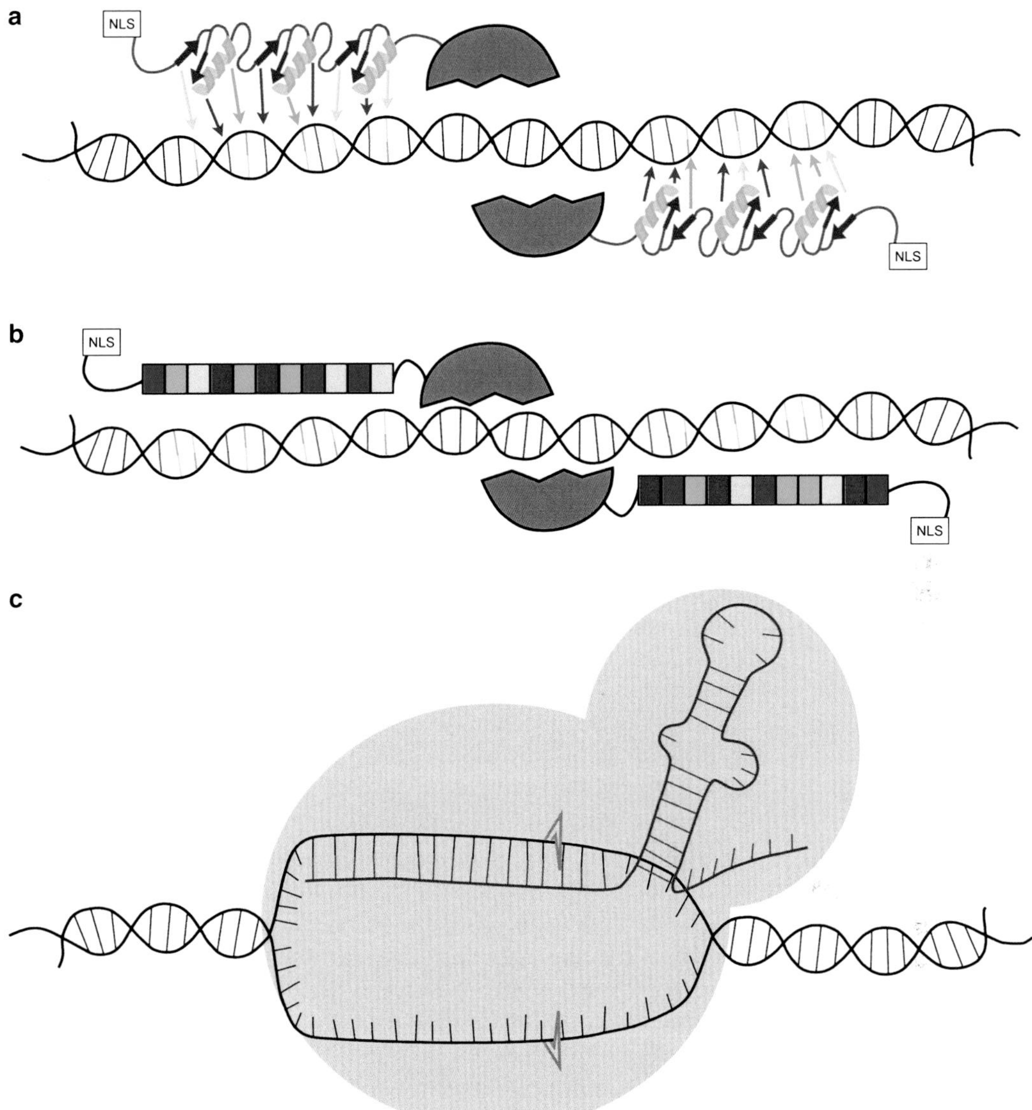

Fig. 1 Diagrams of each of the double strand break technologies. (**a**) ZFNs are made from two different arms that each typically comprises of three binding proteins recognizing three bases each, depicted by the *arrows* pointing to the bases to which the individual motifs would bind. Attached to the N-terminus is the nuclease domain from the FokI restriction enzyme. This dimeric protein creates double strand breaks. On the opposite end is the nuclear localization signal (NLS) (**b**) TALENs are made of repeats that have two amino acids in the center required for sequence-specific DNA binding. Each repeat binds a single DNA base, represented by the *boxes* of the same color as the base to which it binds. (**c**) The CRISPR/Cas9 system requires an RNA (shown in *purple*) that hybridizes to the unwound DNA. The Cas 9 protein (the *green* surrounding the RNA) creates the double strand breaks at the DNA/RNA hybrid

and in vitro [20] systems were created to screen multiple zinc fingers to identify which would best bind a desired DNA sequence. This relatively long and technically challenging process has made ZFNs not very accessible for high-throughput knockout projects or for smaller laboratories. Despite their limitations, including often-modest efficacy at introducing germ line mutations in zebrafish, ZFNs established the modern paradigm for genome engineering techniques using custom restriction enzymes in living animals, and the more than two decades' worth of ZFN research has been heavily used as the basis for subsequent custom restriction enzyme work using other systems [21].

Transcription activator-like effector nucleases (TALENs) were subsequently developed just a few years ago and rapidly used for in vivo targeting in zebrafish (Fig. 1b; [7, 22]). TALEs were first discovered in the plant pathogen *Xanthomonas* [23]. Each TALE domain consists of a ~30 amino acid repetitive motif with two specific amino acids that bind to each DNA base (Fig. 1b; [23, 24]). These repeats are modular and therefore do not influence the binding of the next [24, 25]. This feature makes designing TALENs against a particular sequence relatively simple, with very high success (95 % or higher) in the latest TALEN designs (for review *see* [9]). To further increase the utility of the system, Golden Gate modular assembly systems were deployed using a system of unique restriction nucleases to create TALENs in two steps [26]. Golden Gate assembly has since been expanded and simplified ([27, 28]) for high-throughput assembly [29], making multi-gene targeted knockout projects considerably more feasible and affordable, even for smaller laboratories. Furthermore, the TALEN system has been optimized for rapid, specific cutting that is efficiently transmitted to the germ line [7, 30]. To date, TALENs remain the most versatile custom restriction enzyme system capable of targeting nearly any sequence-specific genomic location and with the lowest off-targeting rates in zebrafish (*see* more, below).

A third site-specific mutagenesis system based on the type II CRISPR/Cas9 system has been very effective for in vivo gene targeting in zebrafish (Fig. 1c; [10, 11, 31, 32]). CRISPRs were discovered in prokaryotes as an adaptive defensive system against viruses [33, 34], and their mechanism for creating DSB is different than ZFN and TALENs. The current artificial system uses a single synthetic guide RNA (sgRNA) that, with the help of the Cas9 protein, hybridizes with the DNA sequence of interest through Watson–Crick base pairing. The sgRNA recognition sequence is downstream of a three base protospacer adjacent motif [35]. The Cas9-RNA-DNA complex creates a double-strand break in the corresponding genomic DNA (Fig. 1c). This technology is relatively simple to use, and the target system easier to create than TALENs. One complication of the CRISPR system is its reduced target specificity relative to ZFNs and TALENs. Using the Cas9 protein, the system will still bind and create DSB despite several mismatched bases. This creates many nonspecific mutations within the genome; in some cases, off-target cutting was at a higher rate than on-target nuclease activity [36]. Recently, this complication has been addressed using in vitro systems by mutating Cas9 to create a protein that only cuts on a single strand, a version of Cas9 called a nickase [37]. This nickase-based Cas9 protein system reduces off-target effects by over 1,000×, thanks to the increased specificity from the use of two guide RNAs. In these cell-based systems, the Cas9 nickase-based off-target rates are, however, still higher than what have been noted for TALENs [37], and their efficacy in zebrafish has yet to be described.

In conjunction with these targeted knockout technologies, zebrafish researchers now have the ability to insert exogenous sequences at DSB target sites. This was first accomplished by co-delivering a high-efficiency TALEN with either single-stranded DNA (ssDNA) [30] or double-stranded DNA (dsDNA) [38]. The ssDNA serves as a template to induce homology directed repair (HDR), enabling changes designed into the ssDNA template to "knock-in" specific sequences at the target site. To date, small changes (such as a new restriction enzyme sequence), as well as longer sequence tags such as a loxP site have been engineered into the zebrafish genome [30]. Using dsDNA, longer sequences, such as enhanced green fluorescent protein has been added to the genome [38]. Germ line transmission of exogenous sequences through HDR has been demonstrated, though at a lower frequency than germ line transmission of NHEJ-created mutations.

The ability to easily create site-specific mutations in vivo and pioneered in zebrafish has opened the world of targeted genome engineering to a large number of cellular and animal model systems [39, 40]. What was once a costly enterprise has now become routine. The ability to genome engineer has been particularly transformative for the zebrafish community. Prior to these technologies, only transient knockdown and overexpression experiments were readily accessible, and generating knockouts was a tedious and challenging exercise. Now, zebrafish researchers can create designer mutations using site-specific genome engineering. This opens up the model to be used for more targeted, and human disease-specific, experimentation [41]. For example, it is now possible to design a TALEN-ssDNA experiment that specifically changes the zebrafish genome to model known human sequence variations, including those associated with disease and those of unknown significance. These fish can be bred to homozygosity, and the phenotype can be defined in this rich in vivo vertebrate. These engineered animal models can consequently be used to elucidate the mechanism of the human disease mutation. Furthermore, these fish can be subsequently used to screen for drugs that would stabilize or attenuate the phenotype and be used for human drug trials. Therefore, the ability to engineer model vertebrate genomes will enhance our abilities to understand basic developmental and disease mechanisms. Furthermore, the combination of a vertebrate model with strong genome engineering capabilities and easily obtaining large numbers of mutated embryos will enable faster and more efficient drug screening.

While these genome editing technologies themselves are interesting and unique, it is the new, functionally limitless opportunities for hypothesis-driven, human disease-focused experimentation that truly excites our imagination. Therefore, with these new tools, we believe the zebrafish has become a terrific bridge between the high-throughput but invertebrate genetic models such as *Drosophila* and the nematode, and the classic human disease-relevant models such as the mouse. With the now highly accessible zebrafish genome and ease of the generation of large numbers of animals makes this aquatic species an outstanding vertebrate biology model with increasing impact on human health science for years to come.

The following protocol is a method using TALENs and ssDNA for creating knockout and knockin zebrafish. We used TALENs because of their high efficiency and low off-target rates. In principle, any of the systems described above can be used to create double-strand breaks and be used with ssDNA to create knockin animals. The most difficult and important part of this protocol is designing the TALENs and ssDNA. The implementation is straightforward and uses largely standard tools in the field such as microinjection and PCR-based genotyping.

2 Materials

2.1 Kits Used in this Protocol

1. RNAeasy MinElute Cleanup Kit (Ambion).
2. TALEN Golden Gate kit (Addgene; from Dan Voytas).
3. pC-GoldyTALEN (Addgene plasmid 38143; from Daniel Carlson).
4. T3 mMessage Kit (Life Technologies, Carlsbad, CA, USA).
5. Spin Miniprep kit.
6. PCR purification kit.
7. Nucleotide removal kit.

2.2 Buffers and Other Reagents

1. 0.4 g/100 ml Tricaine (Ethyl 3-aminobenzoate methanesulfonate): 0.8 g Tricaine, 199 ml of embryo water (or fish water from the zebrafish facility), 1.6 g of sodium bicarbonate to bring the pH to 7.0.
2. 50× TAE Buffer: 242 g Tris base, 57.1 ml 100 % acetic acid, 100 ml 0.5 M sodium EDTA. Bring the volume up to 1 l nanopure water.

3. 0.4 mg/ml 5× Cresol Red loading dye. Dissolve 4 mg of cresol red into 1 ml of nanopure water. Mix until completely dissolved. Dissolve 60 g of sucrose in 80 ml of nanopure water. Add the 4 mg/ml cresol red with the sucrose water and bring the volume up to 100 ml.
4. Restriction enzymes.
   (a) BsmbI.
   (b) SacI.
5. Chemically competent bacteria.
6. 100 mM NaOH.
7. 1 M Tris–HCl, pH 8.0.
8. Agarose.
9. 10 mg/ml ethidium bromide.

2.3 Equipment

1. Razor blades.
2. Centrifuge.
3. Thermocycler.
4. Gel box and comb.
5. UV lightbox and filter.

3 Methods

3.1 TALEN Design (Fig. 2a)

1. Binding sites range from 15–25 bases in length.
2. Spacer length is generally 13–20 bps.
3. Design around a unique restriction enzyme binding site centrally located within the spacer for genotyping (*see* **Notes 1–3**).

3.2 TALEN Constructs

1. TALEN assembly of the repeat-variable di-residues (RVD)-containing repeats was conducted using the Golden Gate kit [26]. This is a 5-day restriction digest/ligation-based protocol that can be performed in any molecular biology laboratory (*see* Chapter 7 by Voytas and colleagues).
2. Confirm appropriate RVD assembly by sequencing.
3. Clone the RVDs into pC-GoldyTALEN backbone using BsmBI restriction enzyme.
4. Transform bacteria, using the standard laboratory procedure, with the completed TALEN.
5. Using a single colony, grow 1–5 ml of the bacteria containing the TALEN.
6. Purify the plasmid using a spin miniprep kit.

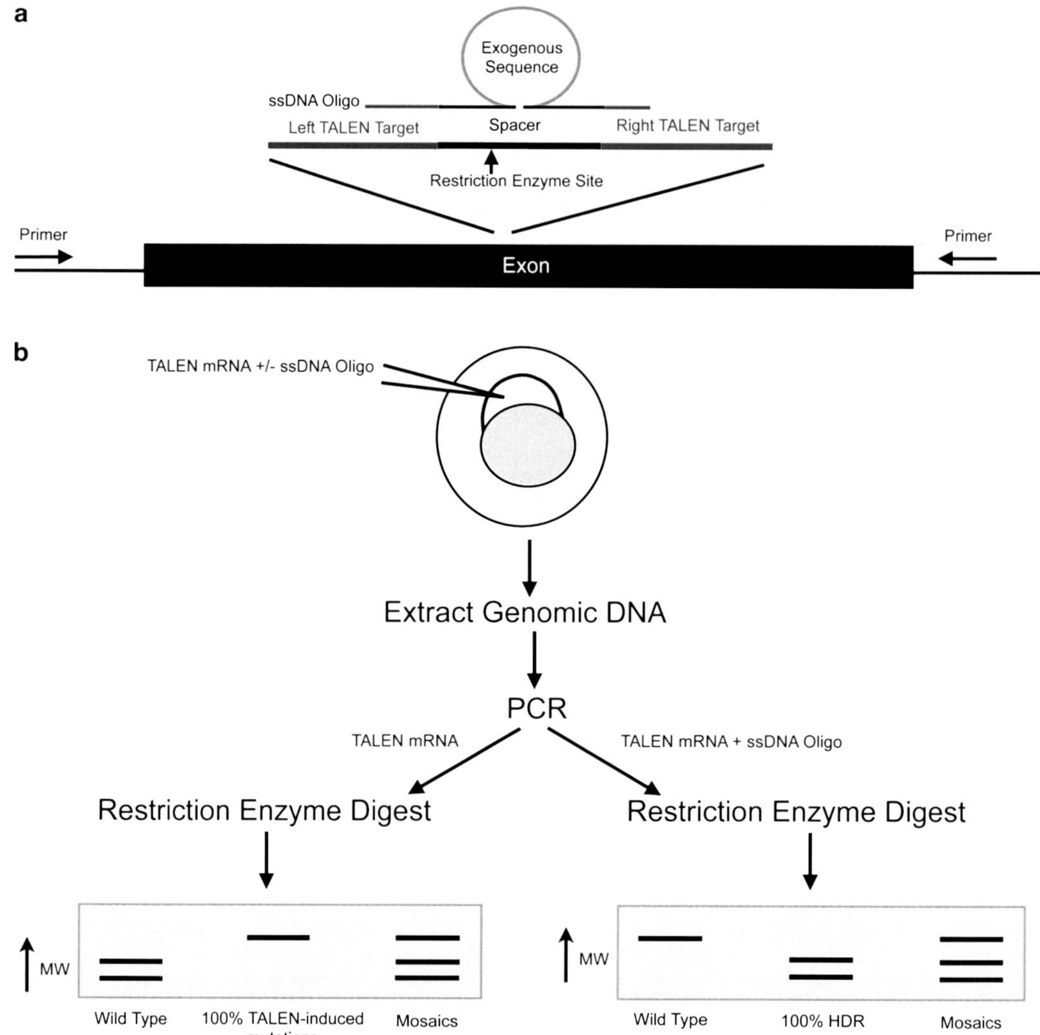

Fig. 2 Experimental design of zebrafish knockin and knockout experiments. (**a**) A diagram of a genetic locus to be targeted. To create a knockout mutation, TALENs should target a conserved exon early exon. Exon one is often an isoform-specific exon, so it is often better to target a downstream exon. Two TALENs (*left* and *right*) bind to the genomic sequence on opposite strands and create a double-strand break. Within the spacer region, we typically design around a local unique restriction enzyme for simplified downstream screening. For knockin experiments, a ssDNA oligo is designed as shown. The homology arms do not normally include the entire TALEN binding site. An exogenous sequence can be added to the center of the spacer region. (**b**) Experimental design for creating knockout or knockin zebrafish. Inject TALEN mRNA to create a knockout or inject TALEN mRNA plus ssDNA oligo for HDR. Extract the genomic DNA from the embryos, perform a PCR and subsequently digest the PCR product. For knockout experiments (injected only with TALEN mRNA), a mutation will result in loss of the restriction enzyme binding site and a larger band on the enzyme analysis on the PCR product. For the knockin experiments such as when adding a new restriction enzyme site to the spacer region, successful HDR results in a digested band and unmodified (wild type) is therefore uncut in the enzymatic assay

7. Digest the plasmid using SacI endonuclease at 37 °C for 2–3 h. Use at least 5–10 mg of plasmid for restriction enzyme digest.
8. Purify the digested plasmid using a PCR purification kit.
9. Use the T3 mMessage Machine kit to create mRNA (*see* **Note 4**).
10. Purify the mRNA using either the phenol–chloroform extraction (T3 mMessage Machine kit user manual protocol) or RNeasy MinElute cleanup kit for injection.

3.3 ssDNA Oligonucleotide (Oligo) Design for Genome Editing

1. Design the oligo to target the spacer sequence between the TALEN cut sites (*see* **Notes 3** and **5**).
2. Add desired insertion sequences to the center of the oligo (*see* **Notes 6–8**).
3. Order oligos from a company such as Integrated DNA Technologies.
4. Purify the oligos using the Nucleotide Removal Kit (*see* **Note 9**).
5. Use RNAse-free water to remove the oligo from the column.

3.4 Creating Knockout or Knockin Zebrafish Embryos (Fig. 2b)

1. Dilute TALEN pair mRNA and/or ssDNA oligo to multiple different concentrations using RNAse-free water. Inject the embryos [1, 42] and create a lethal dose 50 (LD50) curve (*see* **Notes 11** and **12**).
2. Choose the highest concentration at which less than 50 % of the injected embryos are dead.
3. Inject TALEN mRNA alone for knockout experiments using the concentration determined above.
4. Inject TALEN mRNA + ssDNA oligo for knockin experiments (*see* **Notes 13** and **14**). Any experiments with greater than 50 % dead should be disregarded.
5. Isolate genomic DNA to assess for somatic knockout or knockin from 2–5 day-old injected embryos (*see* Subheading 3.5).
6. Raise the remainder of the injected embryos for germ line screening. These fish can be raised, mated and the resulting embryos can be screened for germ line mutation (*see* Subheading 3.7; Fig. 2b).

3.5 Individual Embryo Genomic DNA Isolation (See Note 15)

1. Place a single zebrafish larva (2–5 days old) into 0.2 ml strip tubes using 10 μl.
2. Add 10 μl of 100 mM NaOH, for a final concentration of 50 mM NaOH.
3. Vortex.
4. Incubate at 95 °C for approximately 10 min or until the embryos have dissolved (*see* **Note 16**).

5. Cool to 4 °C in the –20° centigrade freezer.
6. Add 1/10 volume 1 M Tris–HCl pH 8.0 to neutralize the solution.
7. Centrifuge to pellet the debris and use the supernatant for PCR (*see* **Note 17**).

3.6 Isolate Genomic DNA from Fin Clips of Zebrafish at Least 2 Months Old

1. Anesthetize the fish using 200 μg/ml of Tricaine diluted in water from the fish room.
2. Trim the most caudal 2–3 mm of fin using a fresh razor blade for each fish to prevent contamination.
3. Place the tissue in a 0.2 ml tube, which has been placed on ice, until all fin biopsies were collected.
4. Add 150 μl of 50 mM NaOH to the fin clips.
5. Incubate at 95 °C for approximately 10 min or until the tissue has dissolved (*see* **Note 16**).
6. Cool to 4 °C in the –20° centigrade freezer.
7. Adding 1/10 volume 1 M Tris–HCl pH 8.0 to neutralize the solution.
8. Centrifuge to pellet the debris and use the supernatant for PCR.

3.7 Screening for Somatic and Germ Line Knockout Zebrafish (see Note 18)

1. Use the same protocol for screening both somatic and germ line mutations.
2. Design primers to the genomic sequence so that the PCR product is visible when digested with the restriction enzyme (*see* **Notes 3** and **19**).
3. Use 1–2 μl of genomic DNA (*see* Subheadings 3.5 or 3.6) for the PCR (*see* **Note 17**).
4. Digest 1–5 μl of the PCR, depending upon the PCR efficiency, using the unique restriction enzyme (*see* **Note 20**).
5. Run digested production on an agarose gel. The percentage agarose gel depends upon the size of the digested product (*see* Subheading 3.10; *see* **Note 21**).
6. If the TALENs were active in creating mutations, the restriction enzyme site should be lost and the PCR product should not be digested (Fig. 2b).
7. Positive injections should be raised for germ line screening (*see* Subheading 3.9; *see* **Note 22**).

3.8 Screening for Somatic and Germ Line Knockin Zebrafish

1. Screening for sequence knockins depends upon the sequence that is added. In this example a new restriction enzyme is added to the locus (*see* **Note 23**).
2. Design primers to the genomic sequence so that the PCR product is visible when digested with the restriction enzyme (*see* **Notes 3** and **15**).

3. Use 1–2 μl of genomic DNA (*see* Subheadings 3.5 or 3.6) for the PCR.
4. Digest 1–5 μl of the PCR, depending upon the PCR efficiency, using the unique restriction enzyme (*see* **Note 20**).
5. Run digested production on an agarose gel. The percentage agarose gel depends upon the size of the digested product (*see* Subheading 3.10; *see* **Note 21**).
6. If there is HDR the restriction enzyme site should digest the sample creating two unique bands. The wild type sequence will not digest and a single, uncut band is seen (Fig. 2b).
7. Positive injections should be raised for germ line screening (*see* Subheading 3.9).
8. To increase the yield of germ line screening, the F0-injected fish can be screened at 2 months of age for somatic insertion of the sequence (*see* Subheading 3.6; *see* **Note 24**).

3.9 Germ Line Screening

1. Outcross injected F0 fish.
2. Collect the embryos and isolate genomic DNA (*see* Subheading 3.5).
3. Depending upon the efficiency, 10–100 embryos should be screened for germ line mutations (*see* **Notes 25** and **26**).
4. Follow the same screening strategy as somatic mutation analysis (*see* Subheadings 3.7 or 3.8).

3.10 1 % Agarose Gel

1. Measure 1 g of agarose (increase or decrease the amount of agarose depending on the percentage desired) and pour the powder into a flask.
2. Add 100 ml of 1× TAE.
3. Microwave the solution until the agarose is dissolved. Stop every 30 s to 1 min and swirl the solution.
4. Let agarose solution cool down until it is around 60 °C.
5. Add 2–3 μl of 10 mg/ml ethidium bromide (*see* **Note 27**).
6. Pour the agarose into a gel tray with the well comb in place.
7. Let the gel sit at room temperature until it is solid.
8. Fill the gel box with 1× TAE until the gel is completely covered.
9. Remove the comb.
10. Add 5× cresol red loading dye to each samples so that the final concentration is 1× cresol red and load the samples into the well. A ladder should be placed either first or last well for a size comparison.
11. Run the gel at 80–150 V until the dye line is over 2/3 of the way down the gel.
12. Visualize the gel using a UV lightbox and filter (*see* **Note 28**).

4 Notes

1. It is possible to screen for mutations without the restriction enzyme through use of random sequencing or through point-mutation screening methods not discussed here.
2. To simplify the TALEN design process, use an open access software such as Mojo Hand (www.talendesign.org; [43]).
3. The zebrafish genome contains a great deal of sequence diversity within standard laboratory strains, especially within non-exonic regions. Genomic sequences of fish within even an individual laboratory's facility frequently differ from the published genomic sequence. TALENs and primers designed against a published genomic sequence will not bind if an investigator's sequence of interest in the experimental animals differs from the consensus. Confirming the targeted genomic sequence in advance and in the selected specific zebrafish strain can avoid false negative TALEN efficacy due to natural polymorphisms.
4. The enzyme mix was modified from the mMessage kit by using 1.5 μl of enzyme mix added to 0.5 μl of RNAse inhibitor.
5. The tails of the oligo are usually relatively short and do not normally contain the whole TALEN binding site. To date, any sequences tested that contained the whole TALEN binding site has not resulted in HDR.
6. To date, only small insertion sequences have been published, such as EcoRV and LoxP sites.
7. For single base-pair changes, it is possible to use the wobble bases around the desired change to create a restriction enzyme site while leaving the protein sequence unchanged.
8. ssDNA oligos are regularly less than 100. Larger sizes have been shown to have markedly reduced efficiency due to an unknown mechanism.
9. Oligos must be purified prior to injection into the zebrafish. The standard oligo purification leaves chemicals that kill the embryos.
10. Reference [42] is an excellent review for injection set up, calibration and loading embryos.
11. Usually, the TALENs concentration ranged from 25 to 400 pg.
12. The ssDNA oligo ranged from 25 to 100 pg.
13. Do not inject more than 9 nl total of pure water into the zebrafish embryo. Greater than that causes developmental problems.
14. Inject the TALEN mRNA is injected separately from the ssDNA oligo to ensure that the mRNA is not degraded.

15. DNA can also be isolated from pools of ten larval zebrafish using the the DNAeasy Blood and Tissue kit (Qiagen).
16. Using a vortexing incubator increases the DNA yield and ensures the entire embryo is dissolved.
17. When using the supernatant, make sure no cellular debris is pipetted because it can inhibit the PCR reaction.
18. How screening is conducted for knockout efficacy depends upon the TALEN site chosen. If the spacer region contained a unique restriction enzyme site within the nearby genomic locus, restriction enzyme analysis on a locus-specific PCR product can be easily used to screen for mutations. www.talen-design.org [43] is the support software we use to help with TALEN design.
19. It is ideal that the restriction enzyme sequence is asymmetrically located within the PCR product so two bands can be visualized after digest.
20. Most restriction enzymes do not require purification of the PCR product to allow for digestion. However, it is best to test your enzyme in the PCR buffer. Also, have a positive control to ensure the enzyme is active.
21. Most sequences can be resolved using a 1–2 % agarose gel. However, this depends upon the size of the bands needing to be resolved using the specified genotyping assay.
22. Screening embryos somatic mutations is ideal. This ensures that the injection was successful. However, given the efficiency of the GoldyTALENs, it is feasible to skip the somatic screening step and raise the injections.
23. If larger sequences are added, for example a LoxP site, a PCR primer can be made to bind the LoxP site. This can, then, be used to screen for the insertion of the sequence. However, this screening method is significantly more sensitive than PCR and restriction enzyme digestion.
24. If stable somatic HDR activity is detected, there can be an increased likelihood that the insertion is also in the germ line, and these fish can be prioritized for subsequent analyses. Through fin clip screening, the number of fish that need to be mated and screened for germ line mutations is significantly decreased.
25. Genomic DNA isolation for germ line screening can be done either as individual embryos or as groups of ten embryos.
26. For knockout GoldyTALEN injections, the efficiency is high enough that screening 10–20 embryos is sufficient. However, knockin experiments are significantly less efficient. Therefore, screening up to 100 embryos increases the chances of finding the germ line mutation.

27. Ethidium bromide binds to DNA, therefore, it is a carcinogen. Gloves should be worn whenever it is used.
28. There are many gel imaging systems. Currently, the system used in the laboratory is a Fotodyne gel imaging system.

Acknowledgements

Grants: State of Minnesota grant H001274506; NIH GM63904; NIH grant P30DK084567; NIH grant DK083219; Mayo Foundation.

References

1. Hyatt TM, Ekker SC (1999) Vectors and techniques for ectopic gene expression in zebrafish. Methods Cell Biol 59:117–126
2. Nasevicius A, Ekker SC (2000) Effective targeted gene "knockdown" in zebrafish. Nat Genet 26:216–220
3. Sood R, English MA, Jones M et al (2006) Methods for reverse genetic screening in zebrafish by resequencing and TILLING. Methods 39:220–227
4. Amsterdam A, Hopkins N (1999) Retrovirus-mediated insertional mutagenesis in zebrafish. Methods Cell Biol 60:87–98
5. Zhu C, Smith T, McNulty J et al (2011) Evaluation and application of modularly assembled zinc-finger nucleases in zebrafish. Development 138:4555–4564
6. Urnov FD, Rebar EJ, Holmes MC et al (2010) Genome editing with engineered zinc finger nucleases. Nat Rev Genet 11:636–646
7. Huang P, Xiao A, Zhou M et al (2011) Heritable gene targeting in zebrafish using customized TALENs. Nat Biotechnol 29:699–700
8. Sander JD, Cade L, Khayter C et al (2011) Targeted gene disruption in somatic zebrafish cells using engineered TALENs. Nat Biotechnol 29:697–698
9. Campbell JM, Hartjes KA, Nelson TJ et al (2013) New and TALENted genome engineering toolbox. Circ Res 113:571–587
10. Hwang WY, Fu Y, Reyon D et al (2013) Efficient genome editing in zebrafish using a CRISPR-Cas system. Nat Biotechnol 31:227–229
11. Blackburn PR, Campbell JM, Clark KJ et al (2013) The CRISPR system–keeping zebrafish gene targeting fresh. Zebrafish 10:116–118
12. Gaj T, Gersbach CA, Barbas CF III (2013) ZFN, TALEN, and CRISPR/Cas-based methods for genome engineering. Trends Biotechnol 31:397–405
13. Smith J, Bibikova M, Whitby FG et al (2000) Requirements for double-strand cleavage by chimeric restriction enzymes with zinc finger DNA-recognition domains. Nucleic Acids Res 28:3361–3369
14. Vanamee ES, Santagata S, Aggarwal AK (2001) FokI requires two specific DNA sites for cleavage. J Mol Biol 309:69–78
15. Porteus MH, Carroll D (2005) Gene targeting using zinc finger nucleases. Nat Biotechnol 23:967–973
16. Wyman C, Kanaar R (2006) DNA double-strand break repair: all's well that ends well. Annu Rev Genet 40:363–383
17. Doyon Y, McCammon JM, Miller JC et al (2008) Heritable targeted gene disruption in zebrafish using designed zinc-finger nucleases. Nat Biotechnol 26:702–708
18. Meng X, Noyes MB, Zhu LJ et al (2008) Targeted gene inactivation in zebrafish using engineered zinc-finger nucleases. Nat Biotechnol 26:695–701
19. Segal DJ, Crotty JW, Bhakta MS et al (2006) Structure of Aart, a designed six-finger zinc finger peptide, bound to DNA. J Mol Biol 363:405–421
20. Foley JE, Yeh JR, Maeder ML et al (2009) Rapid mutation of endogenous zebrafish genes using zinc finger nucleases made by Oligomerized Pool ENgineering (OPEN). PLoS ONE 4:e4348
21. Lawson ND, Wolfe SA (2011) Forward and reverse genetic approaches for the analysis of vertebrate development in the zebrafish. Dev Cell 21:48–64
22. Liu Y, Luo D, Zhao H et al (2013) Inheritable and precise large genomic deletions of non-

coding RNA genes in zebrafish using TALENs. PLoS ONE 8:e76387

23. Boch J, Scholze H, Schornack S et al (2009) Breaking the code of DNA binding specificity of TAL-type III effectors. Science 326: 1509–1512
24. Moscou MJ, Bogdanove AJ (2009) A simple cipher governs DNA recognition by TAL effectors. Science 326:1501
25. Mahfouz MM, Li L, Shamimuzzaman M et al (2011) De novo-engineered transcription activator-like effector (TALE) hybrid nuclease with novel DNA binding specificity creates double-strand breaks. Proc Natl Acad Sci U S A 108:2623–2628
26. Cermak T, Doyle EL, Christian M et al (2011) Efficient design and assembly of custom TALEN and other TAL effector-based constructs for DNA targeting. Nucleic Acids Res 39:e82
27. Liang J, Chao R, Abil Z et al (2013) FairyTALE: a high-throughput TAL effector synthesis platform. ACS Synth Biol 3:67–73
28. Ma AC, Lee HB, Clark KJ et al (2013) High efficiency In Vivo genome engineering with a simplified 15-RVD GoldyTALEN design. PLoS ONE 8:e65259
29. Zhang Z, Zhang S, Huang X et al (2013) Rapid assembly of customized TALENs into multiple delivery systems. PLoS ONE 8:e80281
30. Bedell VM, Wang Y, Campbell JM et al (2012) In vivo genome editing using a high-efficiency TALEN system. Nature 491:114–118
31. Chang N, Sun C, Gao L et al (2013) Genome editing with RNA-guided Cas9 nuclease in zebrafish embryos. Cell Res 23:465–472
32. Jao L-E, Wente SR, Chen W (2013) Efficient multiplex biallelic zebrafish genome editing using a CRISPR nuclease system. Proc Natl Acad Sci U S A 110:13904–13909
33. Barrangou R, Fremaux C, Deveau H et al (2007) CRISPR provides acquired resistance against viruses in prokaryotes. Science 315:1709–1712
34. Wiedenheft B, Sternberg SH, Doudna JA (2012) RNA-guided genetic silencing systems in bacteria and archaea. Nature 482:331–338
35. Jinek M, Chylinski K, Fonfara I et al (2012) A programmable dual-RNA-guided DNA endonuclease in adaptive bacterial immunity. Science 337:816–821
36. Cho SW, Kim S, Kim Y et al (2014) Analysis of off-target effects of CRISPR/Cas-derived RNA-guided endonucleases and nickases. Genome Res 24:132–141
37. Ran FA, Hsu PD, Lin C-Y et al (2013) Double nicking by RNA-guided CRISPR Cas9 for enhanced genome editing specificity. Cell 154:1380–1389
38. Zu Y, Tong X, Wang Z et al (2013) TALEN-mediated precise genome modification by homologous recombination in zebrafish. Nat Methods 10:329–331
39. Carlson DF, Tan W, Hackett PB et al (2013) Editing livestock genomes with site-specific nucleases. Reprod Fertil Dev 26:74–82
40. Xu L, Zhao P, Mariano A et al (2013) Targeted myostatin gene editing in multiple mammalian species directed by a single pair of TALE nucleases. Mol Ther Nucleic Acids 2:e112
41. Schmid B, Haass C (2013) Genomic editing opens new avenues for zebrafish as a model for neurodegeneration. J Neurochem 127: 461–470
42. Bill BR, Petzold AM, Clark KJ et al (2009) A primer for morpholino use in zebrafish. Zebrafish 6:69–77
43. Neff KL, Argue DP, Ma AC et al (2013) Mojo Hand, a TALEN design tool for genome editing applications. BMC Bioinformatics 14:1

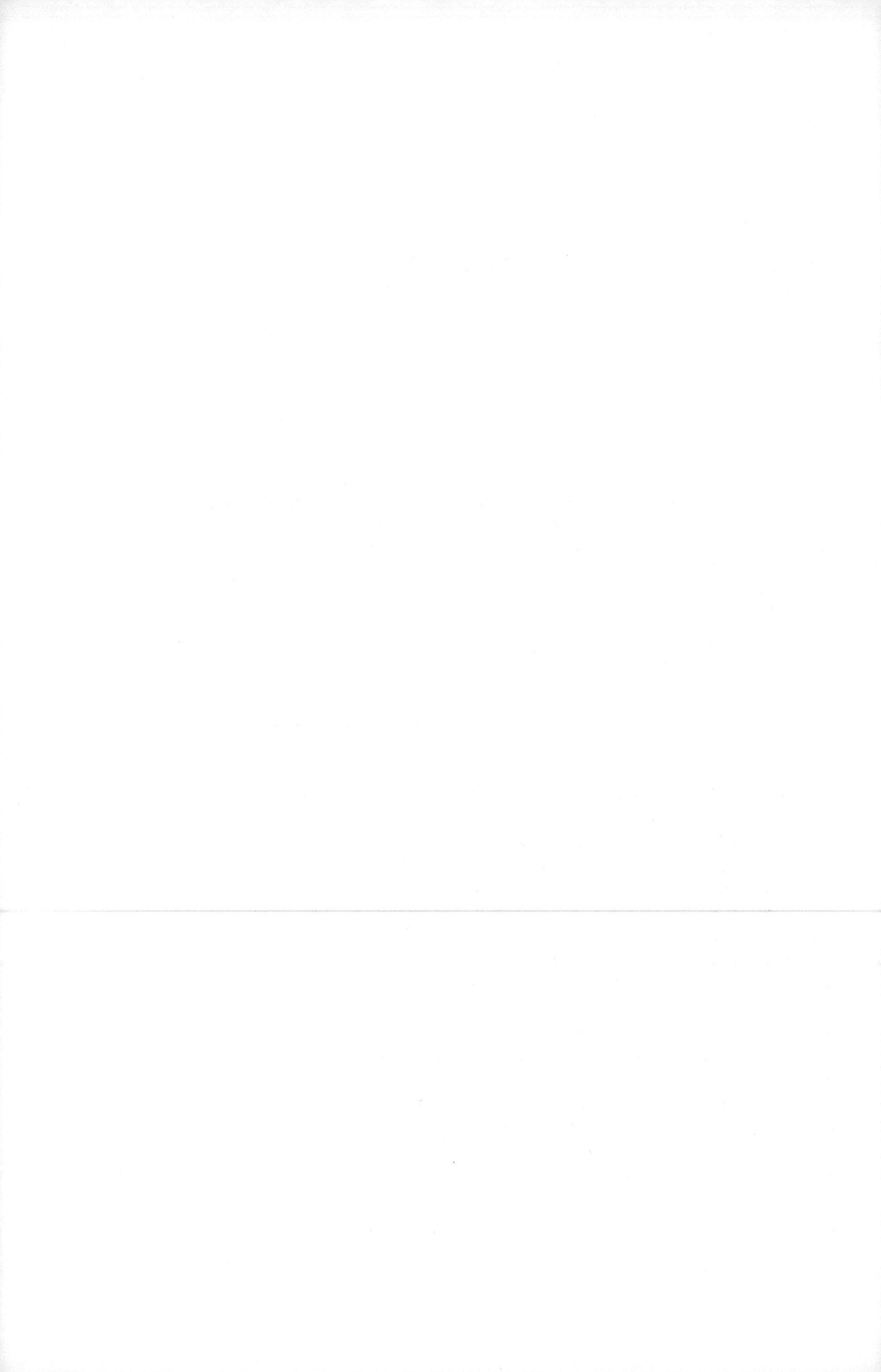

Chapter 18

Creating Knockout and Knockin Rodents Using Engineered Endonucleases via Direct Embryo Injection

Takehito Kaneko and Tomoji Mashimo

Abstract

Genetically engineered rodents have been generated worldwide for biomedical research. Recently, gene targeting techniques have been developed by using engineered endonucleases such as zinc-finger nucleases (ZFN), transcription activator-like effector nucleases (TALEN) and clustered regularly interspaced short palindromic repeats (CRISPR)-Cas9. These endonucleases are useful for simple and rapid production of gene knockout/knockin animals without using embryonic stem (ES) cells. This chapter introduces the latest protocols for producing genetically modified rodents using ZFN, TALEN, and CRISPR/Cas9.

Key words Zinc-finger nucleases (ZFN), Transcription activator-like effector nucleases (TALEN), Clustered regularly interspaced short palindromic repeats (CRISPR), Rat, Mouse, Gene targeting, Microinjection, Embryo transfer

1 Introduction

Transgenic animals can be produced by introducing exogenous DNA fragments into embryos. However, ES cells are required to produce knockout/knockin animals. This means that generation of knockout/knockin animals is limited to those species or strains where ES cells of high quality have been established. In rats, for example, it was very difficult to produce knockout/knockin strains initially because ES cells had not been established. Although knockout rats have now been produced by establishing ES cells [1], it is still difficult to use this protocol routinely in the laboratory. Gene targeting techniques using ZFN, TALEN, and CRISPR/Cas9 provide a new approach for producing knockout/knockin animals [2–6]. Simple and rapid production of genetically modified (GM) animals is now possible even in species and strains where ES cells have not been established. Furthermore, it is a significant advantage that GM animals can be produced by microinjection of these endonucleases into embryos using similar methods to the production of transgenic animals [7–15]. Many successful reports have already

Shondra M. Pruett-Miller (ed.), *Chromosomal Mutagenesis*, Methods in Molecular Biology, vol. 1239, DOI 10.1007/978-1-4939-1862-1_18,

been published on the production of knockout/knockin animals using such engineered endonucleases. These provide powerful tools for the production of GM rodent strains, and these strains will contribute to understanding the etiology and genetics of many human diseases.

2 Materials

2.1 Preparation of ZFN/TALEN/Cas9 mRNA and Guide RNA for Microinjection into Embryos

1. Custom-designed ZFN and TALEN plasmids (*see* **Notes 1** and **2**).
2. Custom-designed CRISPR-Cas9 plasmids.
3. MessageMAX™T7 ARCA-Capped Message Transcription Kit (Cellscript, Madison, WI, USA).
4. MEGAshortscript T7 Kit (Life Technologies Corp., Carlsbad, CA, USA).
5. A-Plus™ Poly (A) polymerase tailing kit (Cellscript).
6. MEGAClear™ kit (Life Technologies Corp.).
7. Distilled water.

2.2 Collection of Pronuclear Stage Embryos

1. Matured male and female mice or rats.
2. HTF medium for mouse embryo manipulation and culture in vitro: *See* Table 1 for individual components. Adjust pH of medium to 7.4 using pH meter. Sterilize medium using a 0.22 μm disposable filter. Store medium at 4 °C until used (*see* **Note 3**).
3. mKRB medium for rat embryo manipulation and culture in vitro: *See* Table 1 for individual components. Adjust pH of medium to 7.4 using pH meter. Sterilize medium using a 0.22 μm disposable filter. Store medium at 4 °C until used (*see* **Note 3**).
4. Sterile mineral oil.
5. Plastic culture dish.
6. CO_2 incubator.
7. Pregnant mare serum gonadotropin (PMSG).
8. Human chorionic gonadotropin (hCG).
9. A pair of small scissors.
10. A sharply pointed forceps.
11. 1 ml syringe.
12. 30 G steel needle.
13. Glass capillary pipettes.

Table 1
Media components

Components	HTF mg/100 ml (mM)	mKRB mg/100 ml (mM)
NaCl	594 (101.6)	553 (94.6)
KCl	35 (4.7)	36 (4.8)
$CaCl_2$	23 (2.0)	19 (1.7)
$MgSO_4 \cdot 7H_2O$	5 (0.2)	29.3 (1.2)
KH_2PO_4	5 (0.4)	16 (1.2)
$NaHCO_3$	210 (25.0)	211 (25.1)
Na-lactate (60 % syrup)	0.34 ml	0.19 ml
Na-pyruvate	4 (0.3)	6 (0.5)
D-Glucose	50 (2.8)	100 (5.6)
Penicillin G	7	7
Streptomycin	5	5
Bovine serum albumin (BSA)	400	400

2.3 Microinjection of mRNA into Pronuclear Stage Embryos

1. Micromanipulator.
2. Micropipette puller (Sutter Instrument, Novato, CA, USA).
3. Microfuge.
4. Glass capillary pipettes.
5. Loading tips.
6. HTF medium.
7. mKRB medium.
8. Sterile mineral oil.
9. Plastic culture dish.
10. CO_2 incubator.

2.4 Embryo Transfer

1. Matured female ICR mice or Wistar rats.
2. Vasectomized male mice or rats.
3. Isoflurane for anesthesia: 1 %, 0.8 L/min for mouse, 2 %, 1 L/min for rat.
4. A pair of small scissors.
5. A sharply pointed forceps.

6. Glass capillary pipettes.
7. 30 G Steel needle.
8. Wound clips.

2.5 Genotyping of Delivered Pups

1. A pair of small scissors.
2. FTA cards.
3. GENEXTRACTOR TA-100 automatic DNA purification system (Takara Bio Inc., Shiga, Japan).
4. PCR system.
5. Electrophoresis system.
6. DNA sequencing system.

3 Methods

3.1 Preparation of ZFN/TALEN mRNA for Microinjection into Embryos

1. Prepare custom-designed ZFN and TALEN plasmids.
2. Transcribe ZFN/TALEN mRNA in vitro using a MessageMAX™T7 ARCA-Capped Message Transcription Kit and carry out polyadenylation using an A-Plus™ Poly (A) polymerase tailing kit.
3. Purify mRNA using a MEGAClear™ kit.
4. Dilute each mRNA with distilled water at 5–10 ng/μl for microinjection into embryos (*see* **Note 4**).

3.2 Preparation of Cas9 mRNA and guide RNA for Microinjection into Embryos

1. Prepare custom-designed CRISPR-Cas9 plasmids.
2. Transcribe Cas9 mRNA and guide (g) RNA in vitro using a MessageMAX™T7 ARCA-Capped Message Transcription Kit and a MEGAshortscript T7 Kit.
3. Polyadenylate Cas9 mRNA using an A-Plus™ Poly (A) polymerase tailing kit.
4. Purify mRNA using a MEGAClear™ kit.
5. Dilute each mRNA with distilled water at 100 ng/μl of Cas9 mRNA and 50 ng/μl of gRNA for microinjection into embryos (*see* **Note 4**).

3.3 Collection of Mouse Pronuclear Stage Embryos Produced by In Vitro Fertilization (IVF)

1. Prepare two culture dishes with 200 μl drops of HTF medium covered with sterile mineral oil for collecting sperm and oocytes.
2. Euthanize a mature male mouse by CO_2 and cervical dislocation.
3. Remove the two epididymal caudae using a small pair of scissors.
4. Remove blood vessels and adipose tissue, and squeeze a dense mass of sperm out of the two epididymides using sharply pointed forceps.

5. Gently place the sperm mass in the 200 μl drops of HTF medium (*see* **Note 5**).
6. Place the culture dish for 60–90 min at 37 °C under 5 % CO_2 and 95 % air.
7. Induce superovulation in female mice by an intraperitoneal injection of PMSG, followed by an intraperitoneal injection of hCG 48 h later (*see* **Note 6**).
8. Euthanize the female mice by CO_2 and cervical dislocation at 13–15 h after hCG injection.
9. Remove the two oviducts using a small pair of scissors.
10. Introduce the oviducts into sterile mineral oil in another culture dish.
11. Hold the oviduct with forceps and puncture the ampulla using a steel needle.
12. Introduce the cumulus–oocyte complexes released from the oviduct into the 200 μl drops of HTF medium (*see* **Note 7**).
13. Collect 3–5 μl of capacitated sperm suspension after incubation.
14. Add the sperm suspension into HTF medium drops with oocytes (*see* **Note 8**).
15. Place the culture dish at 37 °C under 5 % CO_2 and 95 % air.
16. Collect pronuclear stage embryos at 5 h after insemination.

3.4 Collection of Rat Pronuclear Stage Embryos Produced by Natural Mating

1. Induce superovulation in female rats by an intraperitoneal injection of PMSG, followed by an intraperitoneal injection of hCG 48 h later (*see* **Notes 9** and **10**).
2. Check for vaginal plugs in the females that mated with males overnight after hCG injection.
3. Euthanize females with CO_2 and cervical dislocation and remove the two oviducts using a small pair of scissors.
4. Hold the oviduct with forceps and flush ampulla with mKRB medium using a 1 ml syringe with a steel needle.
5. Collect pronuclear stage embryos and transfer them to one of the four 50 μl drops of mKRB medium covered with sterile mineral oil in the culture dish (*see* **Note 11**).
6. Remove cumulus cells and other debris by transferring the embryos to another mKRB medium drop.
7. Place the culture dish at 37 °C under 5 % CO_2 and 95 % air before microinjection.

3.5 Microinjection of mRNA into Pronuclear Stage Embryos

1. Make a holding pipette using a micropipette puller and a microfuge.
2. Introduce mRNA (2–3 μl) into the tip of the injection pipette (*see* **Note 12**).

3. Attach the injection pipette in the right pipette holder of the micromanipulator (Fig. 1).
4. Attach the holding pipette in the left pipette holder of the micromanipulator (Fig. 1).
5. Prepare small drops of HTF medium (mouse) or mKRB medium (rat) covered with sterile mineral oil in the 10 cm manipulation dish (Fig. 1).
6. Stabilize an embryo using the holding pipette.
7. Inject 2–3 pl of mRNA into the male pronucleus or cytoplasm (Fig. 2) (*see* **Note 13**).
8. Culture injected embryos in 100 μl drops of HTF medium (mouse) or mKRB medium (rat) covered with sterile mineral oil at 37 °C under 5 % CO_2 and 95 % air.

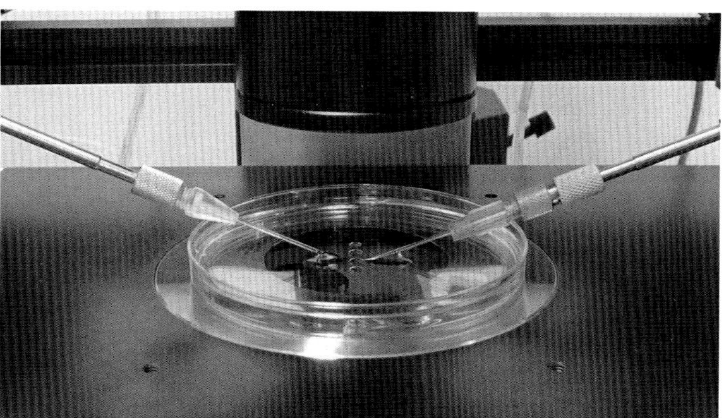

Fig. 1 Set up the micromanipulator

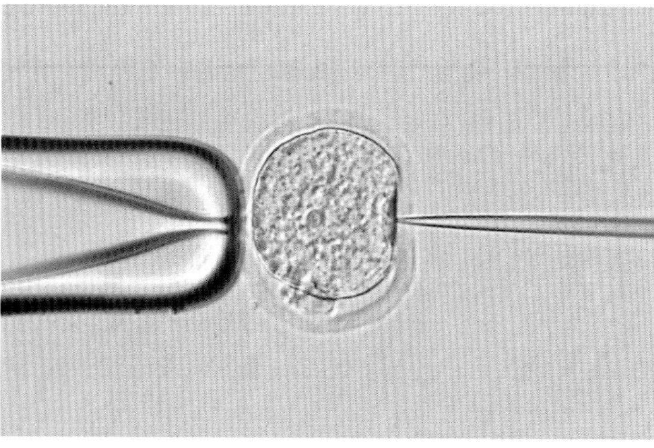

Fig. 2 Injection of RNA into pronuclear stage embryos (rat)

3.6 Embryo Transfer

1. Check that female ICR mice or Wistar rats have been mated with vasectomized males of the same strain on the day before transfer (*see* **Note 14**).
2. Check the vaginal plugs of female mice or rats to ensure they have been mated to induce pseudopregnancy.
3. Anesthetize a pseudopregnant animal.
4. Expose the ovary, oviduct, and part of the uterus through an abdominal incision.
5. Make a small hole in the upper ampulla using a sharp 30G needle.
6. Aspirate 5–10 two-cell embryos into a glass capillary pipette with a few small air bubbles.
7. Insert this capillary pipette into the oviducts and transfer the embryos.
8. Return the ovary, oviduct, and uterus back inside the body cavity and seal the incision with wound clips.
9. Transfer more embryos with the same procedure into another oviduct.
10. Deliver pups at 19 days (mouse) or 21 days (rat) of gestation.

3.7 Genotyping of Delivered Pups

1. Blot blood samples from 3-week-old pups onto the FTA cards.
2. Extract genomic DNA using a GENEXTRACTOR TA-100 automatic DNA purification system.
3. Amplify the target sequences by PCR and electrophorese the PCR products on agarose gels.
4. Sequence the PCR products to confirm the mutations (*see* **Notes 15** and **16**).

4 Notes

1. You can purchase these plasmids as commercial products.
2. Check the activity of the plasmids by transferring them into cultured rat cells for a surveyor (Cel-1) assay before microinjecting mRNA into embryos [11].
3. All media can be stored for up to 1 month.
4. Fresh mRNA should be made up for each experiment.
5. Quick handling leads to high motility of sperm samples.
6. Approximately 5–7.5 IU PMSG/hCG is introduced in each female.
7. To avoid the degeneration of oocytes, cumulus–oocyte complexes should be collected from the oviduct immediately after euthanasia.

Table 2
Development of rat embryos after injection of engineered endonucleases

Targeted genes	Nucleases	No. embryos injected	No. of embryos survived and transferred	No. (%) of offspring	No. of knockout offspring	References
Tyr	TALEN	201	86 (43)	20 (23)	0 (0)	[11]
	TALEN + *ExoI*	68	29 (43)	12 (41)	3 (25)	[11]
	CRISPR/Cas9	19	10 (53)	3 (30)	3 (100)	Unpublished data
Il2rg	ZFN	93	41 (44)	9 (22)	3 (33)	[17]
	Platinum TALEN	52	20 (39)	6 (30)	6 (100)	[17]

8. The final sperm concentration is approximately 1×10^5 cells/ml.
9. Approximately 150–300 IU/kg PMSG/hCG is introduced in each female.
10. F344/Stm rats supplied from the National BioResource Project-Rat (Kyoto, Japan, http://www.anim.med.kyoto-u.ac.jp/NBR/Default.aspx) are suitable for gene modification because many embryos can be obtained [16].
11. Remove cumulus cells using 0.1 % hyaluronidase if they stick to the embryos.
12. It is possible to purchase a commercial products.
13. For generating gene knockin animals, single-strand oligodeoxynucleotides (ssODNs) are coinjected with RNA of the engineered endonucleases into embryos [14, 15].
14. Suitable results are obtained by transfer of embryos at the 2-cell stage.
15. You should sequence any candidate site for off-target effects through the whole genome in founder animals.
16. You should confirm germ line transmission of the mutations by crossing the control strain.
17. Anticipated results are shown in Table 2. Coinjection of Exonuclease I (*Exo I*) with mRNA into embryos increased the rate of producing offspring with the targeted gene edited correctly (Fig.3) [11]. The use of platinum TALEN obtained the expected successful rates of production of knockout rats [17].

Fig. 3 DA rat (founder) with a knockout in the *Tyr* gene using TALEN + *Exo1* (right) and wild type DA rat (left) [11]

References

1. Tong C, Li P, Wu NL et al (2010) Production of p53 gene knockout rats by homologous recombination in embryonic stem cells. Nature 467:211–213
2. Bibikova M, Beumer K, Trautman JK et al (2003) Enhancing gene targeting with designed zinc finger nucleases. Science 300:764
3. Bogdanove AJ, Voytas DF (2011) TAL effectors: customizable proteins for DNA targeting. Science 333:1843–1846
4. Joung JK, Sander JD (2013) TALENs: a widely applicable technology for targeted genome editing. Nat Rev Mol Cell Biol 14:49–55
5. Pennisi E (2013) The CRISPR craze. Science 341:833–836
6. Mashimo T (2013) Gene targeting technologies in rats: zinc finger nucleases, transcription activator-like effector nucleases, and clustered regularly interspaced short palindromic repeats. Dev Growth Differ 56:46–52
7. Cui X, Ji D, Fisher DA et al (2011) Targeted integration in rat and mouse embryos with zinc-finger nucleases. Nat Biotechnol 29:64–67
8. Geurts AM, Cost GJ, Freyvert Y et al (2009) Knockout rats via embryo microinjection of zinc-finger nucleases. Science 325:433
9. Li D, Qiu Z, Shao Y et al (2013) Heritable gene targeting in the mouse and rat using a CRISPR-Cas system. Nat Biotechnol 31:681–683
10. Li W, Teng F, Li T et al (2013) Simultaneous generation and germline transmission of multiple gene mutations in rat using CRISPR-Cas systems. Nat Biotechnol 31:684–686
11. Mashimo T, Kaneko T, Sakuma T et al (2013) Efficient gene targeting by TAL effector nucleases coinjected with exonucleases in zygotes. Sci Rep 3:1253
12. Mashimo T, Takizawa A, Voigt B et al (2010) Generation of knockout rats with X-linked severe combined immunodeficiency (X-SCID) using zinc-finger nucleases. PLoS One 5:e8870
13. Tesson L, Usal C, Menoret S et al (2011) Knockout rats generated by embryo microinjection of TALENs. Nat Biotechnol 29:695–696
14. Wang H, Yang H, Shivalila CS et al (2013) One-step generation of mice carrying mutations in multiple genes by CRISPR/Cas-mediated genome engineering. Cell 153:910–918
15. Yang H, Wang H, Shivalila CS et al (2013) One-step generation of mice carrying reporter and conditional alleles by CRISPR/Cas-mediated genome engineering. Cell 154:1370–1379
16. Taketsuru H, Kaneko T (2013) Efficient collection and cryopreservation of embryos in F344 strain inbred rats. Cryobiology 67:230–234
17. Sakuma T, Ochiai H, Kaneko T et al (2013) Repeating pattern of non-RVD variations in DNA-binding modules enhances TALEN activity. Sci Rep 3:3379

Chapter 19

Simple Sperm Preservation by Freeze-Drying for Conserving Animal Strains

Takehito Kaneko

Abstract

Freeze-drying spermatozoa is the ultimate method for the maintenance of animal strains, in that the gametes can be preserved for a long time in a refrigerator at 4 °C. Furthermore, it is possible to realize easy and safe transportation of spermatozoa at an ambient temperature that requires neither liquid nitrogen nor dry ice. Freeze-drying spermatozoa has been established as a new method for storing genetic resources instead of cryopreservation using liquid nitrogen. This chapter introduces our latest protocols for freeze-drying of mouse and rat spermatozoa, and the anticipated results of the fertilizing ability of these gametes following long-term preservation or transportation.

Key words Sperm, Freeze-drying, Long term preservation, Transportation, Intracytoplasmic sperm injection (ICSI), Fertilization, Mouse, Rat

1 Introduction

Many genetically engineered rodent strains have been generated worldwide. Sperm preservation is an efficient method for storing these strains as genetic resources. Freezing using liquid nitrogen has been used as standard method for storing strains [1–3]. However, continuous supply of liquid nitrogen and mechanical maintenance of the equipment is required. Furthermore, valuable samples might be lost when the supply of liquid nitrogen is interrupted or during disasters such as earthquakes and typhoons [4].

Freeze-drying spermatozoa has been studied as the ultimate method that can overcome these problems in storing animal strains by sperm freezing. Successful freeze-drying attempts have been reported previously in various mammals [5–11]. In the mouse and rat, especially, the methods of sperm freeze-drying have been improved simply and efficiently for increasing success rates [12–16]. Methods for freeze-drying of mouse and rat spermatozoa that allowed long term preservation for 3–5 years at 4 °C have been established using simple solutions containing Tris and ethylenediaminetetraacetic acid

Shondra M. Pruett-Miller (ed.), *Chromosomal Mutagenesis*, Methods in Molecular Biology, vol. 1239, DOI 10.1007/978-1-4939-1862-1_19,

(EDTA) [17–19]. Furthermore, no deterioration of freeze-dried spermatozoa was shown during storage for 3 months and following air transportation at ambient temperature [20, 21].

The sperm freeze-drying protocol described in this chapter is now applicable to the conservation of endangered animals. This method is now a safe and economical tool for the biobanking of valuable animals.

2 Materials

2.1 Freeze-Drying of Spermatozoa

1. Matured male mice or rats.
2. Solution for sperm freeze-drying: Dissolve 10 mM Tris and 1 mM EDTA (TE solution) in 10 ml of distilled water [17]. Adjust pH to 8.0 using a pH meter [14]. Store the freeze-drying solution at 4 °C (*see* **Note 1**). Sterilize the solution using a 0.22 μm disposable filter before use.
3. Freeze-drying machine (Labconco Corporation, Kansas City, MO, USA).
4. 2 ml long-necked glass ampoule (Wheaton, Millville, NJ, USA).
5. 1.5 ml microcentrifuge tube.
6. A pair of small scissors.
7. Sharply pointed forceps.
8. Liquid nitrogen.
9. Gas burner.

2.2 Rehydration of Freeze-Dried Spermatozoa and Collection of Oocytes

1. Sterile distilled water.
2. Matured female mice or rats.
3. Hepes-mCZB medium [22] for mouse oocytes manipulation: *See* Table 1 for individual components. Adjust pH of medium to 7.4 using pH meter. Sterilize medium using a 0.22 μm disposable filter. Store medium at 4 °C until used (*see* **Note 2**).
4. Hepes-mR1ECM medium [23] for rat oocytes manipulation: *See* Table 2 for individual components. Adjust pH of medium to 7.4 using pH meter. Sterilize medium using a 0.22 μm disposable filter. Store medium at 4 °C until used (*see* **Note 2**).
5. Pregnant mare serum gonadotropin (PMSG).
6. Human chorionic gonadotropin (hCG).
7. Plastic culture dish.
8. Sterile mineral oil.
9. A pair of small scissors.
10. Sharply pointed forceps.
11. 30G Steel needle.

Table 1
Media components for mouse embryo manipulation

Components	Hepes-mCZB mg/100 ml (mM)	mCZB mg/100 ml (mM)
NaCl	476 (81.62)	476 (81.62)
KCl	36 (4.83)	36 (4.83)
$CaCl_2 \cdot 2H_2O$	25 (1.70)	25 (1.70)
$MgSO_4 \cdot 7H_2O$	29 (1.18)	29 (1.18)
KH_2PO_4	16 (1.18)	16 (1.18)
$EDTA \cdot Na_2$	4 (0.11)	4 (0.11)
$NaHCO_3$	42 (5.00)	211 (25.00)
Na-lactate (60 % syrup)	0.53 ml	0.53 ml
Na-pyruvate	3 (0.27)	3 (0.27)
D-Glucose	100 (5.55)	100 (5.55)
L-Glutamine	15 (1.00)	15 (1.00)
Penicillin G	5	5
Streptomycin	7	7
Bovine serum albumin (BSA)	–	500
Hepes-Na	520 (20.00)	–
Polyvinylalcohol (PVA)	10	–
Gas phase	Air	5 % CO_2 in air

12. Solution for removal of cumulus cells: Dissolve 0.1 % hyaluronidase in Hepes-mCZB medium for mouse oocytes or Hepes-mR1ECM medium for rat.
13. CO_2 incubator.

2.3 Intracytoplasmic Sperm Injection (ICSI)

1. Micropipette puller (Sutter Instrument, Novato, CA, USA).
2. Microfuge.
3. Micromanipulator with piezo-drive unit.
4. Glass capillary pipettes.
5. Mercury.
6. Plastic culture dish.
7. Hepes-mCZB and mCZB media [24, 25] for mouse embryo manipulation and culture in vitro: *See* Table 1 for individual

components. Adjust pH of Hepes-mCZB medium to 7.4 using pH meter. Sterilize all media using a 0.22 μm disposable filter. Store all media at 4 °C until use (*see* **Note 2**).

8. Hepes-mR1ECM and mKRB media [26] for rat embryo manipulation and culture in vitro: *See* Table 2 for individual components. Adjust pH of Hepes-mR1ECM medium to 7.4 using pH meter. Sterilize all media using a 0.22 μm disposable filter. Store all media at 4 °C until use (*see* **Note 2**).
9. Solution for sperm manipulation: Dissolve 12 % (w/v) of polyvinylpyrrolidone (PVP; Mr 360,000) in Hepes-mCZB medium for mouse spermatozoa or Hepes-mR1ECM for the rat. Sterilize the solution using a 0.45 μm disposable filter. Store the solution at 4 °C until use (*see* **Note 3**).

Table 2
Media components for rat embryo manipulation

	Hepes-mR1ECM	mKRB	mR1ECM
Components	mg/100 ml (mM)	mg/100 ml (mM)	mg/100 ml (mM)
NaCl	448 (76.7)	553 (94.6)	448 (76.7)
KCl	24 (3.2)	36 (4.8)	24 (3.2)
$CaCl_2$	22 (2.0)	19 (1.7)	22 (2.0)
$MgCl_2 \cdot 6H_2O$	10 (0.5)	–	10 (0.5)
$MgSO_4 \cdot 7H_2O$	–	29.3 (1.2)	–
KH_2PO_4	–	16 (1.2)	–
$NaHCO_3$	42 (5.0)	211 (25.1)	210 (25.0)
Na-lactate (60 % syrup)	112 (10.0)	0.19 ml	112 (10.0)
Na-pyruvate	6 (0.5)	6 (0.5)	6 (0.5)
D-Glucose	135 (7.5)	100 (5.6)	135 (7.5)
L-Glutamine	2 (0.1)	–	2 (0.1)
EAA	2 ml	–	2 ml
NEAA	1 ml	–	1 ml
Penicillin G	7	7	7
Streptomycin	5	5	5
Polyvinylalcohol (PVA)	100	–	100
Bovine serum albumin (BSA)	–	400	–
Hepes-Na	573 (22.0)	–	–
Gas phase	Air	5 % CO_2 in air	5 % CO_2 in air

10. Sterile mineral oil.
11. CO_2 incubator.

2.4 Embryo Transfer

1. Female ICR mice or Wistar rats.
2. Vasectomized male mice or rats.
3. 30G steel needle.
4. Isoflurane for anesthesia: 1 %, 0.8 l/min for mouse, 2 %, 1 l/min for rat.
5. A pair of small scissors.
6. Sharply pointed forceps.
7. Glass capillary pipettes.
8. 30 G Steel needle.
9. Wound clips.

3 Methods

3.1 Freeze-Drying of Spermatozoa

1. Prepare 1.5 ml of solution for sperm freeze-drying in a 1.5 ml microcentrifuge tube.
2. Euthanize a mature male mouse or rat by CO_2 and cervical dislocation.
3. Remove both epididymal caudae using a pair of small scissors.
4. Remove blood vessels and adipose tissue, and squeeze a dense mass of spermatozoa out of the two epididymides using sharply pointed forceps.
5. Gently place the sperm mass on the bottom of a 1.5 ml microcentrifuge tube containing 1.5 ml of solution for sperm freeze-drying (*see* **Note 4**).
6. Leave for 10 min at room temperature to allow the spermatozoa to disperse into the solution.
7. Carefully collect 1 ml of supernatant in another tube.
8. Transfer 100 μl aliquots of the sperm suspension into a 2 ml long-necked glass ampoule for freeze-drying.
9. Make up 10 ampoules and dip them into liquid nitrogen for 20 s.
10. Connect ampoules to the manifold of a freeze-drying machine (Fig. 1a) (*see* **Notes 5** and **6**).
11. Dry for 4 h at a pressure of 0.03–0.05 hPa (*see* **Note 7**).
12. Flame-seal each ampoule using a gas burner with the inside pressure of the ampoules kept at 0.03–0.05 hPa (*see* **Note 8**).
13. Store ampoules at 4 °C (Fig. 1b) (*see* **Note 9**).

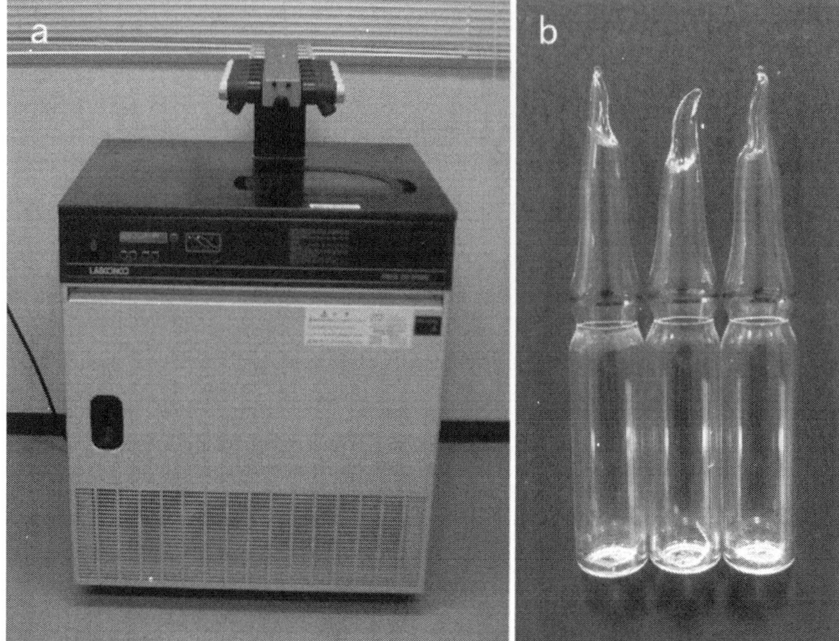

Fig. 1 (**a**) Freeze-drying machine using long-necked glass ampoules. (**b**) Freeze-dried spermatozoa in the glass ampoules

3.2 Rehydration of Freeze-Dried Spermatozoa

1. Carefully open an ampoule.
2. Rehydrate the spermatozoa by adding 100 μl of sterile distilled water.
3. Confirm that the spermatozoa are morphologically normal.

3.3 Collection of Oocytes

1. Prepare culture dishes each with four 50 μl drops of Hepes-mCZB (mouse) or Hepes-mR1ECM (rat) covered with sterile mineral oil.
2. Induce superovulation in female mice or rats by an intraperitoneal injection of pregnant mare serum gonadotropin (PMSG), followed by an intraperitoneal injection of human chorionic gonadotropin (hCG) 48 h later (*see* **Note 10**).
3. Euthanize the animals at 13–15 h after hCG injection by CO_2 and cervical dislocation.
4. Remove the two oviducts using a pair of small scissors.
5. Introduce oviducts into sterile mineral oil in a culture dish.
6. Hold each oviduct using forceps and dissect the ampulla using a steel needle.
7. Introduce the cumulus–oocyte complexes released from the oviduct into one of the medium droplets.
8. Remove cumulus cells from oocytes using 0.1 % hyaluronidase.

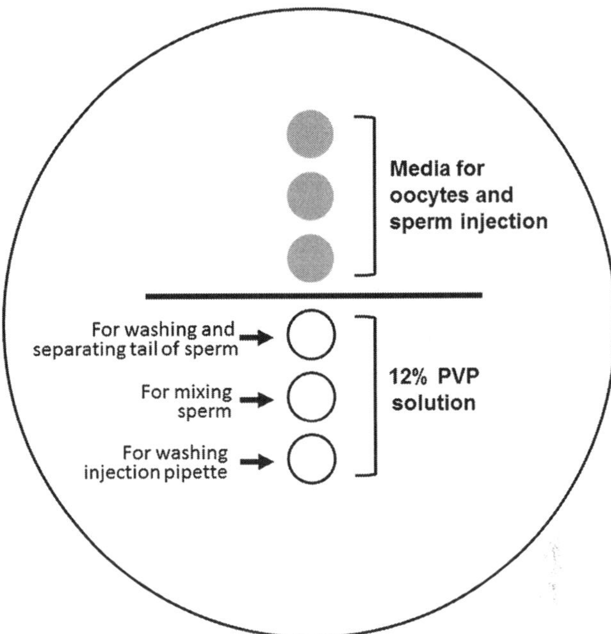

Fig. 2 Manipulation dish for ICSI

9. Introduce cumulus-free oocytes into another medium drop.
10. Place the culture dish at 37 °C under 5 % CO_2 and 95 % air before sperm injection.

3.4 Set Up the Micromanipulator for Intracytoplasmic Sperm Injection (ICSI)

1. Make injection and holding pipettes using a micropipette puller and microfuge (Fig. 3c) (*see* **Notes 11** and **12**).
2. Turn on the power of the micromanipulator with piezo-drive unit (*see* **Note 13**).
3. Inject a small volume of mercury into the injection pipette (*see* **Note 14**).
4. Attach the injection pipette in the right pipette holder of the micromanipulator.
5. Attach the holding pipette in the left pipette holder of the micromanipulator.
6. Prepare small drops of Hepes-mCZB medium (mouse) or Hepes-mR1ECM (rat) in the 10 cm manipulation dish (Fig. 2).
7. Prepare small drops of solution for sperm manipulation (12 % PVP solution) in the 10 cm manipulation dish (Fig. 2).
8. Cover medium and 12 % PVP solution drops with sterile mineral oil (Fig. 2).
9. Carefully transfer oocytes into the medium drops and mix the rehydrated spermatozoa in the drop of 12 % PVP solution.

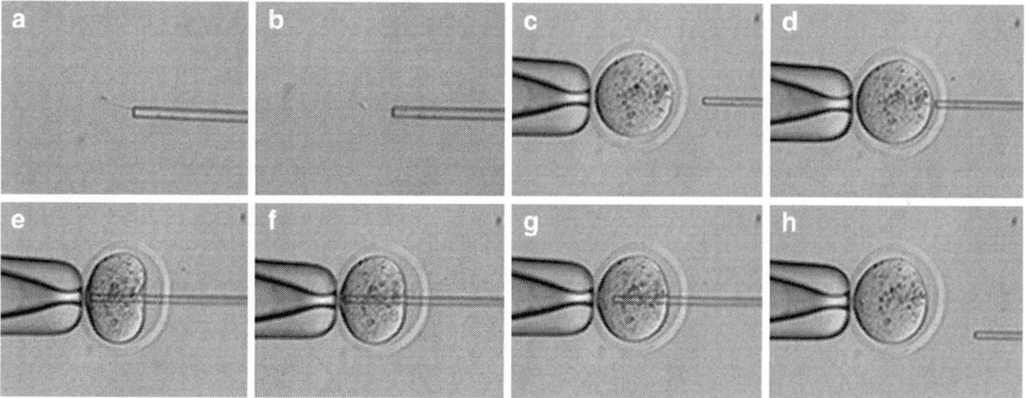

Fig. 3 Injection of a mouse sperm head into oocytes

3.5 Sperm Injection into Oocytes

1. Aspirate a morphologically normal spermatozoon into the injection pipette (Fig. 3a).
2. Separate the tail from the head by piezo-pulse impact with lower power (Fig. 3b) (*see* **Note 15**).
3. Aspirate 5–10 sperm heads into the injection pipette (*see* **Note 16**).
4. Stabilize an oocyte using the holding pipette (Fig. 3c).
5. Make a hole in the zona pellucida using several of piezo-pulse impacts with lower power (Fig. 3d).
6. Push the injection pipette into the oocyte and make a hole on the oolemma by one shot of piezo-pulse impact with the lowest power (Fig. 3e, f).
7. Inject one sperm head into the cytoplasm (Fig. 3g, h) (*see* **Note 17**).
8. Culture ICSI-oocytes in 100 μl drops of mCZB for mouse or mKRB for rat covered with sterile mineral oil at 37 °C under 5 % CO_2 and 95 % air (*see* **Note 18**).

3.6 Embryo Transfer

1. Proestrus female ICR mice or Wistar rats should have been mated with vasectomized males of the same strain on the day before transfer (*see* **Notes 19** and **20**).
2. Check the presence of a vaginal plug in each female to confirm pseudopregnancy.
3. Anesthetize a pseudopregnant female.
4. Expose the ovary, oviduct, and part of the uterus through an abdominal incision.
5. Make a small hole in the upper ampulla using a 30G steel needle.

6. Aspirate 5–10 two-cell embryos into a glass capillary pipette with a few small air bubbles.
7. Insert the glass capillary pipette into the oviducts and transfer the embryos.
8. Return the ovary, oviduct, and uterus back inside the body cavity and seal the incision with wound clips.
9. Transfer embryos with same procedure into another oviduct.
10. Deliver pups at 19 days (mouse) and 21 days (rat) of gestation.

4 Notes

1. It is possible to purchase a commercial product from Ambion [18, 19].
2. All media can be stored at 4 °C for up to 1 month.
3. This solution can be stored at 4 °C for 1 month.
4. Immature spermatozoa collected from the testis can also be freeze-dried by treatment with diamide. This supports DNA condensation in spermatozoa to protect them from damage by freeze-drying [18, 27] (*see* **Note 21**).
5. Quick handling leads to high fertilizing ability of spermatozoa after freeze-drying.
6. You should prepare the freeze-drying machine using long-necked glass ampoules (Fig. 1b). Spermatozoa cannot be preserved at 4 °C for the long term if freeze-dried in a machine using rubber-capped vials [28].
7. The pressure during drying is important to highly maintain fertilizing ability of the spermatozoa. Keep the pressure at less than 0.1 hPa [18, 19].
8. Do not make a hole when flame-sealing to maintain the inside pressure of the ampoules.
9. Spermatozoa that are freeze-dried in glass ampoules can be stored for up to 3 months at room temperature (24 °C) [20].
10. Approximately 5–7.5 IU/female (for mice) or 150–300 IU/kg (for rats) PMSG/hCG is injected into each animal.
11. Tip diameter of injection pipette: mouse, external 10 μm, internal 8 μm; rat, external 6 μm, internal 4 μm.
12. The diameter of the holding pipette: external 80 μm, internal 20 μm.
13. The piezo-drive is necessary to make a hole in the oolemma of mouse and rat oocytes. The oolemma is quite susceptible to physical damage [29].

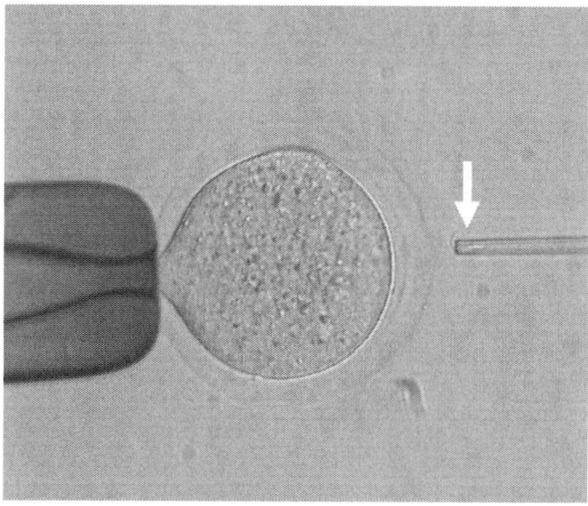

Fig. 4 ICSI in the rat. The *arrow* shows a single sperm head with a normal shape hung on the tip of a narrow injection pipette

14. Mercury is necessary to transmit the impact of the piezo-drive unit. Similar manipulation is possible using Fluorinert instead of mercury.
15. Separate the tail from rat sperm heads for 1 sec using low power output of an ultrasonic homogenizer before mixing with 12 % PVP solution.
16. In rat, a single sperm head with a normal shape is hung on the tip of an injection pipette because the survival rate of oocytes after ICSI was increased when using a pipette with a narrow tip (4 μm) (Fig. 4).
17. Using frozen oocytes and suitable media increase success rates of sperm injection [30, 31].
18. In the rat, 2-cell stage embryos are transferred into mR1ECM medium [32] if cultured to the blastocyst stage in vitro. Individual components of mR1ECM medium are shown in Table 2. Medium are sterilized using a 0.22 μm disposable filter, and can be stored at 4 °C for up to 1 month.
19. Suitable results are obtained by transfer of embryos at the 2-cell stage.
20. You can preserve temporary embryos by freezing if you cannot prepare pseudopregnant females [33–35].
21. The tables show anticipated results for the developmental ability of oocytes fertilized with mouse and rat spermatozoa freeze-dried using the methods described in this chapter. Freeze-dried spermatozoa can be stored for the long term at 4 °C without deterioration (Tables 3 and 4). Although immature spermatozoa collected from testis are damaged by freeze-drying, their tolerance is increased by treatment with diamide before freeze-drying (Table 5). Freeze-dried spermatozoa can be transported worldwide by air at ambient temperatures (Table 6).

Table 3
Development of oocytes fertilized with fresh and freeze-dried mouse spermatozoa stored at 4 °C for various times

Storage term	No. of embryos transferred	No. (%) of embryos implanted	No. (%) of offspring
Fresh	51	35 (67)	20 (39)
1 day	63	42 (67)	16 (25)
5 months	49	40 (82)	18 (37)
1 year	52	29 (56)	20 (39)
3 years	71	30 (42)	22 (31)

C57BL/6J mice were used in this study
The data are reproduced from ref. 19

Table 4
Development of oocytes fertilized with freeze-dried rat spermatozoa stored at 4 °C for various times

Storage term	No. of embryos transferred	No. (%) of embryos implanted	No. (%) of offspring
1–4 days	36	11 (31)	5 (14)
6 months	18	11 (61)	3 (17)
1 year	19	8 (42)	3 (16)
5 years	54	15 (28)	6 (11)

Wistar rats were used in this study
The data are reproduced from refs. 16 and 18

Table 5
Development of oocytes fertilized with freeze-dried mouse and rat testicular spermatozoa

Species	Diamide treatment	No. of embryos transferred	No. (%) of embryos implanted	No. (%) of offspring
Mouse	–	72	2 (3)	2 (3)
	+	67	38 (57)	15 (22)
Rat	–	36	0 (0)	0 (0)
	+	50	4 (8)	4 (8)

B6D2F1 mice and Wistar rats were used in this study
The data are reproduced from refs. 18 and 27

Table 6
Development of oocytes fertilized with freeze-dried mouse spermatozoa after international air transportation at ambient temperature

Transportation	No. of embryos transferred	No. (%) of offspring
Before	25	15 (60)
After	28	15 (54)

B6D2F1 mice were used in this study
Freeze-dried spermatozoa were transported between Japan and USA
The data are reproduced from ref. 21

References

1. Kaneko T, Yamamura A, Ide Y et al (2006) Long-term cryopreservation of mouse sperm. Theriogenology 66:1098–1101
2. Benson JD, Woods EJ, Walters EM et al (2012) The cryobiology of spermatozoa. Theriogenology 78:1682–1699
3. Agca Y (2012) Genome resource banking of biomedically important laboratory animals. Theriogenology 78:1653–1665
4. Dickey RP, Lu PY, Sartor BM et al (2006) Steps taken to protect and rescue cryopreserved embryos during Hurricane Katrina. Fertil Steril 86:732–734
5. Wakayama T, Yanagimachi R (1998) Development of normal mice from oocytes injected with freeze-dried spermatozoa. Nat Biotechnol 16:639–641
6. Keskintepe L, Pacholczyk G, Machnicka A et al (2002) Bovine blastocyst development from oocytes injected with freeze-dried spermatozoa. Biol Reprod 67:409–415
7. Kwon IK, Park KE, Niwa K (2004) Activation, pronuclear formation, and development in vitro of pig oocytes following intracytoplasmic injection of freeze-dried spermatozoa. Biol Reprod 71:1430–1436
8. Liu JL, Kusakabe H, Chang CC et al (2004) Freeze-dried sperm fertilization leads to full-term development in rabbits. Biol Reprod 70: 1776–1781
9. Sánchez-Partida LG, Simerly CR, Ramalho-Santos J (2008) Freeze-dried primate sperm retains early reproductive potential after intracytoplasmic sperm injections. Fertil Steril 89: 742–745
10. Muneto T, Horiuchi T (2011) Full-term development of hamster embryos produced by injecting freeze-dried spermatozoa into oocytes. J Mamm Ova Res 28:32–39
11. Gianaroli L, Magli MC, Stanghellini I et al (2012) DNA integrity is maintained after freeze-drying of human spermatozoa. Fertil Steril 97:1067–1073
12. Kusakabe H, Szczygiel MA, Whittingham DG et al (2001) Maintenance of genetic integrity in frozen and freeze-dried mouse spermatozoa. Proc Natl Acad Sci U S A 98:13501–13506
13. Ward MA, Kaneko T, Kusakabe H et al (2003) Long-term preservation of mouse spermatozoa after freeze-drying and freezing without cryoprotection. Biol Reprod 69: 2100–2108
14. Kaneko T, Whittingham DG, Yanagimachi R (2003) Effect of pH value of freeze-drying solution on the chromosome integrity and developmental ability of mouse spermatozoa. Biol Reprod 68:136–139
15. Kaneko T, Kimura S, Nakagata N (2007) Offspring derived from oocytes injected with rat sperm, frozen or freeze-dried without cryoprotection. Theriogenology 68: 1017–1021
16. Kaneko T, Kimura S, Nakagata N (2009) Importance of primary culture conditions for the development of rat ICSI embryos and long-term preservation of freeze-dried sperm. Cryobiology 58:293–297
17. Kaneko T, Nakagata N (2006) Improvement in the long-term stability of freeze-dried mouse spermatozoa by adding of a chelating agent. Cryobiology 53:279–282
18. Kaneko T, Serikawa T (2012) Successful long-term preservation of rat sperm by freeze-drying. PLoS One 7:e35043
19. Kaneko T, Serikawa T (2012) Long-term preservation of freeze-dried mouse spermatozoa. Cryobiology 64:211–214

20. Kaneko T, Nakagata N (2005) Relation between storage temperature and fertilizing ability of freeze-dried mouse spermatozoa. Comp Med 55:140–144
21. Kaneko T (2014) The latest improvements in the mouse sperm preservation. Mouse molecular embryology. Method Mol Embryol, Springer, USA 1092:357–365
22. Kimura Y, Yanagimachi R (1995) Intracytoplasmic sperm injection in the mouse. Biol Reprod 52:709–720
23. Hirabayashi M, Kato M, Aoto T et al (2012) Offspring derived from intracytoplasmic injection of transgenic rat sperm. Transgenic Res 11:221–228
24. Chatot CL, Ziomek CA, Bavister BD et al (1989) An improved culture medium supports development of random-bred 1-cell mouse embryos in vitro. J Reprod Fertil 86:679–688
25. Chatot CL, Lewis JL, Torres I et al (1990) Development of 1-cell embryos from different strains of mice in CZB medium. Biol Reprod 42:432–440
26. Toyoda Y, Chang MC (1974) Fertilization of rat eggs in vitro by epididymal spermatozoa and the development of eggs following transfer. J Reprod Fertil 36:9–22
27. Kaneko T, Whittingham DG, Overstreet JW et al (2003) Tolerance of the mouse sperm nuclei to freeze-drying depends on their disulfide status. Biol Reprod 69:1859–1862
28. Kawase Y, Araya H, Kamada N et al (2005) Possibility of long-term preservation of freeze-dried mouse spermatozoa. Biol Reprod 72: 568–573
29. Yanagimachi R (2009) Germ cell research: a personal perspective. Biol Reprod 80:204–218
30. Sakamoto W, Kaneko T, Nakagata N (2005) Use of frozen-thawed oocytes for efficient production of normal offspring from cryopreserved mouse spermatozoa showing low fertility. Comp Med 55:136–139
31. Kaneko T, Ohno R (2011) Improvement in the development of oocytes from C57BL/6 mice after sperm injection. J Am Assoc Lab Anim Sci 50:33–36
32. Miyoshi K, Abeydeera LR, Okuda K et al (1995) Effects of osmolarity and amino acids in a chemically defined medium on development of rat one-cell embryos. J Reprod Fertil 103:27–32
33. Whittingham DG, Leibo SP, Mazur P (1972) Survival of mouse embryos frozen to −196 degrees and −269 degrees C. Science 18: 411–414
34. Jin B, Mochida K, Ogura A et al (2010) Equilibrium vitrification of mouse embryos. Biol Reprod 82:444–450
35. Taketsuru H, Kaneko T (2013) Efficient collection and cryopreservation of embryos in F344 strain inbred rats. Cryobiology 67: 230–234

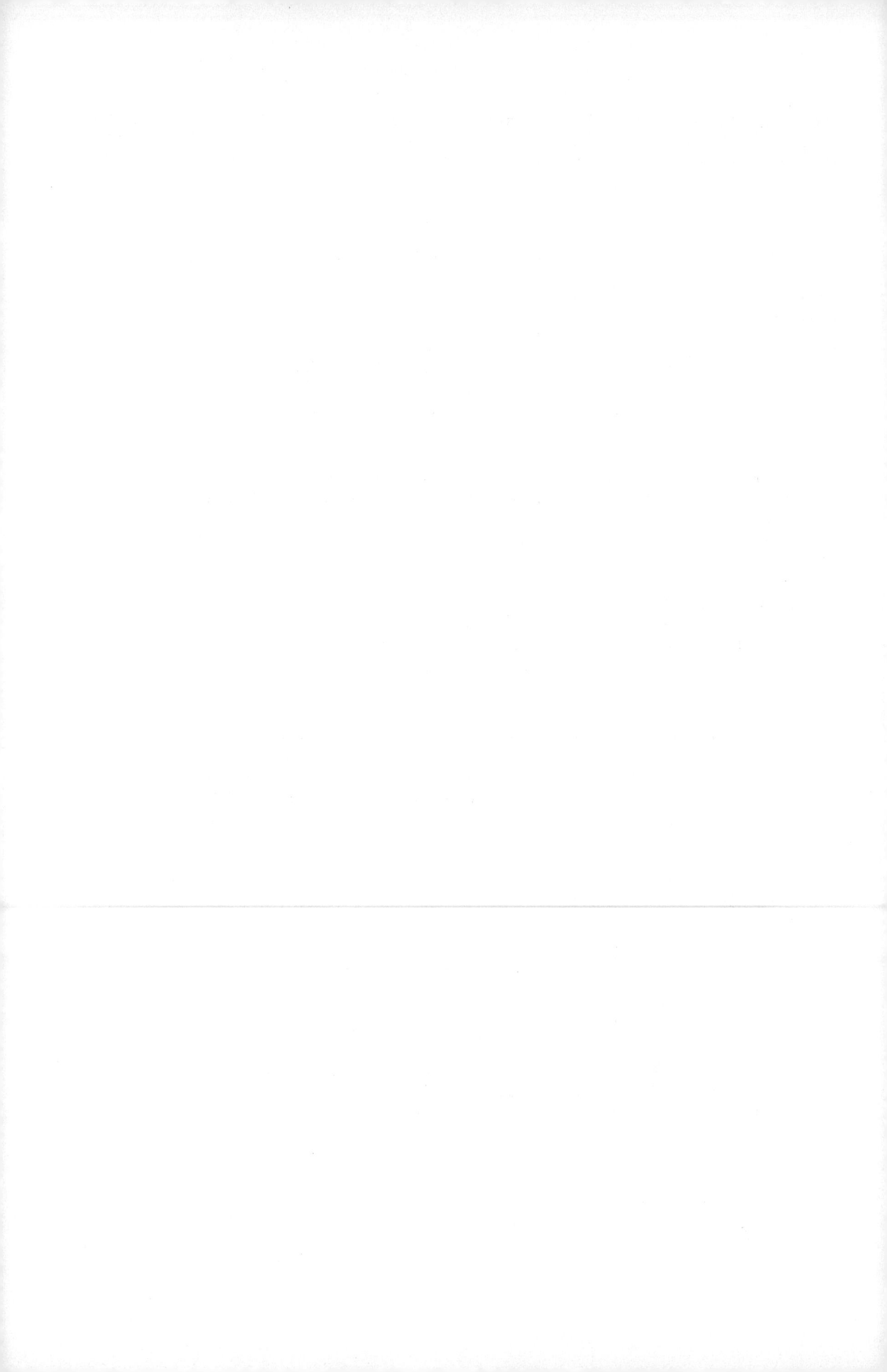

INDEX

Shondra M. Pruett-Miller (ed.), *Chromosomal Mutagenesis*, Methods in Molecular Biology, vol. 1239, DOI 10.1007/978-1-4939-1862-1, © Springer Science+Business Media New York 2015

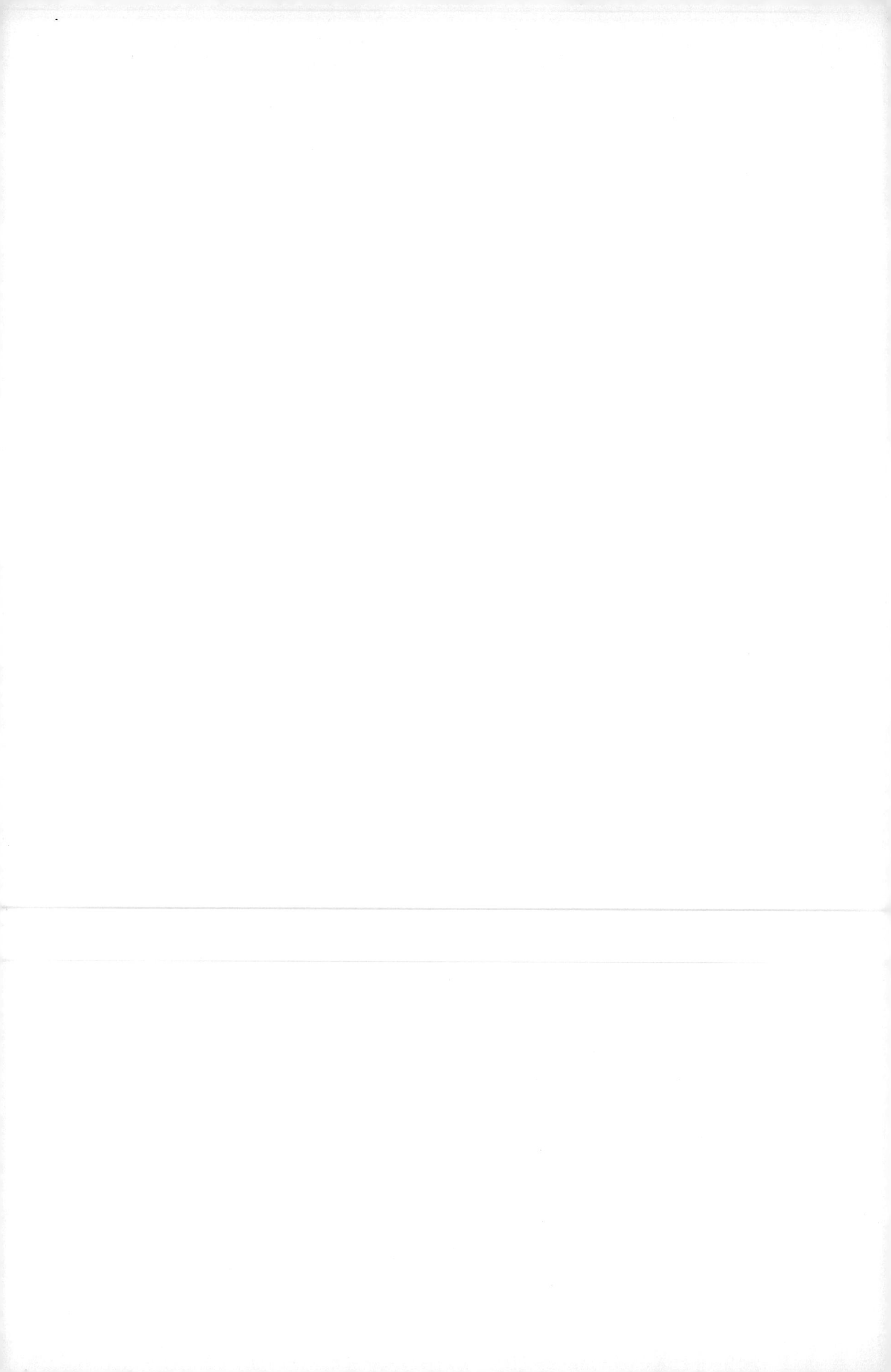